Metamaterials

Beyond Crystals, Noncrystals,
and Quasicrystals

Metamaterials

Beyond Crystals, Noncrystals, and Quasicrystals

Tie Jun Cui
Southeast University, Nanjing, P.R. of China
Wen Xuan Tang
Xin Mi Yang
Zhong Lei Mei
Wei Xiang Jiang

CRC Press is an imprint of the
Taylor & Francis Group, an **informa** business

CRC Press
Taylor & Francis Group
6000 Broken Sound Parkway NW, Suite 300
Boca Raton, FL 33487-2742

© 2016 by Taylor & Francis Group, LLC
CRC Press is an imprint of Taylor & Francis Group, an Informa business

No claim to original U.S. Government works

Printed on acid-free paper
Version Date: 20160211

International Standard Book Number-13: 978-1-4822-2310-1 (Hardback)

This book contains information obtained from authentic and highly regarded sources. Reasonable efforts have been made to publish reliable data and information, but the author and publisher cannot assume responsibility for the validity of all materials or the consequences of their use. The authors and publishers have attempted to trace the copyright holders of all material reproduced in this publication and apologize to copyright holders if permission to publish in this form has not been obtained. If any copyright material has not been acknowledged please write and let us know so we may rectify in any future reprint.

Except as permitted under U.S. Copyright Law, no part of this book may be reprinted, reproduced, transmitted, or utilized in any form by any electronic, mechanical, or other means, now known or hereafter invented, including photocopying, microfilming, and recording, or in any information storage or retrieval system, without written permission from the publishers.

For permission to photocopy or use material electronically from this work, please access www.copyright.com (http://www.copyright.com/) or contact the Copyright Clearance Center, Inc. (CCC), 222 Rosewood Drive, Danvers, MA 01923, 978-750-8400. CCC is a not-for-profit organization that provides licenses and registration for a variety of users. For organizations that have been granted a photocopy license by the CCC, a separate system of payment has been arranged.

Trademark Notice: Product or corporate names may be trademarks or registered trademarks, and are used only for identification and explanation without intent to infringe.

Library of Congress Cataloging-in-Publication Data

Names: Cui, Tie Jun.
Title: Metamaterials : beyond crystals, noncrystals, and quasicrystals / Tie Jun Cui [and four others].
Other titles: Metamaterials
Description: Boca Raton : Taylor & Francis, 2016. | "A CRC title." | Includes bibliographical references and index.
Identifiers: LCCN 2015050548 | ISBN 9781482223101 (alk. paper)
Subjects: LCSH: Metamaterials. | Electronics--Equipment and supplies. | Modulation (Electronics)--Equipment and supplies. | Radio frequency modulation--Equipment and supplies.
Classification: LCC TK7871.15.M48 M484 2016 | DDC 620.1/1--dc23
LC record available at http://lccn.loc.gov/2015050548

Visit the Taylor & Francis Web site at
http://www.taylorandfrancis.com

and the CRC Press Web site at
http://www.crcpress.com

Contents

List of Figures xi
List of Tables xxiii
Preface xxv
Authors xxvii

1 Introduction 1
   1.1 Natural Materials and Metamaterials 1
   1.2 Homogeneous Metamaterials: Several Special Cases 2
       1.2.1 Left-Handed Materials 2
       1.2.2 Zero-Refractive-Index Metamaterials 3
       1.2.3 Negative-Epsilon Materials 4
       1.2.4 Negative-Mu Materials 6
   1.3 Random Metamaterials 7
   1.4 Inhomogeneous Metamaterials 9
       1.4.1 GO Method 9
       1.4.2 Quasi-Conformal Mapping Method 10
       1.4.3 Transformation Optics 11
   1.5 Structure of This Book 12
   Acknowledgments 13
   References 13

2 Effective Medium Theory 17
   2.1 Lorentz–Drude Models 17
   2.2 Retrieval Methods of Effective Medium Parameters 21
   2.3 General Effective Medium Theory 24
   References 28

3 Artificial Particles: "Man-Made Atoms" or "Meta-Atoms" 29
   3.1 Electrically Resonant Particles 30
   3.2 Magnetically Resonant Particles 34
   3.3 Dielectric-Metal Resonant Particles 36
   3.4 Complementary Particles 38
   3.5 Dielectric Particles 43

v

3.6 Nonresonant Particles 48
3.7 LC Particles 51
3.8 D.C. Particles 57
References 62

4 Homogeneous Metamaterials: Super Crystals 67
4.1 Homogeneous Metamaterials: Periodic Arrangements of Particles 68
    4.1.1 SNG Metamaterials 68
    4.1.2 DNG Metamaterials 79
    4.1.3 Zero-Index Metamaterials 85
    4.1.4 DPS Metamaterials 85
4.2 Single-Negative Metamaterials 89
    4.2.1 Evanescent-Wave Amplification in MNG–ENG Bilayer Slabs 89
    4.2.2 Partial Focusing by Anisotropic MNG Metamaterials 97
4.3 Double-Negative Metamaterials 101
    4.3.1 Strong Localization of EM Waves Using Four-Quadrant LHM–RHM Open Cavities 101
    4.3.2 Free-Space LHM Super Lens Based on Fractal-Inspired DNG Metamaterials 107
4.4 Zero-Index Metamaterials 112
    4.4.1 Electromagnetic Tunneling through a Thin Waveguide Channel Filled with ENZ Metamaterials 112
    4.4.2 Highly Directive Radiation by a Line Source in Anisotropic Zero-Index Metamaterials 119
    4.4.3 Spatial Power Combination for Omnidirectional Radiation via Radial AZIM 122
    4.4.4 Directivity Enhancement to Vivaldi Antennas Using Compact AZIMs 125
4.5 Double-Positive Metamaterials 127
    4.5.1 Transmission Polarizer Based on Anisotropic DPS Metamaterials 127
    4.5.2 Increasing Bandwidth of Microstrip Antennas by Magneto-Dielectric Metamaterials Loading 132
Appendix: 2D Near-Field Mapping Apparatus 140
References 141

5 Random Metamaterials: Super Noncrystals 147
5.1 Random Metamaterials: Random Arrangements of Particles 147
    5.1.1 Randomly Gradient Index Metamaterial 147
    5.1.2 Metasurface with Random Distribution of Reflection Phase 150

5.2	Diffuse Reflections by Metamaterial Coating with Randomly Distributed Gradients of Refractive Index 152	
	5.2.1 Role of Amount of Subregions or Length of Coating......... 157	
	5.2.2 Influence of Impedance Mismatch 157	
	5.2.3 Influence of Random Distribution Mode 158	
	5.2.4 Experimental Verification of Diffuse Reflections 159	
5.3	RCS Reduction by Metasurface with Random Distribution of Reflection Phase 163	
References 166		

6 Inhomogeneous Metamaterials: Super Quasicrystals........................ 169
    6.1 Inhomogeneous Metamaterials: Particularly Nonperiodic Arrays of Meta-Atoms 169
    6.2 Geometric Optics Method: Design of Isotropic Metamaterials 171
    6.3 Quasi-Conformal Mapping: Design of Nearly Isotropic Metamaterials 173
    6.4 Optical Transformation: Design of Anisotropic Metamaterials 176
    6.5 Examples 178
        6.5.1 Invisibility Cloaks 178
        6.5.2 Concentrators 180
        6.5.3 High-Performance Antennas....................................... 182
        6.5.4 Illusion-Optics Devices 185
    References 188

7 Gradient-Index Inhomogeneous Metamaterials 191
    7.1 Several Representative GRIN Metamaterials 194
        7.1.1 Hole-Array Metamaterial 194
        7.1.2 I-Shaped Metamaterial 195
        7.1.3 Waveguide Metamaterial 195
    7.2 2D Planar Gradient-Index Lenses 197
        7.2.1 Derivation of the Refractive Index Profile...................... 197
        7.2.2 Full-Wave Simulations (Continuous Medium)................. 198
        7.2.3 Hole-Array Metamaterials 199
        7.2.4 Full-Wave Simulations (Discrete Medium) 200
        7.2.5 Experimental Realization 201
    7.3 2D Luneburg Lens 201
        7.3.1 Refractive Index Profile 202
        7.3.2 Ray Tracing Performance 202
        7.3.3 Full-Wave Simulations (Continuous Medium)................. 204
        7.3.4 Metamaterials Utilized 204
        7.3.5 Experiments 204
    7.4 2D Half Maxwell Fisheye Lens 207
        7.4.1 Refractive Index Profile 207

		7.4.2	Ray Tracing Performance 208
		7.4.3	Full-Wave Simulations (Continuous Medium) 208
		7.4.4	Metamaterials Utilized 209
		7.4.5	Experiments 210
	7.5	3D Planar Gradient-Index Lens 212
		7.5.1	Refractive Index Profile 213
		7.5.2	Full-Wave Simulations (Continuous Medium) 214
		7.5.3	Metamaterials Utilized 215
		7.5.4	Experiments 217
	7.6	3D Half Luneburg Lens 218
		7.6.1	Refractive Index Profile 219
		7.6.2	Ray Tracing Performance 219
		7.6.3	Full-Wave Simulations 220
		7.6.4	Metamaterials Utilized 221
		7.6.5	Experiments 222
	7.7	3D Maxwell Fisheye Lens 223
		7.7.1	Refractive Index Profile 223
		7.7.2	Ray Tracing Performance 223
		7.7.3	Full-Wave Simulations and Experiments 224
	7.8	Electromagnetic Black Hole 225
		7.8.1	Refractive Index Profile 226
		7.8.2	Ray Tracing Performance 227
		7.8.3	Full-Wave Simulations (Continuous Medium) 227
		7.8.4	Metamaterials Utilized 228
		7.8.5	Experiments 228
	References 230
8	**Nearly Isotropic Inhomogeneous Metamaterials** 233
	8.1	2D Ground-Plane Invisibility Cloak 233
	8.2	2D Compact Ground-Plane Invisibility Cloak 241
	8.3	2D Ground-Plane Illusion-Optics Devices 247
	8.4	2D Planar Parabolic Reflector 251
	8.5	3D Ground-Plane Invisibility Cloak 256
	8.6	3D Flattened Luneburg Lens 262
	References 268
9	**Anisotropic Inhomogeneous Metamaterials** 271
	9.1	Spatial Invisibility Cloak 271
	9.2	D.C. Circuit Invisibility Cloak 274
	9.3	Spatial Illusion-Optics Devices 281
		9.3.1	Shrinking Devices 281
		9.3.2	Material Conversion Devices 283
		9.3.3	Virtual Target Generation Devices 286

	9.4	Circuit Illusion-Optics Devices	291
	References		296
10	**Conclusions and Remarks**		**299**
	10.1	Summary of the Book	299
	10.2	New Trends of Metamaterials	300
		10.2.1 Planar Metamaterials: Metasurfaces	300
		10.2.2 Coding Metamaterials and Programmable Metamaterials	301
		10.2.3 Plasmonic Metamaterials	302
	References		303
Index			**305**

List of Figures

Figure 1.1	(a) The electric field, magnetic field, and propagation vector form a left-handed system	3
Figure 1.2	(a) The interface of negative-epsilon material (lower) and dielectric (upper)	5
Figure 1.3	Schematic representations of the specular reflection and Lambertian reflection	8
Figure 1.4	The structure of the book, showing the relationship of the nine chapters	12
Figure 2.1	A very simple model for the derivation of Lorentz–Drude dispersion relation	18
Figure 2.2	A typical Lorentz–Drude dispersion curve for natural materials	19
Figure 2.3	An SRR unit cell (a) and its equivalent circuit (b)	19
Figure 2.4	Lorentz–Drude-like dispersion curve for magnetic permeability of SRR	21
Figure 2.5	Schematic demonstration of the measurement system	22
Figure 2.6	Waveguide method used for the simulation of the metamaterial unit cell	24
Figure 2.7	Retrieved EM parameters for an SRR unit cell	24
Figure 2.8	Schematic diagram for the derivation of discrete Maxwell's equations	25
Figure 2.9	Comparison of the retrieved parameters and those effective ones	27
Figure 3.1	The periodic structure is composed of infinite wires arranged in a simple cubic lattice, joined at the corners of the lattice	31

xi

xii ■ List of Figures

Figure 3.2	Several kinds of electrically resonant particles	32
Figure 3.3	Effective permittivity of I-shaped resonator arrays	33
Figure 3.4	(a) The model of an SRR (side view)	35
Figure 3.5	(a)–(c) Structures serve as the magnetically resonant particles	36
Figure 3.6	Dielectric-metal resonant particles and their arrangement	37
Figure 3.7	Several metamaterial examples composed of dielectric-metal resonant particles	39
Figure 3.8	(a) The SRR and (b) CSRR	40
Figure 3.9	Sketch of the electric and magnetic field lines in (a) the SRR and (b) the CSRR	41
Figure 3.10	(a) Microstrip line with CSRRs etched on the back substrate side	42
Figure 3.11	(a) Dielectric sphere arrays in the air	43
Figure 3.12	An example of the distribution of effective relative permittivity and permeability for the array of lossless magnetodielectric spheres	45
Figure 3.13	(a) Top view of the ground-plane cloak. The blocks with the same number have the same parameters	46
Figure 3.14	Construction of a 3D flatten Luneburg lens fabricated with multilayered dielectric plates by drilling inhomogeneous holes	47
Figure 3.15	Typical Lorentz dispersion curves for electrically or magnetically resonant particles	49
Figure 3.16	Some example applications using nonresonant particles	50
Figure 3.17	(a) The LC unit cell for right-handed materials (RHMs)	53
Figure 3.18	The left-handed planar transmission line lens	54
Figure 3.19	(a) Equivalent circuit model for the ideal CRLH TL	55
Figure 3.20	The principle of D.C. invisibility cloak	58
Figure 3.21	Components of anisotropic conductivity tensor required for the D.C. invisibility cloak and their corresponding resistors	59
Figure 3.22	A conducting material plate, its equivalent resistor network, and the fabricated device	60

Figure 3.23	The negative-resistor model to realize negative conductivity and resistance	61
Figure 4.1	(a) Visualization of metamaterial formed by periodic and symmetric arrangement of EC-SRRs, where a_x, a_y, and a_z are the lattice constants	69
Figure 4.2	(a) Visualization of metamaterial formed by periodic and symmetric arrangement of SRRs, where a_x, a_y, and a_z are the lattice constants	70
Figure 4.3	(a) Visualization of metamaterial formed by periodic arrangement of dielectric disk resonators, where a_x, a_y, and a_z are the lattice constants and r and h are the radius and height of disk, respectively	71
Figure 4.4	(a) Visualization of simple wire media formed by 2D periodic arrangement of parallel and continuous conducting wires, where a_x and a_z are the lattice constants and r is the radius of wire	72
Figure 4.5	(a) Visualization of metamaterial formed by periodic arrangement of ELC resonators (ELCRs), where a_x, a_y, and a_z are the lattice constants	73
Figure 4.6	(a) Visualization of metamaterial formed by periodic arrangement of modified ELCRs, where a_x, a_y, and a_z are the lattice constants	74
Figure 4.7	(a) Visualization of metamaterial formed by periodic arrangement of MLRs, where a_x, a_y, and a_z are the lattice constants	75
Figure 4.8	(a) Visualization of WG metamaterial formed by pattering periodic array of CSRR in the lower plate of a planar waveguide, where a_x and a_y are the periodicity of the array and h is the height of waveguide	76
Figure 4.9	(a) Visualization of WG metamaterial formed by pattering periodic array of CELCR in the lower plate of a planar waveguide, where a_x and a_y are the periodicity of the array and h is the height of waveguide	77
Figure 4.10	(a) The C–C cascaded circuit representing MNG media	77
Figure 4.11	(a) Cubic lattice of DNG metamaterial formed by periodic arrangement of pairs of ELCR and EC-SRR	80

xiv ■ List of Figures

Figure 4.12 (a) Schematic of the planar fractal-meandering particle for DNG metamaterial development, where a_x and a_y are unit cell size in the xoy plane, b, l, w, d_1, d_2, and d_3 are the geometric parameters of the particle.. 82

Figure 4.13 (a) Two-dimensional TL grid loaded with series capacitance and shunt inductance, acting as the unit cell of 2D NRI TL metamaterial... 83

Figure 4.14 (a) Visualization of DPS metamaterial formed by periodic arrangement of I-shaped particles, where a_x, a_y, and a_z are the lattice constants 86

Figure 4.15 (a) Visualization of DPS metamaterial formed by periodic arrangement of Jerusalem cross (or crossed-I) particles 87

Figure 4.16 (a) Visualization of WG DPS metamaterial formed by patterning periodic array of EML in the lower plate of a planar waveguide, where a_x and a_y are the periodicity of the array and h is the height of waveguide 88

Figure 4.17 An MNG–ENG bilayer slab with a source located before the MNG layer.. 90

Figure 4.18 An ideal CC–LL bilayer structure, in which the voltage source represents an effective plane wave incident from free space to the CC circuit or effective MNG layer....................... 92

Figure 4.19 Comparison of transmission coefficients between circuit simulation and theoretical prediction for the ideal CC–LL bilayer structure... 93

Figure 4.20 Comparison of field distribution between circuit simulation and theoretical prediction for the ideal CC–LL bilayer structure... 94

Figure 4.21 Fabricated sample of the realistic CC–LL bilayer structure.. 95

Figure 4.22 Comparison of transmission coefficients between measurement and CST simulation for the realistic CC–LL bilayer structure... 95

Figure 4.23 Comparison of field distribution between measurement and CST simulation for the realistic CC–LL bilayer structure.. 96

Figure 4.24 Theoretical model for partial focusing in a planar waveguide...... 97

List of Figures ■ xv

Figure 4.25	Isofrequency curves for free space (circle) and the anisotropic MNG slab (hyperbolic sheets)	98
Figure 4.26	(a) Experimental setup for partial focusing utilizing 2D mapper	99
Figure 4.27	Retrieval results of effective medium parameters for the CELCR-based WG metamaterial	101
Figure 4.28	Comparison of 2D electric field distribution associated with the partial focusing between measurement and simulation	102
Figure 4.29	An open cavity formed by LHM in the first and third quadrants and RHM in the second and fourth quadrants	103
Figure 4.30	Realization of the four-quadrant LHM–RHM open cavity using NRI and PRI TL metamaterials	104
Figure 4.31	Simulated voltage (a) amplitude and (b) phase distributions in the open cavity, for which the voltage source resides at the node numbered as (24, 27).	106
Figure 4.32	Simulated voltage (a) amplitude and (b) phase distributions in the open cavity, for which the voltage source resides at the node numbered as (13, 38)	106
Figure 4.33	Simulated voltage (a) amplitude and (b) phase distributions in the open cavity, for which the voltage source resides at the node numbered as (10, 41)	107
Figure 4.34	Numerically simulated magnetic field distributions at 5.35 GHz for a super lens-like LHM slab with $\mu_y = -1.006 + i0.132$ and $\epsilon_x = \epsilon_y = -0.995 - i0.049$	109
Figure 4.35	Photograph of a fabricated 3D free-space LHM super lens composed of fractal-meandering particles	110
Figure 4.36	Schematic illustration of free-space focusing and near-field measurement system for the fabricated free-space LHM super lens	110
Figure 4.37	Measured magnetic field distributions at the rear side of the fabricated LHM slab lens at (a) 5.35 and (b) 5.4 GHz	111
Figure 4.38	Two planar waveguides separated by (a) an empty thin channel and (b) a thin channel loaded with CSRR-based WG metamaterial	113

xvi ■ *List of Figures*

Figure 4.39 Simplified two-port network model for the thin waveguide channel 114

Figure 4.40 Retrieved permittivity ϵ_z of the CSRR-based WG metamaterial used for verification of the EM tunneling effect ... 115

Figure 4.41 Simulated, theoretical, and experimental transmittances for both CSRR and control channels.................................. 116

Figure 4.42 Simulated Poynting vector distribution at 8.8 GHz for thin waveguide channel loaded with the CSRR-based WG metamaterial.. 117

Figure 4.43 Two-dimensional electric field distributions at 8.04 GHz measured in the 2D mapper for (a) CSRR and (b) control channels 118

Figure 4.44 Measured phase distributions along the wave-propagation direction (the x direction) for CSRR channel at 8.04 GHz and control channel at 7 GHz 119

Figure 4.45 Sketch of emission patterns of (a) a line source in free space and (b) a line source embedded in an AZIM slab 120

Figure 4.46 Fabricated AZIM lenses with (a) a slab shape and (b) a semicircular shape 121

Figure 4.47 Scanned near electric fields of (a) the slab AZIM lens and (b) the semicircular AZIM lens 122

Figure 4.48 (a) Photograph of a fabricated RAZIM ring 125

Figure 4.49 (a) Photograph of a Vivaldi antenna loaded with AZIM 126

Figure 4.50 A transmission polarizer made from ELCRs and illuminated with normally incident plane waves 129

Figure 4.51 Retrieved anisotropic medium parameters of the optimized ELCR particle from 5 to 15 GHz....................................... 130

Figure 4.52 Theoretical polarization patterns (dashed lines) and measured polarization patterns (plus signs) of the (a) incident waves 131

Figure 4.53 (a) Simulated (solid lines) and measured (dots) transmission of electric field from p1 to p1 (light color) and from p1 to p2 (deep color) for the linear-to-circular polarizer.................. 133

Figure 4.54 (a) Illustration of microstrip patch antenna loaded with EML-based WG magneto-dielectric material (WG-MDM) 135

List of Figures ■ xvii

Figure 4.55	Simulated input conductance (left) and susceptance (right) of both the WG-MDM antenna (the deep color line) and the control antenna (the light color line)	136
Figure 4.56	Simulated and measured reflection coefficients of the WG-MDM antenna (solid black line and dash gray line, respectively) and simulated reflection coefficient of the control antenna (solid gray line)	137
Figure 4.57	Simulated magnitude distribution of magnetic field intensity under the patch at 3.49 GHz for (a) the WG-MDM antenna and (b) the control antenna	138
Figure 4.58	Measured radiation patterns of the WG-MDM antenna in E-plane (left) and H-plane (right) at 3.494 GHz	139
Figure 4A.1	Photograph of the 2D field mapping planar waveguide chamber in its open position	140
Figure 5.1	Schematic plot of a planar slab of 1D randomly gradient index medium	148
Figure 5.2	(a) Surface view of a fabricated sample of 1D randomly gradient index metamaterial composed of I-shaped particles	149
Figure 5.3	Visualization of random surface based on three-layer stacked square patches of variable size	151
Figure 5.4	Reflecting element characterizing three-layer stacked square patches of variable size	151
Figure 5.5	(a) Visualization of a random surface sample composed of reflecting elements based on three-layer stacked patches	153
Figure 5.6	An infinite s-polarized plane wave incident on a 1D randomly gradient index coating backed with a PEC plate	154
Figure 5.7	Comparison of diffuse reflection performances between 1D random coatings matched and unmatched to free space	155
Figure 5.8	Probability density functions of local-beam deflection angles for different random distribution modes	159
Figure 5.9	(a) Visualization of a metamaterial sample of 1D random coating composed of crossed-I particles	161
Figure 5.10	A sketch of experimental setup for measuring the near-field surrounding the coating sample and the metal sheet with a 45° incident wave	162

Figure 5.11	Measured scattered near-field distributions at 10.5 GHz for (a) naked metal sheet and (b) coated metal sheet	163
Figure 5.12	Comparison of measured backscattering patterns between naked and coated metal sheets at 10.5 GHz	164
Figure 5.13	Experimental setup for measuring the reflectance of random surface under normal incidence	165
Figure 5.14	Experimental results of normalized reflectance of random surface with respect to both x- and y-polarized incident waves	166
Figure 6.1	Design of the gradient-index focusing lens which consists of inhomogeneous metamaterials	171
Figure 6.2	Experimental observation of fish with a geometrical-optics-based cloak	173
Figure 6.3	Illustration of ground-plane cloak	174
Figure 6.4	The principle of transformation optics	177
Figure 6.5	The principle of the invisibility cloak	179
Figure 6.6	Near-field distributions of (a) elliptical–cylindrical cloak	180
Figure 6.7	(a) Visualization of the space transformation for concentrator	181
Figure 6.8	(a) Real part of E-field distribution	182
Figure 6.9	(a) Configuration of cylindrical-to-plane-wave conversion	183
Figure 6.10	(a) Photograph of the fabricated broadband transformation optics lens	185
Figure 6.11	The schematic illustration of illusion device	186
Figure 6.12	The scattering field of (a) a metallic square object with an illusion device in real space	187
Figure 7.1	Flowchart for the design, analysis, and realization of the GRIN devices	192
Figure 7.2	Refractive index discretization	193
Figure 7.3	Parallel capacitor model	194
Figure 7.4	Effective refractive index of an I-shaped unit cell structure	196
Figure 7.5	A planar dielectric waveguide	196
Figure 7.6	Schematic diagram of the lenses	197

Figure 7.7	Simulated electric field distributions for the deflection lens (a) and focusing lens (b)	199
Figure 7.8	Geometry of the unit cell (a) and the effective refractive index variation along with the radius of the hole (b)	200
Figure 7.9	Dispersion of the hole-array material	200
Figure 7.10	Simulated electric field distributions for the discrete lens structure	201
Figure 7.11	Measured electric field distributions for the fabricated lens structure at 10 GHz	202
Figure 7.12	Refractive index profile for 2D Luneburg lens	203
Figure 7.13	Ray trajectories of the 2D Luneburg lens	203
Figure 7.14	Simulated electric field distribution of the 2D Luneburg lens	204
Figure 7.15	Effective EM parameters of the I-shaped unit cell	205
Figure 7.16	Near-field distributions of electric fields for the metamaterial Luneburg-like lens antenna	206
Figure 7.17	Far-field pattern for the Luneburg lens at different frequencies	207
Figure 7.18	Ray tracing performance of the HMFE lens	208
Figure 7.19	Full-wave simulation results for HMFE lens antenna	209
Figure 7.20	Variation of retrieved EM parameters with the geometrical size (a), dispersion of the I-shaped unit cell (b), and the cell structure (c)	209
Figure 7.21	Picture of the fabricated HMFE lens antenna	210
Figure 7.22	Near-field distribution of electric fields for the metamaterial HMFE lens antenna	211
Figure 7.23	Normalized far-field radiation patterns for HMFE lens antenna at 12 GHz	212
Figure 7.24	Schematic illustration of the proposed lens, where the cross section along its rotational axis is given	213
Figure 7.25	Refractive index needed for the two lenses	214
Figure 7.26	Near-field distributions inside and outside the GRIN lens B and horn	215

xx ■ List of Figures

Figure 7.27	Far-field pattern of lens antenna B at 10 GHz	216
Figure 7.28	Structure of the closed square ring	216
Figure 7.29	Effective parameters of the CSR structure in terms of the geometrical size	217
Figure 7.30	Measured E plane far-field directivity of the antenna with lens B at 8.2, 9.8, 10.6, and 12.2 GHz	218
Figure 7.31	Ray tracing for the 3D half Luneburg lens	219
Figure 7.32	Simulated far-field radiation patterns for different incident angles	220
Figure 7.33	Unit cells and effective medium parameters for the 3D half Luneburg lens	221
Figure 7.34	Measured far-field pattern for the 3D half Luneburg lens	222
Figure 7.35	Ray tracing performance for the 3D HMFE lens	224
Figure 7.36	Simulated and measured far-field radiation patterns of the half fisheye lens antenna under HPP polarization at 15 GHz	225
Figure 7.37	Dielectric permittivity distribution for the black hole	227
Figure 7.38	Ray tracing for the "black hole" with the dielectric profile given in Equation 7.23	227
Figure 7.39	Simulated electric field distributions when a Gaussian beam is incident on the device on-center (a) and off-center (b)	228
Figure 7.40	Effective material parameters for the I-shaped unit cell in the shell (a) and the ELC unit cell in the center (b)	229
Figure 7.41	Distributions of electric fields $|E_z|$ for the designed omnidirectional absorbing device at the frequency of 18 GHz	230
Figure 8.1	Design of a ground-plane cloak	234
Figure 8.2	2D section view of (a) the physical space and (b) the virtual space	235
Figure 8.3	(a) Metamaterial refractive index distribution	241
Figure 8.4	(a) Photograph of the fabricated metamaterial sample	242
Figure 8.5	Measured field mapping (E field) of the ground, perturbation, and ground-plane cloaked perturbation	242

Figure 8.6	Design for a compact-sized ground-plane cloak in free-space background	243
Figure 8.7	EM wave propagates in the spatial domain at a fixed time	244
Figure 8.8	(a) Dielectric 2D map of a ground-plane cloak composed of 4 × 2 blocks	246
Figure 8.9	(a) Photograph of the fabricated metamaterial ground-plane cloak and the concealed car model	247
Figure 8.10	Measured electric field mapping of (a) the ground plane, (b) triangular metallic bump, and (c) ground-plane cloaked bump when collimated beam is incident at 10 GHz	248
Figure 8.11	Quasi-conformal mappings between (a) the virtual space, (b) the intermediate space, and (c) the physical space	249
Figure 8.12	The electric field distribution under a Gaussian beam incidence, in which the incident angle is $\pi/4$, the waist width is two wavelengths, and the waist center is located at the origin	250
Figure 8.13	Geometry of the parabolic surface	252
Figure 8.14	(a) The virtual space with distorted coordinates	253
Figure 8.15	(a) Relative permittivity map consisting of 64 × 16 blocks	254
Figure 8.16	Real part of the E_z field at 8 GHz	255
Figure 8.17	(a) Photograph of the fabricated planar antenna. (b) Experimental setup for the near-field measurement	256
Figure 8.18	Measured results for the parabolic reflector and planar reflector at different frequencies	257
Figure 8.19	3D microwave ground-plane cloak and its refractive index distribution	259
Figure 8.20	Simulated and measured electric fields in the far region	260
Figure 8.21	The simulated and measured electric fields in the far region	261
Figure 8.22	Ray tracing for a spherical Luneburg lens	263
Figure 8.23	Quasi-conformal mapping of 2D flattened Luneburg lens and generation of 3D flattened Luneburg lens	265

xxii ■ List of Figures

Figure 8.24	Measurement results of near-electric fields outside the 3D lens	266
Figure 8.25	Measured far-field results of the 3D lens	267
Figure 9.1	Photograph of the first free-space invisibility cloak in microwave band	273
Figure 9.2	Simulated and experimented near-field distributions of the first microwave invisibility cloak	274
Figure 9.3	(a) Illustration of a D.C. invisibility cloak	276
Figure 9.4	Simulated and experimented electrical potential distributions of the ultrathin D.C. invisibility cloak	278
Figure 9.5	(a) The negative-resistor model to realize negative conductivity and resistance	279
Figure 9.6	(a) The simulated voltage and current distributions when the designed D.C. exterior cloak is placed close to the objects	280
Figure 9.7	Photograph of the first free-space illusion device, a shrinking device, in microwave band	283
Figure 9.8	Near-field distribution of the shrinking device	284
Figure 9.9	Experimental setup and photograph of the material conversion device	286
Figure 9.10	Simulation and experiment results of the material conversion device	287
Figure 9.11	Schematic of the virtual target generation device	288
Figure 9.12	(a) Experimental setup of the virtual target generation device	290
Figure 9.13	(a) Schematic of the EM invisible gateway	292
Figure 9.14	Measured node voltage distribution of the circuit illusion device	293
Figure 9.15	Design scheme of the localized transformation optics device	294
Figure 9.16	(a) The measured equipotential lines when the metallic cylinder is coated by the localized D.C. cloak	294
Figure 9.17	Illustration of a D.C. illusion device	295

List of Tables

Table 4.1 Comparison of Radiation Performance between the WG-MDM Antenna and the Control Antenna 138

Table 5.1 Statistical Results of Diffusion Degree for 1D Random Coatings with Different Coating Lengths or Amount of Subregions 157

Table 5.2 Statistical Results of Diffusion Degrees for Unmatched 1D Random Coatings with Different Distribution Modes of Index Gradients 160

Preface

Since the realization of artificial wire medium with negative permittivity in 1996, metamaterials have been well developed. Different from the earlier periodic structures (e.g., frequency-selective surfaces and photonic bandgap structures), metamaterials have two unique features: First, metamaterials are composed of subwavelength-scale particles (meta-atoms), which enable them to be equivalent to effective media; second, nonperiodicity (or inhomogeneity) plays the most important role in metamaterials in controlling electromagnetic or light waves. Based on the effective medium theory, metamaterials have good mapping to natural materials. According to different spatial arrangements of atoms, natural materials are classified into three types: crystals (with periodic arrangement of atoms), noncrystals (with random arrangement of atoms), and quasicrystals (with nonperiodic but ordered arrangement of atoms). Similarly, metamaterials are also classified into three types: homogeneous metamaterials, random metamaterials, and inhomogeneous metamaterials based on the spatial arrangements of meta-atoms, which can be regarded as super crystals, super noncrystals, and super quasicrystals because one could create countless meta-atoms and make arbitrary arrangements of meta-atoms. Such unique characteristics result in extreme-value effective parameters (e.g., negative or very large permittivity and permeability, and zero index of refraction) and tailored distribution of inhomogeneous medium parameters, which cannot be realized by natural materials.

The powerful abilities of metamaterials in controlling effective medium parameters make it possible to manipulate electromagnetic waves or lights at will, and hence metamaterials receive intensive attention in both science and engineering communities. Based on metamaterials, many new concepts (e.g., transformation optics and zero refractive index) have been proposed, and a large number of exciting physical phenomena (e.g., negative refraction, invisibility cloaking, electromagnetic black hole, and radar illusion) have been experimentally verified. The topics on metamaterials were twice selected as one of the "Ten Breakthroughs of Science" by *Science Magazine* in the year 2002 (for negative refraction, by using homogeneous metamaterial with negative values of permittivity and permeability) and 2006 (for invisibility cloak, by using anisotropic and inhomogeneous metamaterials governed by transformation optics). In 2010, metamaterials was again listed in the

"Ten Breakthroughs of Science" in the first 10 years of the twenty-first century by *Science Magazine*. Beside the area of basic science, metamaterials have also found wide applications in engineering. In microwave frequencies, wideband and low-loss metamaterials have been fabricated and a series of practical devices have been realized for wireless communications, satellite communications, radar, and imaging.

Metamaterials is a fast-developing subject, and there have been a great number of new discoveries, results, information, and knowledge in this exciting area. Recently, a number of new versions of metamaterials have emerged, such as metasurfaces, coding metamaterials, programmable metamaterials, and planar plasmonic metamaterials. Also, optical metamaterials have attracted much more attention due to their high impacts and challenges in fabrications. Hence, it is impossible to include all excellent advances on metamaterials in a single book. This book focuses only on recent advances of bulk metamaterials based on the effective medium theory in microwave frequencies, most of which were conducted in the State Key Laboratory of Millimeter Waves, Southeast University, China. The book is divided into two parts: fundamentals and applications of metamaterials. The fundamentals section includes the effective medium theory for artificial metamaterial structures and the designs of metamaterial particles (or meta-atoms). Based on two such fundamentals, three kinds of metamaterials—homogeneous metamaterials (super crystals), random metamaterials (super noncrystals), and inhomogeneous metamaterials (super quasicrystals)—are investigated in detail, in which new physical phenomena and practical applications are presented. We hope that this book will be helpful to scientists, engineers, and graduate/undergraduate students in physics and electrical engineering. We also hope that the book will promote the applications of metamaterials.

We would like to thank Ruijun He (editor) and Joselyn Banks, Delroy Lowe, and Laurie Oknowsky (project coordinators at CRC Press and Taylor & Francis Group) for their effort and support in the publication of this book.

Tie Jun Cui
Southeast University, Nanjing, China

Wen Xuan Tang
Southeast University, Nanjing, China

Xin Mi Yang
Soochow University, Suzhou, China

Zhong Lei Mei
Lanzhou University, Lanzhou, China

Wei Xiang Jiang
Southeast University, Nanjing, China

Authors

Tie Jun Cui earned his BSc, MSc, and PhD degrees in electrical engineering from Xidian University, Xi'an, China, in 1987, 1990, and 1993, respectively. In March 1993, he joined the Department of Electromagnetic Engineering, Xidian University, and was promoted to the post of associate professor in November 1993. From 1995 to 1997, he was a research fellow with the Institut fur Hochstfrequenztechnik und Elektronik (IHE) at the University of Karlsruhe, Germany. In July 1997, he joined the Center for Computational Electromagnetics, Department of Electrical and Computer Engineering, at the University of Illinois at Urbana-Champaign, first as a postdoctoral research associate and then as a research scientist. In September 2001, he became a Cheung-Kong Professor with the Department of Radio Engineering, Southeast University, Nanjing, China. Currently, he is the associate dean of the School of Information Science and Engineering, and the associate director of the State Key Laboratory of Millimeter Waves. Since 2013, he has been a representative of People's Congress of China.

Dr. Cui is the coeditor of the book *Metamaterials: Theory, Design, and Applications* (Springer, November 2009) and the author of six book chapters. He has published over 300 peer-reviewed journal articles in *Science, PNAS, Nature Communications, Physical Review Letters, IEEE Transactions*, etc. His research interests include metamaterials, computational electromagnetic, wireless power transfer, and millimeter wave technologies. His research publications have been cited more than 10,000 times. According to Elsevier, he is one of the most-cited Chinese researchers. Dr. Cui was awarded a research fellowship from the Alexander von Humboldt Foundation, Bonn, Germany, in 1995; received the Young Scientist Award from the International Union of Radio Science (URSI) in 1999; was awarded the Cheung Kong Professorship under the Cheung Kong Scholar Program by the Ministry of Education, China, in 2001; received the National Science Foundation of China for Distinguished Young Scholars award in 2002; received a Special Government Allowance from the Department of State, China, in 2008; received the Award of Science and Technology Progress from Shaanxi Province Government in 2009; was awarded a May 1st Labour Medal by Jiangsu Province Government in 2010; received the First Prize of Natural Science from Ministry of Education, China, in 2011; and received the Second Prize of National Natural Science, China, in 2014.

Dr. Cui's research have been selected as one of the 10 Breakthroughs of China Science in 2010, Best of 2010 in *New Journal of Physics*, Research Highlights in *Europhysics News*, *Journal of Physics D: Applied Physics*, *Applied Physics Letters*, and *Nature China*. His work has been reported by *Nature News*, *Science*, *MIT Technology Review*, *Scientific American*, *New Scientist*, etc.

Dr. Cui is an IEEE fellow and an active reviewer for *Science*, *Nature Materials*, *Nature Photonics*, *Nature Physics*, *Nature Communications*, *Physical Review Letters*, *Advanced Materials*, and a series of *IEEE Transactions*. He was an associate editor of *IEEE Transactions on Geoscience and Remote Sensing* and a guest editor of *Science China: Information Sciences*. He served on the editorial staff in *IEEE Antennas and Propagation Magazine*, and is on the editorial boards of *Progress in Electromagnetic Research* (PIER) and *Journal of Electromagnetic Waves and Applications*. He served as general cochair of the International Workshops on Metamaterials (META'2008, META'2012), TPC cochair of Asian Pacific Microwave Conference (APMC'2005), and TPC cochair Progress in Electromagnetic Research Symposium (PIERS'2004).

Wen Xuan Tang earned her BSc degree in electronic engineering and her MSc degree in electromagnetic field and microwave technology from Southeast University, Nanjing, China, in 2006 and 2009, respectively, and her PhD in electromagnetics from Queen Mary University of London, London, United Kingdom in 2012. In November 2012, she joined the School of Information Science and Engineering, Southeast University, Nanjing, China as a lecturer.

Her main research interests include metamaterials and their applications, transformation electromagnetics, and microwave devices and antennas. She has published over 20 technical articles in highly ranked journals, including *IEEE Transactions on Antenna and Propagation*, *New Journal of Physics*, *Optics Express*, *Applied Physics Letters*, *Scientific Reports*, etc.

Xin Mi Yang was born in Suzhou, Jiangsu Province, China in March 1982. He earned his BSc and PhD degrees from Southeast University, Nanjing, China in 2005 and 2010, respectively, both in the School of Information Science and Engineering. Since November 2010, he has been with the School of Electronics and Information Engineering, Soochow University, Suzhou, China. His current research interests include metamaterials, metasurfaces, LTCC technology, and their applications in antennas and microwave engineering.

Zhong Lei Mei was born in Luoyang, Henan Province, China in 1974. He is currently a professor in the School of Information Science and Engineering, Lanzhou University. He is also the deputy dean of the school. He earned his BSc, MSc, and PhD degrees in radio physics from Lanzhou University, China in 1996, 1999, and 2007, respectively. In 2004, Dr. Mei attended a training course on "Teaching Science in English" at Sydney University, Australia. In February 2008, he joined the State Key Laboratory of Millimeter Waves, Southeast University, Nanjing as a postdoctoral research associate. He is a visiting research fellow in the State

Key Laboratory of Millimeter Waves. Dr. Mei's current research interest includes metamaterials and computational electromagnetics. He has published over 30 peer-reviewed journal articles in international journals, including *Physical Review Letters, IEEE Transactions on Antenna and Propagation, New Journal of Physics, Optics Express, Applied Physics Letters*, etc.

Wei Xiang Jiang was born in Jiangsu Province, China, in October 1981. He earned his PhD in electrical engineering from Southeast University, Nanjing, China in October 2010. He joined the State Key Laboratory of Millimeter Waves, Southeast University in November 2010, and was promoted to the post of associate professor in April 2011, and professor in April 2015. He has published more than 60 peer-reviewed journal articles in *Advanced Materials, Advanced Functional Materials, Materials Today, Applied Physics Letters*, etc. His current research interests include electromagnetic theory, illusion optics, and metamaterials. Dr. Jiang's research has been selected for Research Highlights by *Europhysics News* in June 2008, Research Highlights in 2008 by the *Journal of Physics D: Applied Physics*, and Research Highlights by *Applied Physics Letters* in 2011. Dr. Jiang received the First Prize of Natural Science from the Ministry of Education, China in 2011 (Rank 2); the National Excellent Doctoral Dissertation in 2013; the Young Scientist Award from the International Union of Radio Science (URSI) in 2014; and the Second Prize in National Natural Science, China in 2014 (Rank 3). He served as organization committee cochair of the International Workshop on Metamaterials (META'2012, Nanjing). He is an active reviewer for *Nature Communications, Scientific Reports, Journal of Physics D: Applied Physics, Journal of Optics, Chinese Physics Letters, Journal of Electromagnetic Waves and Applications, Progress in Electromagnetics Research* (PIER), etc.

Chapter 1

Introduction

1.1 Natural Materials and Metamaterials

All natural materials are composed of atoms. According to the different arrangements of atoms, natural materials are basically classified into two types: crystals (with periodic arrangement of atoms) and noncrystals (with random arrangement of atoms). In 1985, Daniel Shechtman, Nobel Prize winner in chemistry, discovered an intermediate state between crystals and noncrystals—quasicrystal, which is composed of nonperiodic atoms with certain orders [1], showing excellent performance beyond that of natural materials.

Unlike natural materials, metamaterials are macroscopic composites of periodic or nonperiodic subwavelength-scale artificial particles with certain shapes, materials, and geometries, which are usually called as meta-atoms. The main advantages of metamaterials are as follows. (1) Meta-atoms can be engineered and tailored as required, and hence have countless choices. However, the natural atoms are very limited. (2) Meta-atoms can be arbitrarily arranged due to their flexibilities. But it is difficult to arrange natural atoms to specific orders. Therefore, we can realize homogeneous metamaterials (super crystals), inhomogeneous metamaterials (super quasicrystals), and random metamaterials.

Usually, meta-atoms are electric and/or magnetic resonators. Depending on the resonant status of meta-atoms, metamaterials are classified into resonant metamaterials and nonresonant metamaterials. For the resonant metamaterials, the meta-atoms are in the resonant state with significant changes of electromagnetic responses, which will produce extreme values of effective permittivity and/or permeability (e.g., negative permittivity, negative permeability, and zero index of refraction). Around the resonant frequency, large dynamic range of effective medium parameters can be achieved, from negative to largely positive. Such parameters will result in unusual physics, such as negative refraction [2], perfect imaging [3], and tunneling effect [4].

Due to the resonance nature, however, resonant metamaterials are always accompanied by large loss and narrow bandwidth. For nonresonant metamaterials, the meta-atoms are away from the resonant state, yielding gradient changes of the electromagnetic responses. This feature produces small loss and wideband medium parameters, which makes metamaterials practical in engineering applications. The drawback of nonresonant metamaterials is the small dynamic range of medium parameters. However, the free arrangement of meta-atoms makes it possible to realize arbitrarily inhomogeneous media.

Only when the meta-atoms have strictly circular (or spherical) symmetries, are the resulting two-dimensional (2D) or three-dimensional (3D) metamaterials isotropic. However, most practical meta-atoms do not have such symmetries, and hence most metamaterials are anisotropic. The extreme values of effective medium parameters, high inhomogeneity, and strong anisotropy enable metamaterials to have powerful abilities to control the medium parameters, and further manipulate electromagnetic waves.

1.2 Homogeneous Metamaterials: Several Special Cases

Due to the availability of extreme medium parameters, homogeneous metamaterials can be used to realize many unusual physical phenomena. Here, we provide four simple examples to show the negative refraction, plane-wave radiation, transverse-magnetic (TM) mode of surface plasmon polaritons (SPPs), and transverse-electric (TE) mode of SPPs using double-negative, zero-refractive-index, and single-negative materials.

1.2.1 Left-Handed Materials

Proposed by Veselago in 1965, left-handed material contains negative permittivity and negative permeability simultaneously [2], which is called as double-negative material. From Maxwell's equations,

$$\nabla \times \mathbf{E} = -\partial \mathbf{B}/\partial t, \quad \nabla \times \mathbf{H} = \partial \mathbf{D}/\partial t, \quad (1.1)$$

in which $\mathbf{D} = \bar{\bar{\epsilon}} \cdot \mathbf{E}$ and $\mathbf{B} = \bar{\bar{\mu}} \cdot \mathbf{H}$ are electric displacement and magnetic flux, $\mathbf{E}$ and $\mathbf{H}$ are electric and magnetic fields, and $\bar{\bar{\epsilon}}$ and $\bar{\bar{\mu}}$ are permittivity and permeability tensors, respectively. Under the time-harmonic $\exp(-i\omega t)$, when plane waves are propagating in the left-handed material, Maxwell's equations will be reduced as

$$\mathbf{k} \times \mathbf{E} = \mu\omega\mathbf{H}, \quad \mathbf{k} \times \mathbf{H} = -\epsilon\omega\mathbf{E}, \quad (1.2)$$

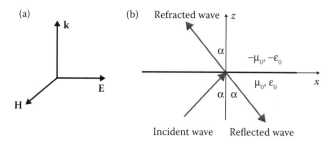

Figure 1.1 (a) The electric field, magnetic field, and propagation vector form a left-handed system. (b) The reflection and refraction features of electromagnetic wave on the interface of right-handed and left-handed materials, in which negative refraction is observed.

in which **k** is the propagating vector, and $\epsilon < 0$ and $\mu < 0$ are scalar permittivity and permeability of the left-handed material, respectively. From Equation 1.2, we clearly note that **E**, **H**, and **k** satisfy a left-handed relation when $\epsilon < 0$ and $\mu < 0$, as shown in Figure 1.1a, which is completely different from the right-handed system in the conventional materials. This is also the reason why Veselago called it as left-handed material.

In the left-handed material, the wave impedance is given by $\eta = \sqrt{|\mu|/|\epsilon|}$, while the index of refraction should be $n = -\sqrt{|\mu\epsilon|/\mu_0\epsilon_0}$ to make the electromagnetic waves inside physical [5], in which ϵ_0 and μ_0 are permittivity and permeability of free space, respectively. Because of this nature, on the interface of free space and left-handed material, there will be negative refraction, as shown in Figure 1.1b, which can be easily derived by using the conventional Snell's law. Unlike the conventional refraction, the incident and refracted waves are in the same side of the interface normal. A special left-handed material slab with $\epsilon = -\epsilon_0$ and $\mu = -\mu_0$ in free space has been proved as a perfect imaging lens [3], breaking the diffraction limit.

1.2.2 Zero-Refractive-Index Metamaterials

Metamaterials with zero index of refraction can be used to enhance the directivity of an antenna [6], since the refracted rays radiated from the antenna in a zero-refractive-index metamaterial are always parallel to the normal direction of the metamaterial surface. Zero-index metamaterials (ZIMs) could be realized by either zero permittivity or zero permeability, or both zero permittivity and zero permeability. In the latter situation, ZIM also has the ability to tunnel electromagnetic waves through a narrow channel [4]. Usually, ZIM with both zero permittivity and zero permeability is narrowband to reach the electric and magnetic resonances simultaneously, while ZIM with either zero permittivity or zero permeability encounters severe impedance mismatch.

4 ■ *Metamaterials*

To solve the impedance mismatch problem, anisotropic zero-index metamaterial (AZIM) was proposed, in which one component of the permittivity or permeability tensor is zero while other components are properly designed to make the impedance matching with free space [7,8]. As an example, we consider a 2D AZIM, in which $\mu_y = 0$ and $\mu_x = \epsilon_z \neq 0$. When an infinitely long electric current filament along the z direction (2D point source) is radiated in free space, cylindrical waves with TE polarization are generated. In AZIM, however, the radiation of 2D point source will be governed by the dispersion relation for the TE polarization:

$$\frac{k_x^2}{\mu_y} + \frac{k_y^2}{\mu_x} = k^2 \epsilon_z, \tag{1.3}$$

in which k is the wavenumber in free space. Since $\mu_y = 0$, we must enforce $k_x = 0$ to guarantee the wave propagation in AZIM physical. As a consequence, the wavenumber in the y direction has the form of $k_y = k\sqrt{\epsilon_z \mu_x}$. Based on Maxwell's equation, we easily have

$$\frac{\partial E_z}{\partial x} = -i\omega\mu_y H_y = 0. \tag{1.4}$$

Therefore, E_z is a constant with respect to the x direction. The above analysis indicates that the 2D point source will radiate a perfect plane wave propagating along the y direction with uniform field distribution along the x direction in AZIM with $\mu_y = 0$, which is significantly different from that in free space. This property enables an AZIM slab with arbitrary shape to enhance the directivity and gain of an antenna, as presented in detail in Chapter 4.

1.2.3 Negative-Epsilon Materials

Negative-epsilon materials contain negative permittivity and positive permeability. They cannot support propagating waves and, therefore, fields inside them are attenuated. One of the most important features of negative-epsilon materials is the guidance of SPPs on their interfaces with dielectrics [9]. Although SPPs are beyond the scope of this book, we show how SPPs emerge on the surface of this special material.

Consider a half space problem, in which the upper space is dielectric (or air) and the lower space is the negative-epsilon material, as illustrated in Figure 1.2a. In the 2D coordinate system with the TM polarization, there exist only three nonzero field components: E_x, E_z, and H_y, which have no variations along the y direction ($\partial/\partial y = 0$). In this case, all field components have the following form: $F(x, y, z) = F(z)e^{i\beta x}$, and Maxwell's equations are simplified as [9]

$$E_x = -\frac{i}{\omega\epsilon_0\epsilon}\frac{\partial H_y}{\partial z}, \quad E_z = -\frac{\beta}{\omega\epsilon_0\epsilon}H_y, \tag{1.5}$$

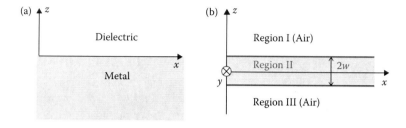

Figure 1.2 (a) The interface of negative-epsilon material (lower) and dielectric (upper). (b) A thin slab of anisotropic negative-mu material embedded in the air.

in which β is the propagation constant along the x direction. Since the propagating waves cannot be supported by the negative-epsilon material, the fields along the z direction must be decayed. Then the fields in the upper space ($z > 0$) are written as

$$H_y(z) = A_2\, e^{i\beta x}\, e^{-k_2 z} \qquad (1.6)$$

$$E_x(z) = iA_2 \frac{1}{\omega \epsilon_0 \epsilon_2} k_2\, e^{i\beta x}\, e^{-k_2 z}$$

$$E_z(z) = -A_1 \frac{\beta}{\omega \epsilon_0 \epsilon_2}\, e^{i\beta x}\, e^{-k_2 z}.$$

In the lower space, we have

$$H_y(z) = A_1\, e^{i\beta x}\, e^{-k_1 z} \qquad (1.7)$$

$$E_x(z) = -iA_1 \frac{1}{\omega \epsilon_0 \epsilon_1} k_1\, e^{i\beta x}\, e^{-k_1 z}$$

$$E_z(z) = -A_1 \frac{\beta}{\omega \epsilon_0 \epsilon_1}\, e^{i\beta x}\, e^{k_1 z}.$$

Here, k_1 and k_2 are attenuation factors in upper and lower spaces along the z direction, respectively, which must be positive to make the fields physical. Using the boundary condition that tangential electric and magnetic fields are continuous on the interface, we further get the relations $A_1 = A_2$ and

$$\frac{k_1}{k_2} = -\frac{\epsilon_1}{\epsilon_2}. \qquad (1.8)$$

The above requirement can only be satisfied when the permittivity in upper and lower spaces have opposite signs since k_1 and k_2 must be positive. That is to say, on the interface of negative-epsilon material (e.g., metal in the visible regime) and dielectric, the waves are propagating along the interface with the fields decayed exponentially on the normal direction, which are the TM-mode SPPs.

1.2.4 Negative-Mu Materials

Negative-mu materials have negative permeability and positive permittivity, which do not exist in nature but can be realized by artificial structures (e.g., split-ring resonators [SRRs]). Based on the duality principle of Maxwell's equations, TE-mode SPPs could be generated and then propagated along the interface of two materials with positive and negative magnetic permeability. Here, we illustrate that an anisotropic negative-mu material slab can also guide TE-mode SPPs [10].

As shown in Figure 1.2b, a 2D metamaterial slab with width $2w$ is residing in the air. The anisotropic metamaterial slab is modeled as a homogeneous medium described by diagonal permittivity and permeability tensors $\bar{\bar{\epsilon}} = \text{diag}(\epsilon_x \, \epsilon_y \, \epsilon_z)$ and $\bar{\bar{\mu}} = \text{diag}(\mu_x \, \mu_y \, \mu_z)$. Under the assumption that the metamaterial slab is very thin ($2w \ll \lambda$) and the surface modes are propagating along the x direction with fields exponentially decaying along the z direction, we derive the electric fields in the air (Regions I and III) and metamaterial slab (Region II) as

$$E_{1y} = A_1 \, e^{-k_1 z} \, e^{i\beta_1 x} \quad (1.9)$$
$$E_{2y} = (A_2 \, e^{-k_2 z} + B_2 \, e^{-k_2 z}) \, e^{i\beta_2 x}$$
$$E_{3y} = B_3 \, e^{k_3 z} \, e^{i\beta_3 x},$$

in which k_1, k_2, and k_3 are attenuation constants along $+z$ and $-z$ directions in Regions I, II, and III; and β_1, β_2, and $\beta 3$ are propagation constants along $+x$ direction in Regions I, II, and III, respectively; all of which are positive numbers. A_1, A_2, B_2, and B_3 are constants. The phase-matching and boundary conditions of E_y and H_x on the interfaces of air and metamaterial require that $k_3 = k_1$; $\beta_1 = \beta_2 = \beta_3 = \beta$; while $A_1 = B_3$ and $A_2 = B_2$ for even modes; and $A_1 = -B_3$ and $A_2 = -B_2$ for odd modes. We further obtain the dispersion equations as

$$\beta^2 - k_1^2 = k^2, \quad (1.10)$$

$$\mu_x \beta^2 = \mu_z k_2^2 + \mu_x \mu_z \epsilon_y k^2, \quad (1.11)$$

$$\tanh(k_2 w) = -\frac{\mu_x k_1}{k_2} \quad \text{(for even modes)}, \quad (1.12)$$

$$\tanh(k_2 w) = -\frac{k_2}{k_1 \mu_x} \quad \text{(for odd modes)},$$

where k is the wavenumber in free space, β is the propagation constant of the surface modes, and k_1 and k_2 are attenuation constants in air and metamaterial, respectively.

From Equation 1.12, we notice that μ_x must be negative in order to support the propagation of TE-mode SPPs.

The electric fields of the TE-mode SPPs propagating in the negative-mu material slab can be then written as

$$E_{2y} = 2A_2 \cosh(k_2 z)e^{i\beta x} \quad \text{(for even modes),} \quad (1.13)$$

$$E_{2y} = -2A_2 \sinh(k_2 z)e^{i\beta x} \quad \text{(for odd modes),}$$

which are spatially even and odd functions along the z direction at fixed x considering the fact that z is small. Hence the SPP modes can propagate along the metamaterial slab ($\mu_x < 0$) with the TE polarization, in which the fields are highly confined in both z and y directions.

1.3 Random Metamaterials

In nature materials, atoms are distributed either periodically to produce crystals, or randomly to generate noncrystals. Similarly, meta-atoms can also be packed randomly to create random metamaterials or metasurfaces, which provide an important route toward controlling the radiations, reflections, and scattering of the electromagnetic waves. One of the most important applications of random metamaterial or metasurface is to generate electromagnetic diffusions (or Lambertian reflections) [11–13], in which the geometrical parameters of meta-atoms are varied in space so as to result in spatially random distributions of the effective medium parameters (e.g., electric permittivity, magnetic permeability, index of refraction, and wave impedance) [11].

If metamaterial or metasurface is large enough, the ideally random distribution of meta-atoms with different geometrical parameters will certainly generate diffusions. In most cases, however, the metamaterial does not have sufficiently large area to allow real randomness. Then some special rules should be invoked in designing the spatial distributions of meta-atoms [12–16], so that they attribute to destructive interferences of reflected or scattered waves from each other, resulting in the diffusion phenomena. The optimization methods and coding approaches have been presented to reach this goal in both microwave and terahertz frequencies [12–16].

In the above designs, the phases of meta-atoms are enforced to be distributed randomly. Then the reflected waves from the meta-atoms will experience destructive inference due to the random phase difference, leading to Lambertian reflections or diffusions of scattered waves. Usually, a large phase shift of available meta-atoms is required to redistribute the scattered energy to all possible directions. As an example, Figure 1.3 illustrates a comparison of specular reflection by a perfectly conducting plane (a) and Lambertian reflection by a random metasurface (b), in which a terahertz wave is incident at an oblique angle of 30°. From the numerical simulation

8 ■ *Metamaterials*

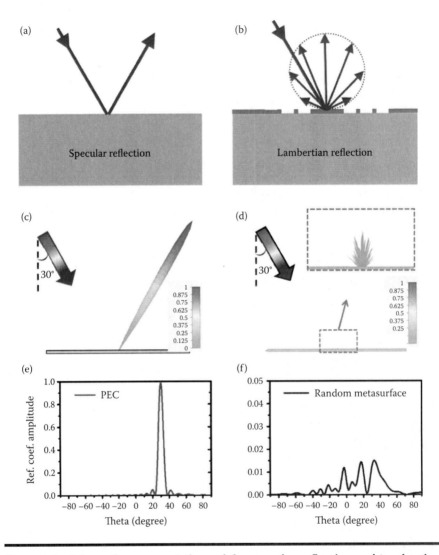

Figure 1.3 Schematic representations of the specular reflection and Lambertian reflection. (a) The specular reflection on a conducting surface at 30° incidence angle. (b) The Lambertian reflection on a random metasurface. (c, d) The far-field scattering patterns of the conducting plane and random metasurface. (e, f) The reflected power distributions of the conducting plane and random metasurface.

results of far-field scattering patterns shown in Figure 1.3c and d, we note that the incident wave is totally reflected to the specular direction by the conducting plane (consistent with the Snell's law), while the fields incident upon the random metasurface are redirected to random directions with the intensity outline performing like Lambertian reflection. Such phenomena are quantitatively observed from the reflected power distributions demonstrated in Figure 1.3e and f.

Therefore, random metamaterials or metasurfaces can be used to reduce the radar cross sections of metallic targets in broad frequency bands in both microwave and terahertz regimes [11–17], which will be discussed in detail in Chapter 5. They have also helped realize high-performance imaging using single sensor with the aid of signal processing methods [18].

1.4 Inhomogeneous Metamaterials

Inhomogeneous metamaterials indicate artificial structures with neither periodic nor random distributions of meta-atoms. Instead, the meta-atoms are distributed with certain rules and can be described by effective inhomogeneous media. Due to the flexibility in designing the meta-atoms and their spatial arrangements, inhomogeneous metamaterials can be regarded as super quasicrystals, which have more capabilities to control the electromagnetic waves, resulting in many novel physical phenomena and engineering applications.

The inhomogeneous metamaterials must be well designed to make them useful. There are a number of designing methods relevant to physical principles: geometrical optics (GO) method to design isotropic inhomogeneous metamaterials, quasi-conformal mapping method to design nearly isotropic inhomogeneous metamaterials, and transformation optics to design anisotropic inhomogeneous metamaterials [19].

1.4.1 GO Method

The traditional GO method plays an important role in designing inhomogeneous metamaterials. As examples, planar lenses with gradient refractive indexes have been designed [20,21] using the GO and ray-tracing method, which could transform the cylindrical waves of line source or spherical waves of point source to planar waves to achieve high-gain and high-directivity antennas. In the 2D case, a high-performance beam-scanning antenna was experimentally demonstrated using the gradient-index planar lens and horn antenna [20], in which metamaterials are carefully designed to obtain different refractive indices while keeping good impedance match. 3D planar lens was presented using inhomogeneous metamaterials composed of squarering arrays [21], showing much better performance than the traditional antenna with broad bandwidth, small return loss, dual polarization, and high directivity. Recently, a very large 3D planar lens has been designed using the GO method [22], which has a diameter of 100 cm and a thickness of 15 cm, operating in the X band (8–12 GHz). To build up this antenna, nearly 1 million inhomogeneous meta-atoms (square rings and drilling-hole dielectrics) have been used.

The GO method has also been used to design curved lenses, such as Luneburg and Maxwell's fisheye lenses [23–25], which have gradient-index distributions in the spherical region. Such lenses can be easily realized using inhomogeneous meta-atoms. For example, 2D Luneburg lens and half Maxwell's fisheye lens were fabricated using inhomogeneous nonresonant I-shaped structures [23,24]; 3D Luneburg lens and Maxwell'afs fisheye lens were realized by multiple stacked dielectric boards by drilling inhomogeneous air holes [25]. Such metamaterial lenses have good performance of high gain, broad bandwidth, and low return loss due to better impedance matching design.

One of the most attractive designs of metamaterials using the GO method is the electromagnetic black hole [26–28], which can simulate the trapping and absorbing behaviors of real black hole because of the similar propagation properties of optical or electromagnetic waves in inhomogeneous metamaterials. The electromagnetic black hole was first demonstrated experimentally in the microwave frequencies using the inhomogeneous I-shaped structures [27], and later was presented in the optical regime [28]. It was shown that the artificial black holes could trap and absorb electromagnetic waves from all directions spirally inward without any reflections.

1.4.2 Quasi-Conformal Mapping Method

Quasi-conformal mapping is a simplified version of transformation optics, which can be used to design more complicated inhomogeneous metamaterial devices than the GO method. Proposed by Li and Pendry [29], the quasi-conformal mapping was first applied in designing ground-plane cloaks (or carpet cloaks). The invisible cloaks are dedicated to the unique feature of inhomogeneous metamaterials in discovering novel phenomena that had not been found earlier. The first realization of ground-plane cloak was made in the microwave frequencies in 2D version, which can hide objects on conducting ground in a broad frequency band with very good cloaking performance [30]. To reduce the size of ground-plane cloak, a simplified design was presented to achieve equivalent cloaking performance in experiments [31], resulting in a size reduction by 6 times. Around the same time, 2D ground-plane cloaks in the optical frequencies have been fabricated independently in two groups [32,33]. To generate practical metamaterial devices, the quasi-conformal mapping has been extended to 3D, case approximately. In 2010, the first 3D ground-plane cloaks were demonstrated in both microwave [34] and optical frequencies [35], independently. The 3D optical cloak is the extruding extension of the 2D version, which is valid in certain incident angles and vertical polarization [35], while the 3D microwave cloak is the revolution of the 2D version, which is valid in all incident angles and both vertical and horizontal polarizations [34]. Excellent cloaking performance was observed in experiments in both 3D cloaks.

The quasi-conformal mapping method has also been applied to achieve new kinds of metamaterial lenses. Based on the quasi-conformal mapping, a 2D planar focusing lens made of gradient-index metamaterials was presented [36], achieving the similar performance to the traditional parabolic antenna in a planar manner.

To realize novel lenses beyond the tradition, the quasi-conformal mapping was used to design flattened Luneburg lenses with planar focal surfaces in the microwave frequency in both 2D [37] and 3D [38] cases, which have great advantages over the conventional lenses with no aberration, zero focal distance, and the ability to form images at extremely large angles. In the 3D case, the flattened Luneburg lens was fabricated by multilayered dielectric plates with inhomogeneous air holes, which has excellent performance with high gain and large scanning angles for different polarizations in broadband [38], and hence can be directly used in engineering to radiate or receive narrow electromagnetic waves.

1.4.3 Transformation Optics

The GO method is good at designing isotropic inhomogeneous metamaterials, and the quasi-conformal mapping is suitable for weakly anisotropic metamaterials. In order to design highly anisotropic inhomogeneous metamaterials, transformation optics has to be used. The transformation optics is a powerful tool in generating new physics and novel engineering devices [39,40], from which the exciting "invisibility cloak" has been realized in both microwave frequency [41] and electrostatics [42]. Besides making objects invisible, the transformation optics was also applied in designing the illusion-optics devices with the aid of anisotropic and inhomogeneous metamaterials, which make an object look exactly like another with different shapes and material makeup [43] (e.g., generating multiple virtual objects [44], moving object virtually from one place to another place [45], and changing metal object to virtual dielectric object [46]).

Recently, several important experiments have been conducted in the microwave frequencies to illustrate the illusion-optics phenomena using the inhomogeneous and anisotropic metamaterials made of SRR structures [47–49], including a shrinking device that can make a large metal or dielectric object with arbitrary shape smaller [47], a material-convention device that can transform a metal object to dielectric [48], and a camouflage device that can create ghost illusions of a real object [49]. In the electrostatic case, a light-controlled transformation direct current (D.C.) illusion device was presented [50], in which the anisotropic inhomogeneous conducting materials composed of light-sensitive semiconductor resistors are dynamically controlled by tuning the intensity of illumination light. Then the functionalities of the transformation-D.C. device can be switchable from a D.C. invisibility cloak to D.C. illusion devices. Such experiments verified the strong abilities of anisotropic inhomogeneous metamaterials to control the electromagnetic waves.

Inhomogeneous metamaterial has been emphasized in this book due to its special role, which will be discussed in detail in Chapters 6 through 9 on the cases of isotropic, nearly isotropic (or weakly anisotropic), and highly anisotropic, respectively.

1.5 Structure of This Book

In this book, we will concentrate on metamaterials described by the effective medium parameters. The book comprises nine chapters, introducing homogeneous, random, and inhomogeneous metamaterials. The structure of the book is as follows:

- Chapter 1 is an overall introduction to metamaterials and their recent advances.
- Chapter 2 describes the effective medium theory of metamaterials, which is the basis of the whole book.
- Chapter 3 gives different types of unit cells of metamaterials—the meta-atoms (or man-made atoms), which are the building blocks to construct metamaterials.
- Chapter 4 discusses in detail homogeneous metamaterials, which can be regarded as super crystals, and their special characteristics in physics.
- Chapter 5 focuses on random metamaterials, revealing their principles, designs, realizations, and applications.
- Chapter 6 presents the overall discussion on inhomogeneous metamaterials, introduces the typical methods to design inhomogeneous metamaterials, and gives several examples.
- Chapter 7 investigates the inhomogeneous metamaterials designed by GO method—the gradient-refractive-index metamaterials—in detail, and their engineering applications.
- Chapter 8 studies the nearly isotropic inhomogeneous metamaterials designed by the quasi-conformal mapping method and their engineering applications.

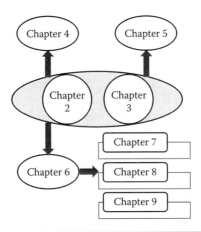

Figure 1.4 The structure of the book, showing the relationship of the nine chapters.

- Chapter 9 provides a detailed discussion on the anisotropic inhomogeneous metamaterials designed by the transformation optics.

For a clear understanding, Figure 1.4 illustrates the flowchart of the chapter organization.

Acknowledgments

This book is authored by Tie Jun Cui, Wen Xuan Tang, Xin Mi Yang, Zhong Lei Mei, and Wei Xiang Jiang. All authors have contributed equally. TJC designed the structure of the book, including all chapters, sections, and subsections, wrote Chapters 1 and 10, and supervised the writing of other chapters. WXT wrote Chapters 3 and 8; XMY wrote Chapters 4 and 5; ZLM wrote Chapters 2 and 7; and WXJ wrote Chapters 6 and 9.

TJC acknowledges support from the National Science Foundation of China (61171024, 61138001, 61571117, 61501112, and 61501117), and the 111 Project (111-2-05); WXT acknowledges the supports from the National Science Foundation of China (61302018 and 61401089); XMY acknowledges support from the National Science Foundation of China (61301076), the Natural Science Foundation of Jiangsu Province (BK20130326), the Open Research Program of State Key Laboratory of Millimeter Waves in China (K201417), the Natural Science Foundation of the Higher Education Institutions of Jiangsu Province (12KJB510030), and Suzhou Key Laboratory for Radio and Microwave/Millimeter Wave (SZS201110); ZLM acknowledges support from the Open Research Funds of State Key Laboratory of Millimeter Waves (K201409), and the Fundamental Research Funds for the Central Universities (LZUJBKY-2015-k07, LZUJBKY-2014-43, and LZUJBKY-2014-237); and WXJ acknowledges support from the National Science Foundation of China (61171026 and 61522106), and the Foundation of National Excellent Doctoral Dissertation of China.

References

1. L. Swartzendruber, D. Shechtman, L. Bendersky, and J. Cahn. Nuclear γ-ray resonance observations in an aluminum-based icosahedral quasicrystal. *Phys. Rev. B*, 32: 1383, 1985.
2. V. G. Veselago. The electrodynamics of substances with simultaneously negative values of ϵ and μ. *Sov. Phys. Usp.*, 10: 509, 1968.
3. J. B. Pendry. Negative refraction makes a perfect lens. *Phys. Rev. Lett.*, 85: 3966, 2000.
4. R. Liu, Q. Cheng, T. Hand, J. J. Mock, T. J. Cui, S. A. Cummer, and D. R. Smith. Experimental demonstration of electromagnetic tunneling through an epsilon near zero metamaterial at microwave frequencies. *Phys. Rev. Lett.*, 100: 023903, 2008.

5. T. J. Cui, Z. C. Hao, X. X. Yin, W. Hong, and J. A. Kong. Study of lossy effects on the propagation of propagating and evanescent waves in left-handed materials. *Phys. Lett. A*, 323: 484, 2004.
6. S. Enoch, G. Tayeb, P. Sabouroux, N. Guerin, and P. Vincent. A metamaterial for directive emission. *Phys. Rev. Lett.*, 89: 213902, 2002.
7. Y. G. Ma, P. Wang, X. Chen, and C. K. Ong. Near-field plane-wave-like beam emitting antenna fabricated by anisotropic metamaterial. *Appl. Phys. Letts.*, 94: 044107, 2009.
8. Q. Cheng, W. X. Jiang, and T. J. Cui. Radiation of planar electromagnetic waves by a line source in anisotropic metamaterials. *J. Phys. D: Appl. Phys.*, 43: 335406, 2010.
9. S. A. Maier. *Plasmonics: Fundamentals and Applications*. Springer, New York, 2007.
10. X. Gu. Manipulation of Surface Plasmon by Metamaterials, Chapter 3. B.S. Dissertation, Southeast University, China, 2011.
11. X. M. Yang, X. Y. Zhou, Q. Cheng, H. F. Ma, and T. J. Cui. Diffuse reflections by randomly gradient index metamaterials. *Opt. Lett.*, 35: 808, 2010.
12. K. Wang, J. Zhao, Q. Cheng, D. Dong, and T. J. Cui. Broadband and broad-angle low-scattering metasurface based on hybrid optimization algorithm. *Sci. Rep.*, 4: 5935, 2014.
13. D. S. Dong, J. Yang, Q. Cheng, J. Zhao, L. H. Gao, S. J. Ma, S. Liu et al. Terahertz broadband low-reflection metasurface by controlling phase distributions. *Adv. Opt. Mater.*, 3(10): 1405, 2015.
14. T. J. Cui, M. Q. Qi, X. Wan, J. Zhao, and Q. Cheng. Coding metamaterials, digital metamaterials, and programmable metamaterials. *Light Sci. Appl.*, 3: e218, 2014.
15. L. Gao, Q. Cheng, J. Yang, S. Ma, J. Zhao, S. Liu, H. Chen et al. Broadband diffusion of terahertz waves by multi-bit coding metasurfaces. *Light Sci. Appl.*, 4: e324, 2015.
16. L. Liang, M. Qi, J. Yang et al. Anomalous terahertz reflection and scattering by flexible and conformal coding metamaterial. *Adv. Opt. Mater.*, 3(10): 1373, 2015.
17. X. Shen and T. J. Cui. THz broadband antireflection coating based on diffusion metasurface. Unpublished manuscript, 2015.
18. J. Hunt, T. Driscoll, A. Mrozack, G. Lipworth, M. Reynolds, D. Brady, and D. R. Smith. Metamaterial apertures for computational imaging. *Science*, 339: 310, 2013.
19. T. J. Cui. Metamaterials—Beyond crystals, noncrystals, and quasicrystals: Microwave applications. In *Proceedings of the Sixth International Congress on Advanced Electromagnetic Materials in Microwaves and Optics*, St. Petersburg, Russia, September 17–22, 2012.
20. H. F. Ma, X. Chen, H. S. Xu, X. M. Yang, W. X. Jiang, and T. J. Cui. Experiments on high-performance beam-scanning antennas made of gradient- index metamaterials. *Appl. Phys. Lett.*, 95: 094107, 2009.
21. X. Chen, H. F. Ma, X. Y. Zou, W. X. Jiang, and T. J. Cui. Three-dimensional broadband and high-directivity lens antenna made of metamaterials. *J. Appl. Phys.*, 110: 044904, 2011.
22. X. Y. Zhou, X. Y. Zou, Y. Yang, H. F. Ma, and T. J. Cui. Three-dimensional large-aperture lens antennas with gradient refractive index. *Sci. China Inform. Sci.*, 56: 120410, 2013.
23. H. F. Ma, X. Chen, X. M. Yang, H. S. Xu, Q. Cheng, and T. J. Cui. A broadband metamaterial cylindrical lens antenna. *Chinese Sci. Bull.*, 55: 2066, 2010.
24. Z. L. Mei, J. Bai, T. M. Niu, and T. J. Cui. A half Maxwell fish-eye lens antenna based on gradient-index metamaterials. *IEEE Trans. Antenn. Propag.*, 60: 398, 2012.

25. H. F. Ma, B. G. Cai, T. X. Zhang, Y. Yang, W. X. Jiang, and T. J. Cui. Three-dimensional inhomogeneous microwave metamaterials and their applications in lens antennas. *IEEE Trans. Antenn. Propag.*, 61: 2560, 2013.
26. E. E. Narimanov and A. V. Kildishev. Optical black hole: Broadband omnidirectional light absorber. *Appl. Phys. Lett.*, 95: 041106, 2009.
27. Q. Cheng, T. J. Cui, W. X. Jiang, and B. G. Cai. An omnidirectional electromagnetic absorber made of metamaterials. *New J. Phys.*, 12: 063006, 2010.
28. C. Sheng, H. Liu, Y. Wang, S. N. Zhu, and D. A. Genov. Trapping light by mimicking gravitational lensing. *Nat. Photon.*, 7: 902, 2013.
29. J. Li and J. B. Pendry. Hiding under the carpet: A new strategy for cloaking. *Phys. Rev. Lett.*, 101: 203901, 2008.
30. R. Liu, C. Ji, J. J. Mock, J. Y. Chin, T. J. Cui, and D. R. Smith. Experimental verification of a broadband ground-plane cloak. *Science*, 323: 366, 2009.
31. H. F. Ma, W. X. Jiang, X. M. Yang, X. Y. Zhou, and T. J. Cui. Compact-sized and broadband carpet cloak and free-space cloak. *Opt. Express*, 17: 19947, 2009.
32. J. Valentine, J. Li, T. Zentgraf, G. Bartal, and X. Zhang. An optical cloak made of dielectrics. *Nat. Mater.*, 8: 568, 2009.
33. L. H. Gabrielli, J. Cardenas, C. B. Poitras, and M. Lipson. Silicon nanostructure cloak operating at optical frequencies. *Nat. Photon.*, 3: 461, 2009.
34. H. F. Ma and T. J. Cui. Three-dimensional broadband ground-plane cloak made of dielectrics. *Nat. Commun.*, 1: 21, 2010.
35. T. Ergin, N. Stenger, P. Brenner, J. B. Pendry, and M. Wegener. Three-dimensional invisibility cloak at optical wavelengths. *Science*, 328: 337, 2010.
36. Z. L. Mei, J. Bai, and T. J. Cui. Experimental verification of a broadband planar focusing antenna based on transformation optics. *New J. Phys.*, 13: 063028, 2011.
37. N. Kundtz and D. R. Smith. Extreme-angle broadband metamaterial lens. *Nat. Mater.*, 9: 129, 2010.
38. H. F. Ma and T. J. Cui. Three-dimensional broadband and broad-angle transformation-optics lens. *Nat. Commun.*, 1: 24, 2010.
39. J. B. Pendry, D. Schurig, and D. R. Smith. Controlling electromagnetic fields. *Science*, 312: 1780, 2006.
40. U. Leonhardt. Optical conformal mapping. *Science*, 312: 1777, 2006.
41. D. Schurig, J. J. Mock, B. J. Justice, S. A. Cummer, J. B. Pendry, A. F. Starr, and D. R. Smith. Metamaterial electromagnetic cloak at microwave frequencies. *Science*, 314: 977, 2006.
42. F. Yang, Z. L. Mei, T. Y. Jin, and T. J. Cui. D. C. electric invisibility cloak. *Phys. Rev. Lett.*, 109: 053902, 2012.
43. Y. Lai, J. Ng, H. Y. Chen, D. Z. Han, J. J. Xiao, Z. Q. Zhang, and C. T. Chan. Illusion optics: The optical transformation of an object into another object. *Phys. Rev. Lett.*, 102: 253902, 2009.
44. W. X. Jiang, H. F. Ma, Q. Cheng, and T. J. Cui. Illusion media: Generating virtual objects using realizable metamaterials. *Appl. Phys. Lett.*, 96: 121910, 2010.
45. W. X. Jiang and T. J. Cui. Moving targets virtually via composite optical transformation. *Opt. Express*, 18: 5161, 2010.
46. W. X. Jiang, H. F. Ma, Q. Cheng, and T. J. Cui. Virtual conversion from metal to dielectric objects using metamaterials. *Opt. Express*, 18: 11276, 2010.
47. W. X. Jiang, T. J. Cui, H. F. Ma, X. M. Yang, and Q. Cheng. Shrinking an arbitrarily-shaped object as desired using metamaterials. *Appl. Phys. Lett.*, 98: 204101, 2011.

48. W. X. Jiang and T. J. Cui. Radar illusion via metamaterials. *Phys. Rev. E*, 83: 026601, 2011.
49. W. X. Jiang, C. W. Qiu, T. Han, S. Zhang, and T. J. Cui. Creation of ghost illusions using wave dynamics in metamaterials. *Adv. Funct. Mater.*, 23: 4028, 2013.
50. W. X. Jiang, C. Y. Luo, S. Ge, C. W. Qiu, and T. J. Cui. Optically controllable transformation-dc illusion device. *Adv. Mater.*, 27(31): 4628, 2015.

Chapter 2
Effective Medium Theory

In some natural materials, macroscopic electromagnetic (EM) parameters can be theoretically derived from atomic or molecular structures, which usually involves tedious work. As has been pointed out in Chapter 1, metamaterials use man-made atoms, that is, meta-atoms, to mimic real materials for the flexible control of EM waves. Then it is quite natural that we can borrow the same idea for the calculation. Sometimes, rigorous analysis may be impossible; however, the classical model (e.g., Lorentz model) can be used for numerical optimizations. Moreover, a deep understanding of natural materials helps to improve the metamaterial design and creates new controllable materials. In this chapter, we provide a few theories frequently used in metamaterials and introduce the detailed parameter-retrieval process. It hence lays the theoretical foundation for the later chapters.

2.1 Lorentz–Drude Models

Natural materials consist of large quantities of fundamental particles, atoms or molecules. Let us look at one of these particles, for example, an atom, very carefully. In a very coarse way, we can treat such a particle as a harmonic oscillator, as illustrated in Figure 2.1. Here, the spiral represents the restoring force exerted on the charge when it leaves the position of balance. When an external electric field is turned on, the charge will be forced to vibrate near the balanced position. Using Newton's second law of motion, we have

$$m\ddot{r} = qE - \omega_0^2 mr - dm\dot{r}, \qquad (2.1)$$

where the three terms on the right-hand side of the equation represent the electric force, the restoring force, and the friction force, respectively, and m is the mass of the

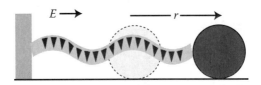

Figure 2.1 A very simple model for the derivation of Lorentz–Drude dispersion relation.

charge. Since the external field is time harmonic, it is convenient to use the phasor form, and the above equation becomes

$$-m\omega^2 r = qE - \omega_0^2 m r - i\omega d m r. \quad (2.2)$$

Hence, we have, at the stable situation

$$r = \frac{q}{m} \frac{E}{-\omega^2 + \omega_0^2 + i\omega d}. \quad (2.3)$$

For this particular case, since the positive and negative particle is separated by a distance r, the corresponding electric dipole must be

$$p = \frac{q^2}{m} \frac{E}{-\omega^2 + \omega_0^2 + i\omega d}. \quad (2.4)$$

Considering that the particle number density is N, the magnitude of the polarization vector, which means the induced electric dipoles in a unit volume, is then written as

$$P = Np = \frac{Nq^2}{m} \frac{E}{-\omega^2 + \omega_0^2 + i\omega d}. \quad (2.5)$$

When the external field is not too strong, we know that **P**, in the first-order approximation, should be proportional to **E**, that is, $\mathbf{P} = \epsilon_0 \chi_e \mathbf{E}$, where χ_e is the electric susceptibility. As a consequence, we have

$$\chi_e = \frac{P}{\epsilon_0 E} = \frac{Nq^2}{\epsilon_0 m} \frac{1}{-\omega^2 + \omega_0^2 + i\omega d}. \quad (2.6)$$

Then, it is easy to see that the relative permittivity of the material should be

$$\epsilon_r = 1 + \chi_e = 1 + \frac{\omega_p^2}{\omega_0^2 - \omega^2 + i\omega d}, \quad (2.7)$$

Effective Medium Theory ■ 19

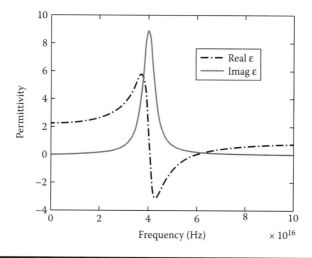

Figure 2.2 A typical Lorentz–Drude dispersion curve for natural materials.

where $\omega_p = \sqrt{Nq^2/\epsilon_0 m}$ is the plasma frequency. This is the well-known Lorentz–Drude dispersion model for ordinary dielectric materials. In Figure 2.2, we show a very typical Lorentz–Drude dispersion curve for material parameters chosen by Brillouin [1]. Similar model holds for the magnetic permeability.

Like natural materials, metamaterials also consist of a large number of man-made particles, that is, electric or magnetic resonators, or unit cells, which are usually arranged in a 3D and periodic way. Let us take the most well-known unit cell, that is, SRR as an example. Figure 2.3a shows the structure of SRR, and Figure 2.3b shows its equivalent circuit, which helps understanding its magnetic resonance nature.

Suppose that the external magnetic fields **H** penetrate the ring surface; then it is easy to get the induced electromotive force (emf) using Faraday's law, which is

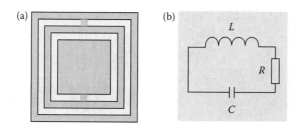

Figure 2.3 An SRR unit cell (a) and its equivalent circuit (b).

emf $= -i\omega\mu_0 HS$. Then, using the Kirchoff's voltage law, we have

$$-i\omega\mu_0 HS = I(R + i\omega L + 1/(i\omega C)). \quad (2.8)$$

So,

$$I = \frac{-i\omega\mu_0 HS}{R + i\omega L + 1/(i\omega C)}. \quad (2.9)$$

Then, the magnetic moment for this unit cell is obtained,

$$m = IS = \frac{-i\omega\mu_0 HS^2}{R + i\omega L + 1/(i\omega C)}. \quad (2.10)$$

If we take the SRR array as a continuous medium, which is magnetized under the external field, then the magnetization (magnetic moment per volume) is

$$M = Nm = \frac{-i\omega\mu_0 NHS^2}{R + i\omega L + 1/(i\omega C)}, \quad (2.11)$$

where N is the number density. Now, it is easy to get the relative magnetic permeability as

$$\mu_r = 1 + \chi_m = 1 + M/H = 1 - \frac{i\omega\mu_0 NS^2}{R + i\omega L + 1/(i\omega C)}. \quad (2.12)$$

Or, in another way

$$\mu_r = 1 + \frac{(\mu_0 NS^2/L)\omega^2}{1/(LC) - \omega^2 + i\omega R/L}. \quad (2.13)$$

Then it is obvious that the magnetic permittivity also has the Lorentz–Drude form. In his seminal paper on metamaterials, Pendry et al. analyzed various unit cell structures, one of which is SRR. And the analytical result for the effective permeability is [2]

$$\mu_{\mathit{eff}} = 1 - \frac{\pi r^2/a^2}{1 + 2l\sigma_1/\omega r\mu_0 i - 3lc_0^2/\pi\omega^2 \ln 2c/dr^3}. \quad (2.14)$$

Then, it is shown that metamaterials using SRR arrays have a Lorentz–Drude-like dispersion properties. Figure 2.4 shows its typical magnetic response. More work on electric or magnetic resonators also confirms the observations [3,4].

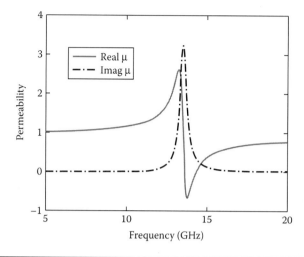

Figure 2.4 Lorentz–Drude-like dispersion curve for magnetic permeability of SRR.

2.2 Retrieval Methods of Effective Medium Parameters

As has been discussed in the previous section, using the analogy between natural atoms and man-made atoms, it is possible to study the EM responses of metamaterial unit cells, and then get the macroscopic parameters of the corresponding materials, whose building blocks are those man-made units. Actually, the method has been used in the chemical sectors along with molecular dynamics to study the EM property of a certain material. And in the physical sector, there are reports on various analytical formulas for the determination of material parameters based on the unit cell response. However, considering the complexity of the unit cell structure and practical realization of the material, novel engineering-oriented method should be developed along with the more theoretical one. In this section, we will show how to retrieve effective medium parameters from the engineer's point of view.

Compared with the more theory-involved method, this retrieval method is rather simple and versatile, and it treats the metamaterial as a black box. Hence, no matter what kind of unit cells are utilized, the method works well. The equivalent material parameters are obtained by measuring the S scattering parameters from the following system, which is schematically illustrated in Figure 2.5.

In this figure, the transmission horn antenna is connected to Port One of the vector network analyzer (VNA), and the receiving antenna is connected to Port Two. The metamaterial slab with thickness d is placed between the two antennas. The core task is to measure the reflection and transmission coefficients on two surfaces of the metamaterial slab. To fulfill the task, two more steps are needed to "calibrate"

22 ■ Metamaterials

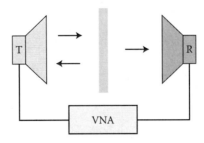

Figure 2.5 Schematic demonstration of the measurement system.

the network analyzer (in addition to the analyzer's own calibration) in order to get the correct phases (the magnitude of the measured data are correct).

Step 1: The "cut-through" measurement. The slab is removed and the transmission coefficient, $S_{12}(s)$, is obtained.

Step 2: The "short" measurement. The slab is replaced by a perfectly electrical conducting (PEC) plate and the reflection coefficient, $S_{11}(s)$, is obtained.

Step 3: The "DUT" measurement. The metamaterial slab is put at the same position, and the reflection and transmission coefficients are measured, that is, $S_{11}(m)$ and $S_{12}(m)$. Note that phases of the two coefficients need adjustments while their magnitudes do not.

Based on the above three steps, phases of the reflection and transmission coefficients on the two surfaces of the slab can be determined, which read as

$$\angle r = \angle S_{11}(m) - \angle S_{11}(s) + \pi, \tag{2.15}$$

$$\angle t = \angle S_{12}(m) - \angle S_{12}(c) - kd, \tag{2.16}$$

and $|r| = |S_{11}(m)|$, $|t| = |S_{12}(m)|$.

Now, we get the "real" reflection and transmission coefficients through the measurements, that is, r and t. As we have mentioned, the metamaterial slab is considered to be in a black box. This suggests that one knows nothing about the box except for the measured S parameters. As a result, we may think an equivalent bulk material, which is homogeneous, isotropic, and may be magnetic, is put inside the box. No one can tell the differences between these two cases if they produce the same S parameters. And for this "reference" slab, suppose its EM parameters are n, z, ϵ, and μ, respectively, we have the following equations [6]:

$$t^{-1} = \left[\cos(nkd) - \frac{i}{2}\left(z + \frac{1}{z}\right)\sin(nkd)\right]e^{ikd}, \tag{2.17}$$

$$\frac{r}{t'} = -\frac{1}{2}i\left(z - \frac{1}{z}\right)\sin(nkd), \tag{2.18}$$

where $t' = \exp(ikd)t$ and k is the wavenumber for the incident waves.

Note that in the above equations, the time factor exp($-i\omega t$) is implicitly used, which is different from most VNA measured results. Special attentions should be paid when one uses measured data to retrieve effective parameters. Through some not too difficult deductions, we get the following results:

$$z = \pm\sqrt{\frac{(1+r)^2 - t'^2}{(1-r)^2 - t'^2}}, \tag{2.19}$$

$$\text{Im}(n) = \pm\text{Im}\left(\frac{\cos^{-1}(1/2t'[1-(r^2-t'^2)])}{kd}\right), \tag{2.20}$$

$$\text{Re}(n) = \pm\text{Re}\left(\frac{\cos^{-1}(1/2t'[1-(r^2-t'^2)])}{kd}\right) + \frac{2\pi m}{kd}, \tag{2.21}$$

where m is an integer. Note that the expressions for n and z are complex functions with multiple branches, the interpretation of which can lead to ambiguities in determining the final effective parameters. We can resolve these ambiguities by making use of additional knowledge about the material. If the material is passive, Re(z) should be greater than zero. Likewise, Im(n) > 0 leads to an unambiguous result for Im(n). The most difficult part will be the determination of Re(n), since it is complicated by the ranches of the arccosine function. To make things easier, the smallest possible thickness of sample should be adopted in the measurement, as has commonly been known in the analysis of continuous materials. Even with a small sample, more than one thickness must be measured to identify the correct branches of the solution, which yields consistently the same values for n.

When the impedance z and refractive index n is determined, electric permittivity and magnetic permeability can be easily obtained by the following equations:

$$\epsilon = n/z, \quad \mu = nz. \tag{2.22}$$

Though we can get the effective EM parameters through the measurement of S parameters, as mentioned above, another way to get the effective parameters, which is often used in theoretical analysis, is to use full-wave simulations. In this case, the so-called "waveguide" method is often utilized. The method is depicted in Figure 2.6. In the simulation, an air box is used to cover one unit cell structure, where PEC and perfectly magnetic conducting (PMC) boundary conditions are set for four surfaces facing each other, and the other two faces are set to wave ports. This configuration will produce a uniform plane wave propagating from one port to the other inside the box, where it impinges the cell, reflected, and transmitted. The S parameters can be easily obtained through numerical calculations.

Figure 2.7 shows the effective EM parameters for an SRR unit cell, which is obtained using the above-mentioned method. Note that even for this typical magnetic resonator, whose equivalent permeability shows a Lorentz–Drude response,

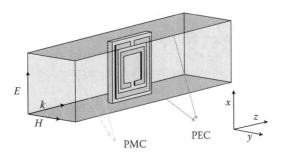

Figure 2.6 Waveguide method used for the simulation of the metamaterial unit cell.

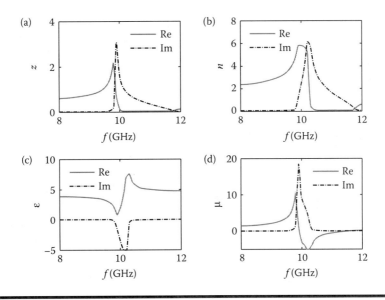

Figure 2.7 Retrieved EM parameters for an SRR unit cell. (a) Impedance; (b) refractive index; (c) permittivity; (d) permeability.

the equivalent permittivity does change with the frequency too. And this is called the antiresonance phenomenon.

2.3 General Effective Medium Theory

As has been described in the previous sections, most metamaterials are composed of electric and/or magnetic resonant particles, for example, SRR [2] or the electric-field-coupled resonator (ELC) [3,4]. The resonances lead to responses to EM fields characterized by dispersive permittivity or permeability. As has been discussed in the previous section, the dispersion is very similar to a Lorentz–Drude model, which can be obtained by local field averaging technique [5]. Meanwhile, the standard retrieval technique [6] provides an easier approach to retrieve the effective permittivity and permeability from simulated reflection and transmission coefficients. It is expected that these two results should agree with each other. However, careful examinations between them show that they do not coincide exactly. Take SRR as an example. Since SRR is a magnetic resonator, its magnetic permeability shows a Lorentz–Drude-like dispersion, while the permittivity is expected to be constant due to the lack of electric resonance at the same band. However, the retrieved parameters show that it is not the case (see Figure 2.7).

From Figure 2.7, it is very clear when one of the effective parameters is resonant, the other one will have antiresonance spontaneously, no matter whether in the retrieval or the theoretical result. This is called the antiresonant phenomenon. Then, it seems that a more accurate theory is needed for the clear explanation and deep understanding of the phenomenon.

As will be shown in this sector, the discrepancy is because metamaterial particles are not infinitely small compared to the wavelength, which generates the effect of spatial dispersion. However, the distortion between the two sets of constitutive parameters can be bridged by the general effective medium theory [7]. In our nomenclature, $\bar{\epsilon}$ and $\bar{\mu}$ are termed as the average permittivity and permeability while ϵ_{eff} and μ_{eff} are termed as the effective permittivity and permeability, respectively.

Let us begin our analysis here from the discrete form of Maxwell's equations. To this end, Yee's grids can be borrowed, which are frequently used in finite difference time domain (FDTD) numerical techniques [8]. The periodicities in the x, y, and z directions are equal to p. Suppose that an incident EM wave with transverse electric and magnetic fields propagates along the z direction, as shown in Figure 2.8. The electric field is x polarized, and the magnetic field is y polarized.

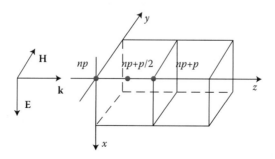

Figure 2.8 Schematic diagram for the derivation of discrete Maxwell's equations.

Periodic boundary conditions are assumed in the x and y directions. Applying Faraday's law of induction to an arbitrary cell in the x–z plane, we obtain for the electric field,

$$\int_0^p E(x, 0, np + p/2)dx - \int_0^p E(x, 0, np - p/2)dx = i\omega \iint B(x, 0, np)ds \quad (2.23)$$

In the field averaging method, we define the following average field:

$$\bar{E}_x(z) = \frac{1}{p} \int_0^p E(x, 0, z)dx, \quad (2.24)$$

$$\bar{B}_y(z) = \frac{1}{p^2} \iint B(0, y, z)ds. \quad (2.25)$$

Then,

$$\bar{E}_x(np + p/2) - \bar{E}_x(np - p/2) = i\omega p \bar{B}_y(np) = i\omega p \bar{\mu} \bar{H}_y(np), \quad (2.26)$$

where $B_y(z) = \mu H_y(z)$, and μ is the ambient magnetic permeability. If we inspect the right-hand side of Equation 2.23, we see that the averaged magnetic permeability emerges naturally:

$$\bar{\mu} = \bar{B}_y(np)/\bar{H}_y(np). \quad (2.27)$$

Similarly, the other Maxwell equation in the discrete form can be obtained:

$$\bar{H}_y(np + p) - \bar{H}_y(np) = i\omega p \bar{\epsilon} \bar{E}_x(np + p/2). \quad (2.28)$$

Since p may not be very small compared with the working wavelength, the Bloch theory must be adopted for the accurate analysis of the structure. Substitute the Bloch boundary conditions into the discrete Maxwell equations, we have

$$\bar{H}_y(np + p) = \bar{H}_y(p)\exp(in\theta), \quad \bar{E}_x(np + p/2) = \bar{E}_x(p/2)\exp(in\theta), \quad (2.29)$$

where θ is the phase advance in each cell. We then obtain the dispersion equation based on the above equations

$$2 - 2\cos\theta = \omega^2 p^2 \bar{\epsilon}\bar{\mu}. \quad (2.30)$$

And the phase advance is

$$\theta = \arccos(1 - \omega^2 p^2 \bar{\epsilon}\bar{\mu}/2). \quad (2.31)$$

To obtain a complete description of wave propagation in a medium, it is also necessary to determine the wave impedance in the medium. According to Liu et al., the following two definitions are usually adopted [7]:

$$\eta(np + p/2) = \bar{E}_x(np + p/2)/\bar{H}_y(np + p/2) = \sqrt{\bar{\mu}/\bar{\epsilon}}\frac{1}{\cos(\theta/2)}, \quad (2.32)$$

$$\eta(np) = \bar{E}_x(np)/\bar{H}_y(np) = \sqrt{\bar{\mu}/\bar{\epsilon}}\cos(\theta/2). \quad (2.33)$$

For magnetic resonators, the first definition will be approximately correct, in which the magnetic field is nearly uniform within one unit cell. Likewise, for electric resonators, the second equation will be approximately correct since the electric field is nearly uniform within one unit cell. Using Liu and Cui's expression, the equations can be combined together to yield a general form:

$$\eta = \sqrt{\bar{\mu}/\bar{\epsilon}}[\cos(\theta/2)]^{S_b}. \quad (2.34)$$

In this expression, $S_b = 1$ is for electric resonators while $S_b = -1$ for magnetic resonators.

Now, we can derive two effective material parameters based on the spatial dispersion and the wave impedance. Let us denote the effective permittivity and permeability as ϵ_{eff} and μ_{eff}, then the phase shift θ and wave impedance η can be expressed in terms of ϵ_{eff} and μ_{eff}, since $\theta = \omega p\sqrt{\epsilon_{eff}\mu_{eff}}$ and $\eta = \sqrt{\mu_{eff}/\epsilon_{eff}}$. The general solution for the effective permittivity and permeability are given as

$$\epsilon_{eff} = \bar{\epsilon}\frac{\theta/2}{\sin(\theta/2)}[\cos(\theta/2)]^{-S_b}, \quad (2.35)$$

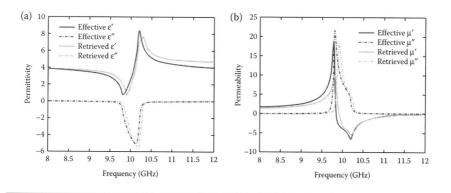

Figure 2.9 Comparison of the retrieved parameters and those effective ones. (a) Permittivity; (b) permeability.

$$\mu_{\mathit{eff}} = \bar{\mu}\frac{\theta/2}{\sin(\theta/2)}[\cos(\theta/2)]^{S_b}. \qquad (2.36)$$

These equations provide a more accurate solution for electrically or magnetically resonant metamaterials, setting up the relationship between the particle response and the system behavior.

Let us again take SRR as an example. As has been discussed in the previous section, its magnetic permeability shows a very typical Lorentz–Drude response, which has been demonstrated in Figure 2.4 and proved in this section too. However, our retrieved results do not exactly coincide with the theoretical/averaged data. Now, by using the general effective medium theory and considering the spatial dispersion effects (which is ignored in the theoretical model), we can get more accurate parameters for the structure and eliminate the disparity. In Figure 2.9, we show the comparison between the retrieved parameters and the effective parameters given by Equations 2.35 and 2.36. Clearly, the two data sets agree very well with each other.

References

1. L. Brillouin. *Wave Propagation and Group Velocity*. New York: Academic Press, 1960.
2. J. B. Pendry, A. J. Holden, D. J. Robbins, and W. J. Stewart. Magnetism from conductors and enhanced nonlinear phenomena. *IEEE Trans. Microw. Theory Tech.*, 47: 2075, 1999.
3. W. J. Padilla, M. T. Aronsson, C. Highstrete, M. Lee, A. J. Taylor, and R. D. Averitt. Electrically resonant terahertz metamaterials: Theoretical and experimental investigations. *Phys. Rev. B*, 75: 041102, 2007.
4. D. Schurig, J. J. Mock, and D. R. Smith. Electric-field-coupled resonators for negative permittivity metamaterials. *Appl. Phys. Lett.*, 88: 041109, 2006.
5. D. R. Smith and J. B. Pendry. Homogenization of metamaterials by field averaging. *J. Opt. Soc. Am. B*, 23: 391, 2006.
6. D. R. Smith, S. Schultz, P. Markos, and C. M. Soukoulis. Determination of effective permittivity and permeability of metamaterials from reflection and transmission coefficients. *Phys. Rev. B*, 65: 195104, 2002.
7. R. Liu, T. J. Cui, D. Huang, B. Zhao, and D. R. Smith. Description and explanation of electromagnetic behaviors in artificial metamaterials based on effective medium theory. *Phys. Rev. E*, 76: 026606, 2007.
8. K. S. Yee. Numerical solution of initial boundary value problems involving Maxwell's equations in isotropic media. *IEEE Trans. Antenn. Propag.*, 14: 302, 1966.

Chapter 3

Artificial Particles: "Man-Made Atoms" or "Meta-Atoms"

As is well known, a crystal is a solid natural material whose constituents, such as atoms, molecules, or ions, are arranged periodically. The crystal lattice extends in all directions, and results in homogeneous property of the solid. A crystal can be either isotropic or anisotropic in different directions. In contrast, a noncrystal is a solid in which the atoms inside it form a random arrangement; whereas a quasicrystal is a solid in the ordered state between crystalline and noncrystalline. The quasicrystal consists of arrays of atoms that are ordered but not strictly periodic. Therefore, a quasicrystal is inhomogeneous, either isotropic or anisotropic. The man-made metamaterials can be classified in a similar way. When the artificial particles, the electrically resonant particles and/or the magnetically resonant particles, are arranged periodically, they form a metamaterial termed as the "super crystal." When they are arranged randomly, they form a metamaterial termed as the "super noncrystal." When they are arranged quasiperiodically, varying in a specific manner, they form a metamaterial termed as the "super quasicrystal."

The electrically resonant particles (Section 3.1) and magnetically resonant particles (Section 3.2) are the fundamental composing unit cells for metamaterials. To date, several different methods have been presented to design and fabricate particle arrays to realize metamaterials at microwave frequencies. Among them, the dielectric-metal resonant particles (Sections 3.3 and 3.4) have been widely reported in literatures and applied for engineering. Besides, the dielectric subwavelength particles (Section 3.5) have been utilized in many 2D and 3D metamaterial designs; while the inductor-capacitor (LC) particles, for example, the composite right-/left-handed (CRLH) structures, have been applied in many planar construction designs (Section 3.6). In addition, the nonresonant particles (Section 3.7) have been developed for broadband novel functional devices and antennas. Most recently, the D.C. particles (Section 3.8) have been created for the static fields, as an extension of the time-varying EM fields.

3.1 Electrically Resonant Particles

The homogeneous view of EM properties of a medium is usually presented by the electric permittivity ε and the magnetic permeability μ. The permittivity is the measure of the resistance that is encountered when forming an electric field in a medium. It is determined by the ability of a medium to polarize in response to the electric field. The permeability, on the other hand, is the degree of magnetization of a medium in response to the magnetic field. In general, permittivity and permeability are not constants, as they can vary with the position in the medium, frequency of the field applied, humidity, temperature, and other parameters. If the medium is anisotropic, the permittivity and permeability are tensors.

For natural media, permittivity and permeability are determined by the composing atoms and their arrangement. And for artificial ones, they are decided by the constructing "man-made atoms" or "meta-atoms," the artificial particles. We mentioned in Chapter 2 that natural atoms can be treated as harmonic oscillators. Similarly, the "man-made atoms" respond to the external EM field as electric or magnetic resonators, resulting in induced field that reduces or even reverses the total field around them. From the point of view of the effective medium theory, as we discussed in Chapter 2, when such resonators are of subwavelength and are periodically placed, they perform as effective media, the metamaterials. Due to the flexibility of the composing particles, as well as their arrangement, metamaterials are able to possess negative, near-zero, positive, and extremely high permittivities and/or permeability, which are rarely found in natural materials.

Strictly speaking, the first metamaterial sample with controllable permittivity was not composed of "man-made atoms" but constructed by "man-made" infinite elements. Inspired by the celebrated concept that the EM response of metals in the visible region and near ultraviolet is dominated by the negative epsilon, in 1996, Pendry et al. proposed a periodic structure with infinite wires to realize negative permittivity in the far infrared or even GHz band [1]. The building blocks of the new material are very thin metallic wires of the order of 1 μm in radius, and are assembled into a periodic cubic lattice, as shown in Figure 3.1.

This design originates from the well-established theory that the plasmons have a profound impact on properties of metals and produces a dielectric function of

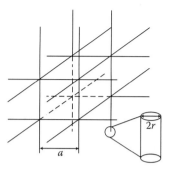

Figure 3.1 The periodic structure is composed of infinite wires arranged in a simple cubic lattice, joined at the corners of the lattice. The large self-inductance of a thin wire delays the onset of current mimicking the effect of electron mass. (Reprinted with permission from J. B. Pendry et al., *Phys. Rev. Lett.*, 76: 4773. Copyright 1996 by the American Physical Society.)

the form

$$\epsilon(\omega) = 1 - \frac{\omega_p^2}{\omega(\omega + i\gamma)}, \quad (3.1)$$

where $\omega(p)$ is the plasma frequency and γ is a damping term representing dissipation of the plasmon's energy into the system. $\omega(p)$ can be calculated by

$$\omega_p^2 = \frac{ne^2}{\epsilon_0 m_{\text{eff}}}, \quad (3.2)$$

where n is the density of electrons and m_{eff} is the effective mass of the electrons. For typical metals, $\omega(p)$ is in the ultraviolet region of the spectrum and γ is small relative to $\omega(p)$. According to Equation 3.1, ϵ is negative below the plasma frequency, and at least down to frequencies comparable to γ. It is true that the EM response of metals is dominated by the negative epsilon concept in the visible region and near ultraviolet. However, from the near infrared downwards, the dielectric function in Equation 3.1 is essentially imaginary. Therefore, metals cannot be directly applied to produce negative permittivity at terahertz, millimeter-wave, and microwave frequencies.

To solve this problem, Pendry et al. used periodic wires to depress the plasma frequency of metals. In Figure 3.1, consider a displacement of electrons along one of the cubic axes. In this circumstance, the active wires are those directed along that axis. If the density of electrons in these wires is n, then the density of these active electrons in the structure is effectively reduced. The reduction is decided by the fraction of space occupied by the wire, and hence the effective density is written as

$$n_{eff} = n\frac{\pi r^2}{a^2}. \qquad (3.3)$$

Clearly, when the radius of the wires, r, is much smaller than the period of the lattice, a, the effective density of the active electrons decreases significantly. On the other hand, from classical mechanics, the new effective mass of the electrons becomes

$$m_{eff} = \frac{\mu_0 \pi r^2 e^2 n}{2\pi} \ln(a/r). \qquad (3.4)$$

Having both the effective density, n_{eff}, and the effective mass, m_{eff}, the new plasma frequency can be easily calculated using Equation 3.2. Reference 1 presents an example that the aluminum wires with $r = 1.0 \times 10^{-6}$ m and $a = 5 \times 10^{-3}$ m obtain a plasma frequency at approximately 8.2 GHz, which is depressed by up to 6 orders of magnitude.

In 2001, the periodically positioned wires were successfully applied to realize negative index metamaterials (NIM) at microwave frequencies for the first time [2]. This demonstration has brought about great expectation for metamaterials. However, arguments arose that wires are not independent artificial particles because they require continuous connections between unit cells. When attempting to construct finite metamaterial objects, however, terminations of the wires are necessary and significant change of the electric property is brought in inevitably. As a result, the infinite wire is not a reliable particle for artificial materials because its electric property changes when it is cut in half.

A straightforward way to solve this problem is to use finite wires, for example, the cut wire in Figure 3.2a. When the external electric field is parallel to the cut wire, currents resonate on the wire as illustrated by the dashed lines in Figure 3.2a, and the induced electric field thereby reduces the total electric field. One serious

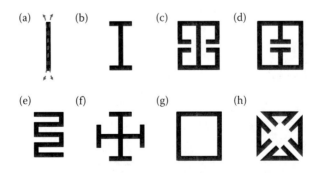

Figure 3.2 Several kinds of electrically resonant particles.

drawback is that the cut wire requires sufficient length that is usually much longer than a subwavelength unit cell. In addition, note that for a cut wire the real part of effective permittivity follows the Lorentz model, while for an infinite wire it follows the Drude model.

A developed electric resonator is the I-shaped structure in Figure 3.2b. Vertical bars are added to the cut wire, increasing the series capacitance, depressing the resonance frequency, and consequently reducing the dimension of the resonator. Figure 3.3 illustrates the typical distribution of its effective permittivity. The real part of permittivity (solid curve) follows the Lorentz model, goes extremely large around the resonance frequency (f_r), and changes rapidly from positive to negative. The imaginary part of permittivity (dashed curve) represents loss of the particle. Far away from the resonance frequency, the permittivity is relatively constant. Besides, instead of increasing the series capacitance, an alternative, the meander-line structure, has been proposed to increase the series inductance and thereby reduce the electric length of the resonator (see Figure 3.2e) [3]. One problem of the I-shaped structure and the meander-line structure is that they are not local and self-contained resonators. Mutual coupling happens between neighboring unit cells. As a result, they are not robust with regard to maintaining the bulk properties close to a boundary or interface.

The resonance frequency of the I-shaped structure can be further reduced when the two bars extend and even bend, as shown in Figure 3.2c. This structure has two capacitor-like structures connected in parallel, coupling to the electric field. This is a self-contained resonator. However, most of the energy is concentrated in the capacitors, close to the left and right boundaries of the unit cell. Therefore, this structure also has couplings with its neighboring ones. A further developed resonator is the electric-LC (ELC) resonator in Figure 3.2d [4]. One capacitor-like structure couples to the electric field and is connected in parallel to two loops, which provides inductance to the circuit. This is a local and self-contained oscillator and most of the energy is concentrated in the center of the unit cell. Therefore, the ELC resonator

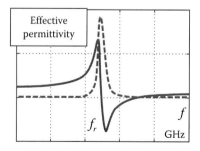

Figure 3.3 Effective permittivity of I-shaped resonator arrays. Solid curve: the real part. Dashed curve: the imaginary part.

can be used to construct artificial metamaterials with the benefit of not requiring continuous current paths between unit cells.

In addition, it should be pointed out that anisotropic property of electrically resonant particles has not been characterized yet. Due to the asymmetry in the particle, differently polarized fields will drive different resonant modes. Note that we define an electrically resonant particle because its first resonant mode is electric, although it may have high-order magnetic resonant modes. Figure 3.2f–h lists some more isotropic electrical resonators. They could be periodically arranged in cubic unit cells to construct near-isotropic 3D metamaterials. Furthermore, in practice, neighboring particles may be oriented "symmetrically" in order to suppress the magnetoelectric coupling that could give rise to bianisotropic response for the composite structure. Arrangement of particles will be further discussed in Chapter 4.

3.2 Magnetically Resonant Particles

In nature, the magnetic response of most materials does not exist in the gigahertz range and, therefore, it is extremely difficult to find the magnetic analog of a good electrical conductor that we discussed in Section 3.1. Instead, microstructured metamaterials with magnetic activity have been proposed since 1999 [5]. They are composed of magnetically resonant particles that are much shorter than the wavelength, and possess controllable permeability ranging from negative to positive values.

The SRR depicted in Figure 3.4a is a widely applied magnetically resonant particle. It contains two layers of conducting cylinders and in each layer there is a gap to prevent currents from flowing around. When the external magnetic field H_0 is parallel to the cylinders, induced currents are cut off at the two gaps and opposite charges assemble on the two conducting layers, as shown in Figure 3.4b. Consequently, capacitive elements have been introduced between the internal and external cylinders, which is represented by C_{SRR} in the equivalent circuit model in Figure 3.4c. In addition to this, there exists a natural magnetic inductance of the cylinders, which is remarked as L_{SRR} in Figure 3.4c. Therefore, the two-layered split ring works as a magnetic LC resonator at microwave frequencies, and the effective permeability can be calculated by Equation 2.14, following the Lorentz–Drude model sketched in Figure 2.4. At the resonance frequency, the structure resonates intensely, and the induced magnetic field is opposite and comparable to the external filed. In this way, negative permeability is achieved. Besides, it is noted that there are extra capacitances to the top and bottom boundaries of the SRR structure, represented by C in Figure 3.4c. They are in nature capacitances between neighboring unit cells.

The cylindrical SRR has a huge stereo volume and therefore is not suitable for constructing artificial materials. Instead, a more compact alternative structure, the 2D SRR printed on high-frequency dielectric substrates has been widely employed

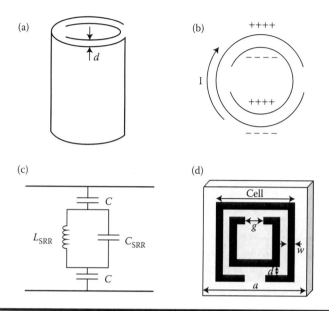

Figure 3.4 (a) The model of an SRR (side view). (b) Currents flowing in the split rings (top view). (c) The equivalent circuit model of a split ring unit cell. (d) The two-dimensional SRR on printed circuit board.

in practice. Figure 3.4d depicts a square SRR unit cell as an example. Since the metallic wires portraying the square rings are very thin, the capacitance at the two splits becomes comparable to the one between the external and internal rings. Therefore, the dimensions of *cell*, *d* and *g* in Figure 3.4d primarily determine the resonance frequency. Broad-side coupled SRR (BC-SRR) has also been developed, as shown in Figure 3.5a, to enhance the capacitance. Note that the permittivity of substrate also tunes the resonance frequency of SRR because it influences the EM property of the background where the particle is embedded.

It has been observed that at the first resonance frequency, the magnetically resonant particles essentially perform as magnetic dipoles, and all of the structures in Figure 3.5a–c contain "loop" elements along which the induced currents could travel, shown as the arrowed curves in the figures. In contrast, the electrically resonant particles are essentially electric dipoles at the first resonance frequency, with induced "line" currents presented by the arrowed lines in Figure 3.5d–f. Note that a structure may serve as the magnetically resonant particle at one frequency and the electrically resonant particle at another frequency. For example, the structure in Figure 3.5b and e is a magnetically resonant particle at a lower frequency and an electrically resonant particle at a higher frequency, while the structure in Figure 3.5c and f is a magnetically resonant particle at a higher frequency and an electrically resonant particle at a lower frequency.

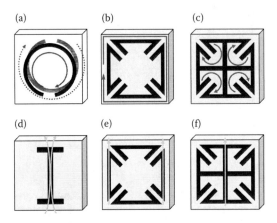

Figure 3.5 (a)–(c) Structures serve as the magnetically resonant particles. (d)–(f) Structures serve as the electrically resonant particles.

Furthermore, it is easy to understand that SRR has magnetic response only when the external magnetic field is aligned along the axes of the cylinders or the rings. More generally, a magnetically resonant particle works only when the external magnetic field is aligned along the axes of the "loop" currents and an electrically resonant particle works only when the external electric field is aligned along the "line" currents. The property to exhibit anisotropy along with cross-polarization is called bianisotropy.

Aside from the conventional SRRs, new magnetically resonant particles have been developed for different applications. For example, to reduce the electrical size, spiral resonators have been proposed [6]; to improve the restricted magnetic response in other directions, nonbianisotropic SRRs (NB-SRRs) have been developed [7]; to simultaneously achieve negative permittivity and permeability, S-shaped resonators have been created [8].

3.3 Dielectric-Metal Resonant Particles

The dielectric-metal resonant particles are essentially printed circuit board (PCB) with patterned metallic structures on it. This realization method has the advantage of low cost and readiness for mass production. At microwave frequencies, commercial PCB substrates such as the Duroid series and the F4R series possess low loss, flexible permittivity, controllable solidity and thickness, and stable dispersive property, and therefore are great choices for metamaterials. For example, structures depicted in Figures 3.2 and 3.5 can be printed on PCB substrate to make a dielectric-metal particle. Obviously, they are planar particles. Following the effective medium theory in Chapter 2, each particle should be of subwavelength. In practice, people construct 3D metamaterials by extending the subwavelength planar particles to three orthogonal directions, as shown in Figure 3.6. The PCBs with printed single split rings on them construct the six surfaces of a unit cell. Low loss and nonconductive glue or brackets are usually used to fix the boards and locate unit cells periodically or nonperiodically, so as to compose homogeneous or inhomogeneous metamaterials, respectively. Note that the property of PCB substrate effects the parameters of metamaterials. For example, loss of the substrate brings in loss for the metamaterials, and permittivity of the substrate alters the resonance frequency of the bare metallic particle.

When the particle is resonant to the electric field alone, the composed metamaterial has negative permittivity and positive permeability near the resonance frequency. This kind of metamaterial is called the ε-negative (ENG) metamaterial. In contrast, when the particle is resonant to magnetic field alone, the composed metamaterial has positive permittivity and negative permeability near the resonance frequency. This kind of metamaterial is called the μ-negative (MNG) metamaterial. Both the ENG metamaterial and the MNG metamaterial are defined as single-negative (SNG) metamaterial because only one of the two parameters, ε and μ,

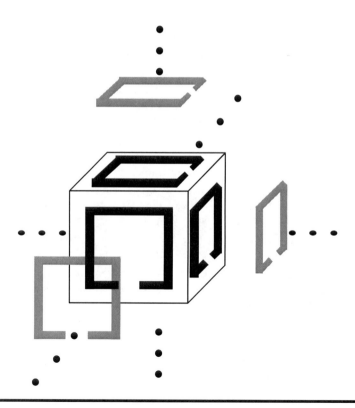

Figure 3.6 Dielectric-metal resonant particles and their arrangement.

has negative real part. Furthermore, when the particle is resonant to both the electric field and the magnetic field (e.g., when the particle includes both electrically resonant structures and magnetically resonant structures, or includes structures resonate to the electric field and the magnetic field at almost the same frequency), the composed metamaterial has negative permittivity and negative permeability near the resonance frequency. This kind of metamaterial is called as the double-negative (DNG) metamaterial, the NIM, or the left-handed materials (LHM) since the electric field, magnetic field, and wavevector of an EM wave form a left-handed triad inside the media. Detailed study on ENG metamaterials, MNG metamaterials, and DNG metamaterials will be carried out in Chapter 4.

The dielectric-metal resonant particles have been successfully applied to build SNG and DNG metamaterials. Figure 3.7 depicts some examples. Figure 3.7a shows the first prototype of DNG metamaterial at Duke University [2]. The particles use SRRs to produce negative magnetic permeability and wire elements to produce negative electric permittivity at some specific frequencies. The resonant structures are printed on fiber glass circuit board and are periodically extended in two dimensions. Therefore, this is a 2D isotropic and homogeneous model. Negative refractive index has been experimentally validated for the first time by measuring the scattering angle of the transmitted beam through a prism fabricated from this material. The model in Figure 3.7b is a ground-plane cloak (or called as carpet cloak) [9]. The particles use I-shaped structures to provide spatially dispersive positive permittivity values and constant positive permeability. This is a nearly isotropic and inhomogeneous double-positive (DPS) metamaterial model. Due to the merit of broadband performance, DPS metamaterials have been employed in a lot of applications such as the gradient-index (GRIN) lenses, microwave antennas, and EM black holes. Figure 3.7c depicts a ZIM lens [10]. Due to the anisotropy of SRRs, this metamaterial model approaches zero permeability and zero refractive index in one direction. According to Snell's law, a line source embedded in ZIM radiates plane waves instead of cylindrical waves when one component of the permeability tensor approaches zero. Therefore, this metamaterial lens creates highly directive beams for antennas. In addition, nonlinear circuit components, such as diodes, also have the potential to further control the EM properties of metamaterials and form nonlinear metamaterial systems [11–13]. In Figure 3.7d, positive-intrinsic-negative (PIN) diodes are loaded to SRRs to create a switchable anisotropic ZIM [14]. When the PIN diodes are switched on, the sample is ZIM which serves to enhance the radiation directivity. When the PIN diodes are switched off, it is transparent to EM waves.

3.4 Complementary Particles

The dielectric-metal resonant particles have been successfully introduced into the construction of metamaterials in one-dimensional (1D), 2D, and 3D circumstances.

Artificial Particles ■ 39

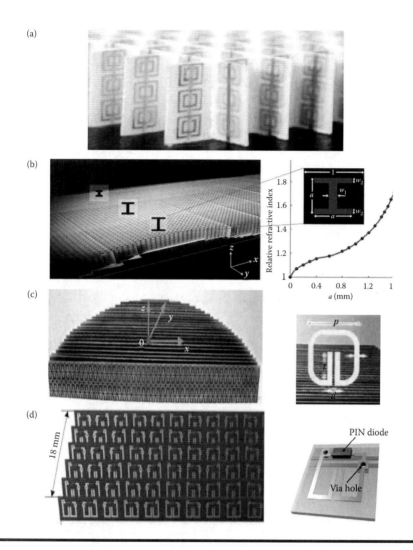

Figure 3.7 Several metamaterial examples composed of dielectric-metal resonant particles. (a) The first prototype of DNG metamaterial. (From R. Shelby, D. R. Smith, and S. Schultz. Experimental verification of a negative index of refraction. *Science*, 292: 77, 2001. Reprinted with permission of AAAS.) (b) A ground-plane cloak (or carpet cloak). (From R. Liu et al. Broadband ground-plane cloak. *Science*, 323: 366, 2009. Reprinted with permission of AAAS.) (c) A zero refractive index metamaterial lens. (The source of the material Radiation of planar EM waves by a line source in anisotropic metamaterials. *J. Phys. D: Appl. Phys.*, 43: 335406. Copyright 2010 IOP Publishing is acknowledged.) (d) A switchable ZIM by loading PIN diodes. (Reprinted with permission from N. Xiang et al. Switchable zero-index metamaterials by loading positive-intrinsic-negative diodes. *Appl. Phys. Lett.*, 104: 053504, Copyright 2014, American Institute of Physics.)

In the 2D and 3D cases, bulk metamaterials have been engineered for left-handed materials [2], invisibility cloaks [9,16], directive lenses [37], etc. In the 1D case, planar metamaterials or metasurfaces, have also been developed. Magnetically resonant particles, for example, SRRs, have been etched on the top or back substrate side of microstrip lines or coplanar waveguides (CPWs) for new functional devices such as the filters [18,19]. The loaded magnetic resonators bring in pass-band and stop-band to the microstrip model, and alter the permeability of the substrate.

One question may arise, that is, one can introduce electric resonance to a microstrip model, and create alterable dispersive permittivity? A straightforward way is to etch electrically resonant particles on the substrate. As is known, in planar transmission lines, most magnetic field assembles surrounding the microstrip. Therefore, magnetically resonant particles etched on the substrate close to the microstrip are easily driven by the vertical magnetic field. However, in contrast, electrically resonant particles on the substrate are not easy to be driven by the external field because most electric field are normal to and right beneath the microstrip. One solution is to vertically embed electrically resonant particles in the substrate. Nevertheless, the height for the embedded particles is extremely limited.

Aiming at this issue, an impactful scheme has been developed. A planar electrically resonant structure, the complementary split-ring resonator (CSRR) has been proposed as the dual counterpart of SRR [7,15]. Figure 3.8a depicts a metallic SRR structure and Figure 3.8b depicts the corresponding CSRR, which is a metallic screen with the negative image of SRR. Assume that the external EM field ($\mathbf{E}^0$, $\mathbf{B}^0$) is launched from $z < 0$, z defined in Figure 3.8a. At the resonance frequency ω_0, an infinitely thin SRR performs as a magnetic dipole with [15]

$$\mathbf{m} = \frac{\alpha_0}{\omega_0^2/\omega^2 - 1}\mathbf{B}^0 \cdot \mathbf{zz}, \tag{3.5}$$

where α_0 is a geometrical factor. Based on the Babinet principle, there is a duality for the complementary structure. Let us assume that the incident EM fields for the

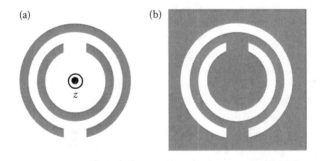

Figure 3.8 (a) The SRR and (b) CSRR.

metallic sheet with CSRR are $\mathbf{E}_c^0$, $\mathbf{B}_c^0$ and those for the metallic SRR are $\mathbf{E}^0$ and $\mathbf{B}^0$. A restriction is applied that $\mathbf{E}^0 = c\mathbf{B}_c^0$, $\mathbf{B}^0 = -(1/c)\mathbf{E}_c^0$.

It has been proved that in the shadowed region of $z > 0$, the CSRR scatters the EM waves in the same manner as an electric dipole with $\mathbf{p} = (1/c)\mathbf{m}$. This property is observed from the electric and magnetic field lines in Figure 3.9. In the region of $z > 0$, the electric field lines in Figure 3.9b have the same pattern as the magnetic field lines in Figure 3.9a. Meanwhile, in the nonshadowed region of $z < 0$, the sign of this dipole must change. This is due to the fact that the scattered fields are caused

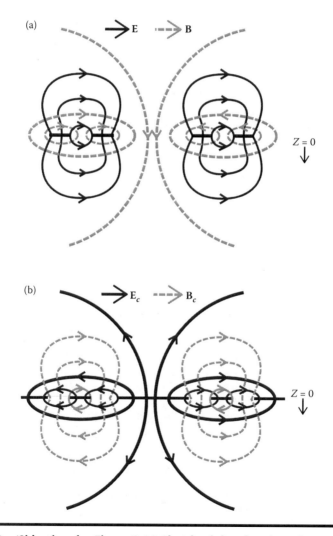

Figure 3.9 (Side view for Figure 3.8.) Sketch of the electric and magnetic field lines in (a) the SRR and (b) the CSRR. The horizontal black lines represent metal.

by the currents in the $z = 0$ plane. As a result, the electric field lines should be symmetric to the $z = 0$ plane and the magnetic field lines should be surrounding the currents. This property can be observed in Figure 3.9. In the region of $z < 0$, the electric field lines in Figure 3.9b have the same pattern as the magnetic field lines in Figure 3.9a, but is in the opposite direction. In this region, CSRR performs as an electric dipole with $\mathbf{p} = -(1/c)\mathbf{m}$.

CSRR exhibits resonant behavior for vertically polarized (with respect to the CSRR's plane) electric fields. The real part of its permittivity follows the Lorentz model. The corresponding SRR exhibits resonant behavior for vertically polarized magnetic fields. The real part of its permeability follows the same Lorentz model. Note that if the metal is assumed to be infinitely thin and lossless, as well as the dielectric substrate, CSRR and SRR resonate at the same frequency.

The complementary resonators have been added to microwave devices and antennas for better performance and novel properties. Figure 3.10 sketches two example applications. The first one is a CSRR-coupled microstrip lines. CSRRs are etched on the back substrate side. This configuration has been used to compress propagating modes at designated frequencies [20], and to realize a left-handed transmission line [7]. The second one is the CSRR-etched planar waveguide. This special configuration is also termed as the "waveguided-metamaterial (WG-MTM)" in some literatures. It consists of two parallel metallic plates constituting a planar waveguide, with etched resonator patterns in the lower plate. The WG-MTMs have been reported to demonstrate microwave tunneling between two planar waveguides separated by a thin permittivity(ϵ)-near-zero (ENZ) channel [21]. In addition, the complementary resonant structures have also been involved in antenna designs. For example, CSRRs have been etched on the ground plane between two coplanar microstrip patch antennas to effectively reduce mutual couplings [22]. This design has the merit of compact size, controllable frequency band, and easy fabrication. More studies on the complementary resonant structures will be carried out in Chapter 4.

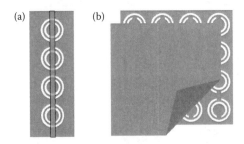

Figure 3.10 (a) Microstrip line with CSRRs etched on the back substrate side. (b) Planar waveguide with CSRRs etched in the lower plate.

3.5 Dielectric Particles

Dielectric particles present electric-dipole response at certain frequencies. They have also been adopted as "atoms" for metamaterials due to the resonant property. Dielectric particle arrays embedded in the air or other host media have been investigated with the effective medium theory for years, first presented in the pioneering works by Maxwell and Rayleigh, and later developed in the notable work of Lewin [23].

Figure 3.11a sketches the unit cell model when a huge number of dielectric spheres are embedded in the air. Size of the spheres, as well as the spacing between neighboring spheres, are small compared to the wavelength in the air. Under the condition, the sphere array as a whole becomes an artificial material bulk. The dielectric particles behave in a way similar to that of the dielectric-metal resonant particles we studied in previous sections, and provide us with permittivity and/or permeability ranging from negative to positive.

It has been studied in References 23 and 24 that for the array of lossless magnetodielectric spheres with ϵ_2 and μ_2, as shown in Figure 3.11a, the effective relative permittivity is real, that

$$\epsilon_{\mathit{eff}} = \epsilon'_{\mathit{eff}} = \epsilon_0 \left(1 + \frac{3A}{(F(\theta) + 2b_e)/(F(\theta) - b_e) - A}\right), \quad (3.6)$$

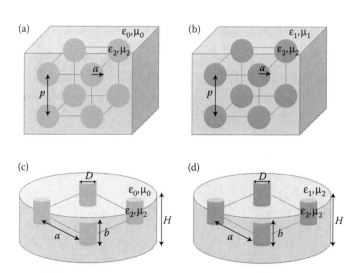

Figure 3.11 (a) Dielectric sphere arrays in the air. (b) Dielectric sphere arrays embedded in host media with ϵ_1 and μ_1. (c) Dielectric cylinder arrays in the air. (d) Dielectric cylinder arrays embedded in host media with ϵ_1 and μ_1.

where A is the volume fraction of the spherical inclusions as

$$A = \frac{4\pi a^3}{3p^3}, \quad (3.7)$$

a, p defined in Figure 3.11a, and

$$b_e = \frac{\epsilon_0}{\epsilon_2}. \quad (3.8)$$

The function $F(\theta)$ is

$$F(\theta) = \frac{2(\sin\theta - \theta\cos\theta)}{(\theta^2 - 1)\sin\theta + \theta\cos\theta}. \quad (3.9)$$

Here,

$$\theta = k_0 a \sqrt{\epsilon_2' \mu_2'}, \quad k_0 = \frac{2\pi}{\lambda}. \quad (3.10)$$

Equations 3.6, 3.9, and 3.10 imply that with fixed b_e and A, the effective permittivity can be negative when

$$-b_e \frac{2+A}{1-A} < F(\theta) < -2b_e \frac{1-A}{1+2A}. \quad (3.11)$$

Since $F(\theta)$ is a function of the frequency for a specific model, the effective permittivity is dispersive in frequency, following the Lorentz distribution. Figure 3.12 [24] describes the dispersion of the effective relative permittivity for an example model when $A = 0.5$, $\epsilon_2 = 40$, and $\mu_2 = 200$. The dispersion curve distributes in the same manner as the one in Figure 2.2.

Similarly, the effective relative permeability is calculated for the array of lossless magnetodielectric spheres as

$$\mu_{\text{eff}} = \mu_{\text{eff}}' = \mu_0 \left(1 + \frac{3A}{(F(\theta) + 2b_m)/(F(\theta) - b_m) - A}\right), \quad (3.12)$$

where

$$b_m = \frac{\mu_0}{\mu_2}. \quad (3.13)$$

Figure 3.12 also describes the Lorentz dispersion of the effective relative permeability. The dispersion curve is in the same manner as the one in Figure 2.4. Therefore, the dielectric particles have similar resonant property as the dielectric-metal particles, and hence are competent for constructing metamaterials.

More generally, the dielectric particles can be embedded in host media, or, in other words, in matrix, other than the air. Figure 3.11b illustrates the model when

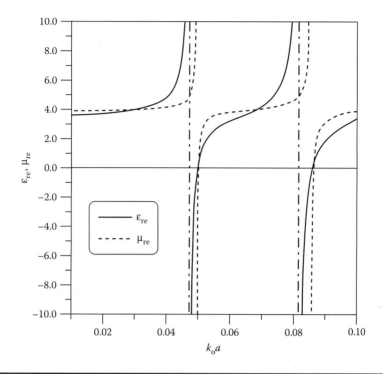

Figure 3.12 An example of the distribution of effective relative permittivity and permeability for the array of lossless magnetodielectric spheres. (C. L. Holloway et al., A double negative (DNG) composite medium composed of magnetodielectric spherical particles embedded in a matrix, *IEEE Trans. Ant. Propag.*, 51(10): 2596 © 2003 IEEE.)

dielectric spheres with permittivity ϵ_2 and permeability μ_2 are embedded in the host media with permittivity ϵ_1 and permeability μ_1. Instead of ϵ_0 and μ_0, ϵ_1 and μ_1 are applied to Equations 3.6 through 3.13, respectively, to calculate the effective relative permittivity and permeability for the mixture. Once the host media is decided, by changing the volume fraction A, one is able to vary the resonance frequency and consequently control the permittivity and/or permeability value at an assigned frequency.

Metamaterials constructed by embedding dielectric spheres in host media can be isotropic, thanks to the symmetric geometry of the sphere. However, in the microwave frequency band, it tends to be expensive and time consuming to fabricate 3D spherical arrays with gradually varied dimensions. An easier substitute for 3D nearly isotropic metamaterials is to use dielectric cylinder arrays, as shown in Figure 3.11c. Cylinders are located in the air, with diameter D, height b, and spacing a. They can also produce effective permittivity and permeability ranging from negative to positive. Nevertheless, when the cylinders operate far away from resonance, the effective permittivity of the mixture is positive and can be simply calculated using the volume average of the dielectric and the air as

$$\epsilon_{\mathit{eff}} = A\epsilon_2 + (1-A)\epsilon_0, \tag{3.14}$$

where

$$A = \frac{\pi D^2 b}{4a^2 H}. \tag{3.15}$$

For most microwave applications, metamaterials realized in this way are considered as approximately isotropic [26]. Figure 3.13 shows a simplified 2D ground-plane cloak realized by nonmagnetic dielectric cylinders. The ground-plane cloak is an exotic device to make objects on the ground "invisible." Inhomogeneous and isotropic metamaterials are required so as to build a space where incident waves cannot reach the concealed objects. The concept of ground-plane cloak will be explained in detail in the following chapters. In Figure 3.13a, the space of ground-plane cloak is simplified to eight blocks with four different refractive indices (numbered by 1, 2, 3, and 4). The dielectric cylinders are mainly made of $(Z_r, S_n)TiO_4$ ceramics, and possess a relative permittivity of about 36.7 and a radius of 1.5 mm. They are located periodically to construct near-isotropic metamaterial blocks. By varying the height of cylinders between 0.1 and 3 mm, one can obtain required refractive indices between 1.02 and 1.49 at microwave frequencies. Performance of the ground-plane cloak has been validated from 4 to 10 GHz in experiment [25].

The dielectric cylinders in Figure 3.11c can also be embedded in host media with permittivity ϵ_1 and permeability μ_1, as shown in Figure 3.11d. For dielectric cylinders working off the resonance, the effective permittivity becomes

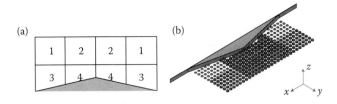

Figure 3.13 (a) Top view of the ground-plane cloak. The blocks with the same number have the same parameters. n1 = 1.08, n2 = 1.14, n3 = 1.01, and n4 = 1.21. (b) Perspective view of the metallic object and the cover of cylinder arrays. The lattice period is 5 mm. (With kind permission from Springer Science+Business Media: *Frontiers of Physics in China*, A broadband simplified free space cloak realized by nonmagnetic dielectric cylinders, Vol. 5, No. 3, 2010, pp. 319–323, D. Bao et al., Figure 4.)

$$\epsilon_{eff} = A\epsilon_2 + (1-A)\epsilon_1 \qquad (3.16)$$

in this model. This design is extremely adaptive at microwave frequencies using the PCB technology. In a very special case, differently sized subwavelength air hole arrays can be drilled on ordinary dielectric substrates for different effective permittivities. The effective permittivity of the mixture becomes

$$\epsilon_{eff} = A\epsilon_0 + (1-A)\epsilon_D, \qquad (3.17)$$

where ϵ_D is the dielectric constant of the substrate. Figure 3.14 shows a 3D flattened Luneburg lens fabricated by drilling inhomogeneous holes in multilayered dielectric plates [27]. Required refractive index varies from 0.5 to 2.2 for the lens design. Two kinds of dielectric substrates, the FR4 dielectric with a higher relative permittivity 4.4 (with loss tangent 0.025) and the F4B dielectric with a lower relative permittivity 2.65 (with loss tangent 0.001), are involved in the lens' realization. Three different thicknesses of the substrates are also applied. The continuous variation of

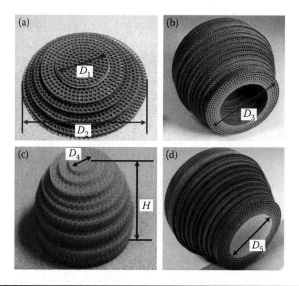

Figure 3.14 Construction of a 3D flatten Luneburg lens fabricated with multilayered dielectric plates by drilling inhomogeneous holes. (a) The first region (the top part) was fabricated using 2 × 2 × 1 mm³ F4B unit cells. (b) The second region (the bottom shell) was fabricated using 2 × 2 × 2 mm³ F4B unit cells. (c) The third region (the bottom core) was fabricated using 2 × 2 × 2 mm³ FR4 unit cells. (d) Construction of the whole lens. (Reprinted by permission from Macmillan Publishers Ltd. *Nat. Commun.* H. F. Ma and T. J. Cui. Three-dimensional broadband and broad-angle transformation-optics lens. 1: 124. Copyright 2010.)

the refractive index is obtained by changing the diameter of the drilled hole in each unit cell. More details about this design can be found in Section 8.6.

It should be pointed out that designs in Figures 3.13 and 3.14 operate in wide frequency bands. In fact, as long as the working frequency is far away from the resonance frequency of the dielectric particles, Equations 3.6 and 3.7 are applicable. In these circumstances, the dielectric particles are actually nonresonant particles whose effective permittivity slowly changes. Nonresonant particles will be dealt with in detail in the next section.

3.6 Nonresonant Particles

In the early stage, research on metamaterials were devoted to the realization of LHMs [28]. Pendry and other scientists have proposed several schemes to realize metamaterials with negative permittivity and permeability using electrically and magnetically resonant particles, respectively [1,5,6,8,29], along with a series of novel applications of LHMs [30–35]. These contributions, accompanied with the successful experimental demonstration of negative refraction at microwave frequencies in 2001 [2], have brought about great expectation for this new kind of artificial material. However, most realized LHMs rely upon resonant structures and, therefore, have the distinct disadvantages of being lossy and narrow-band [36]. In 2005, Smith et al. proposed the idea of "gradient index metamaterials" [37]. A gradient index lens was designed using metamaterials with a constant gradient in the refractive index along one axis. Ever since, the concept of MTMs has been extended. It is no longer restricted to the material with negative permittivity and/or permeability, but can also be referred to as materials with novel properties which are rarely found in natural materials. For example, metamaterials can possess near-zero or extremely high refractive indices, and become inhomogeneous with engineered positive index distribution.

Let us look at Figure 3.15, which depicts the typical dispersion curves for electrically or magnetically resonant particles, including both the real part and the imaginary part. Effective permittivity or permeability follows the Lorentz model, with the real part beginning with a no-less-than-unity value and increases slowly as the working frequency goes higher. This frequency range is marked as Region I in Figure 3.15. In this region, the particle works off the resonance, and therefore can be considered as a relatively "nonresonant" particle that produces nearly constant permittivity or permeability. When the working frequency approaches the resonance frequency (f_r), the real part increases rapidly, and then falls promptly to be negative, as shown in Region II. In this region, very high permittivity or permeability can be achieved. As the working frequency goes further higher, the real part becomes negative around f_r, as indicated in Region III. In this region, negative permittivity or permeability can be achieved. Region IV indicates a special frequency range where the real part of permittivity or permeability is near zero. On the right-hand side of

Artificial Particles ■ 49

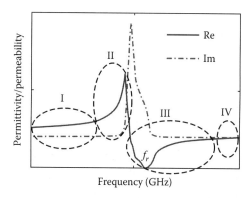

Figure 3.15 Typical Lorentz dispersion curves for electrically or magnetically resonant particles.

Region IV, the real part gradually changes back to the original value and becomes nearly constant again. In addition, the imaginary part represents loss of materials. Clearly, it becomes extremely high around the resonance frequency.

Based on the Lorentz model in Figure 3.15, we can conclude that the electrically or magnetically resonant particles are "resonant" only near the resonance frequency (Regions II, III, and IV here). In fact, for metamaterial design, resonant particles have been applied to realize exotic properties such as the negative permittivities and/or permeabilities, near-zero refractive indices, and extremely high refractive indices. Note that the Lorentz model indicates that resonant metamaterials have essential disadvantages of narrow-band performance and high loss.

In contrast, these particles operate in another way off the resonance (e.g., in Regions I here), being "nonresonant." By varying the geometry of the structure, one is able to control the resonance frequency of a particle, and consequently vary the permittivity or permeability at frequencies of interest. The nonresonant particles take advantage of the broadband flat dispersion region before resonance, and hence possess broadband performance and low loss. To date, nonresonant particles have been developed for inhomogeneous metamaterials with designed index distribution to control the propagation of waves. According to Fermat's principle, light rays passing between two spatial points choose the optically shortest path. Therefore, the path may be curved if the refractive index varies in space. This theory implies that one is able to control the traveling path of EM waves by engineering the refractive index distribution in a space. A series of novel functional devices have been proposed and realized based on this idea, such as the invisibility cloak, the illusion devices, and metamaterial lenses. In Chapter 6, we will explain in detail how to control the propagation of waves using inhomogeneous metamaterials.

Nonresonant particles are especially competent when the required relative permittivities or permeabilities are not much larger or much less than unity. Figure 3.16

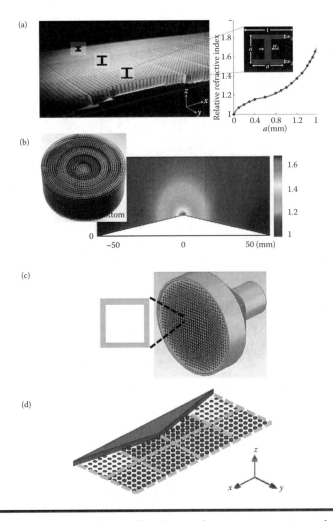

Figure 3.16 Some example applications using nonresonant particles. (a) A ground-plane cloak using I-shaped structures. (From R. Liu et al. Broadband ground-plane cloak. *Science*, 323: 366, 2009. Reprinted with permission of AAAS.) (b) A 3D ground-plane cloak made of drilled-hole dielectric plates. (Reprinted by permission from Macmillan Publishers Ltd. *Nat. Commun.* H. F. Ma and T. J. Cui. Three-dimensional broadband ground-plane cloak made of metamaterials. 1: 21. Copyright 2010.) (c) The sketch of a metamaterial gradient index lens antenna. (Reprinted by permission from Macmillan Publishers Ltd. *Nat. Commun.* H. F. Ma and T. J. Cui. Three-dimensional broadband ground-plane cloak made of metamaterials. 1: 21. Copyright 2010.) (d) A 2D ground-plane cloak made of dielectric cylinder arrays. (With kind permission from Springer Science+Business Media: *Frontiers of Physics in China*, A broadband simplified free space cloak realized by nonmagnetic dielectric cylinders, Vol. 5, No. 3, 2010, pp. 319–323, D. Bao et al., Figure 4.)

lists some example applications. In Figure 3.16a, a broadband ground-plane cloak is fabricated and demonstrated in experiments [9]. In this design, the required refractive index ranges from 1.08 to 1.67. This index span can be achieved using nonresonant I-shaped particles printed on FR4 substrate. Concealing performance of the ground-plane cloak has been validated from 13 to 16 GHz, far away from the resonance frequency of the I-shaped particles. The ground-plane cloak has also been realized in 3D circumstance. Figure 3.16b shows a 3D broadband ground-plane cloak [38]. For this design, the required refractive index ranges from 1 to 1.63. Drilled-hole F4B plates have been employed to fulfill the index map, which is depicted in this figure. In the previous section, we described the technique of drilling inhomogeneous holes in multilayered dielectric plates. The concealing performance has been observed in experiment from 9 to 12 GHz. Nonresonant particles have also been applied to antenna applications. In Reference 38, to test the 3D ground-plane cloak, a metamaterial lens antenna has been used to transmit a narrow-beam plane wave in the near-field region. This lens is made of inhomogeneous closed-square rings (CSR), which are nonresonant from 8 to 12 GHz. By adjusting the size of the CSR, gradient index has been achieved on the aperture of the horn antenna so as to increase the directivity. Owing to the nonresonant nature, the highly directive lens antenna operates in a broad band covering the whole X-band. The simplified 2D ground-plane cloak realized by nonmagnetic dielectric cylinders in Figure 3.16d has been discussed in the previous section [25]. By varying the height of the cylinders between 0.1 and 3 mm, the required refractive indices with values between 1.02 and 1.49 have been achieved. The periodic arrays of high-index dielectric cylinders operate off resonance from 4 to 10 GHz. Therefore, the ground-plane cloak obtains advantages of the nondispersive nature, the broad bandwidth, the low loss, and the ease of fabrication.

Note that resonant particles and nonresonant ones are sometimes needed simultaneously, for example, in the design of an EM black hole. We will study this interesting device in Section 7.7.

3.7 LC Particles

In previous sections, we introduced dielectric-metal resonant particles and dielectric ones. They have been primarily used to construct bulky 3D metamaterials. In addition, for the specific circumstance in waveguides, complementary resonant particles have been developed to construct "waveguided-metamaterials." The question that may arise is how one can apply metamaterials to planar RF/microwave devices and planar circuit applications? The answer is to use a new category of metamaterial particles, the LC particles.

In the year of 2002, Eleftheriades et al. studied the LC distributed network and proposed a method to realize planar negative refractive index media using periodically LC loaded transmission lines [39]. This idea is based on the well-known theory that dielectric properties such as the permittivity and the permeability can be modeled using distributed LC networks. Let us assume a classic model for positive refractive index: a network of series impedances Z and shunt admittances Y. According to the 2D telegrapher's equations, and the relation between field components and the voltages and currents in the medium, one is able to access the effective material parameters as [39]

$$j\omega\mu_e = Z, \quad \mu_e = \frac{Z}{j\omega} \quad (3.18)$$

and

$$j\omega\epsilon_e = Y, \quad \epsilon_e = \frac{Y}{j\omega}. \quad (3.19)$$

For a right-handed material (RHM), refractive index (n), effective permittivity (ϵ_e), and permeability (μ_e) are all positive. Therefore, such a medium should be represented in a low-pass topology with $L = \mu_e(H/m)$ and $C = \epsilon_e(F/m)$ being positive. The unit cell of the distributed LC network is plotted in Figure 3.17a. The corresponding propagation constant reduces to that of a standard transmission line as

$$\beta = \sqrt{-ZY} = \omega\sqrt{LC} = \omega\sqrt{\mu_e\epsilon_e}. \quad (3.20)$$

Consequently, the phase and group velocities are calculated by

$$v_\phi = \frac{\omega}{\beta} = \frac{1}{LC} = \frac{1}{\sqrt{\mu_e\epsilon_e}} = \left(\frac{\partial\beta}{\partial\omega}\right)^{-1} = v_g. \quad (3.21)$$

Equation 3.21 indicates that the phase and group velocities are both positive if L and C are simultaneously chosen to be positive. So is the refractive index, since

$$n = \frac{c}{v_\phi} = \frac{\sqrt{LC}}{\sqrt{\mu_e\epsilon_e}}. \quad (3.22)$$

Next, the scope of L and C parameters in a network are examined. From an impedance perspective, imposing a negative L and C essentially exchanges their inductive and capacitive roles so that the series inductor becomes a series capacitor ($C' > 0$), and the shunt capacitor becomes a shunt inductor ($L' > 0$), as shown in Figure 3.17b. Now, the effective medium is in a high-pass topology with

$$j\omega\mu_e = Z = \frac{1}{j\omega C'}, \quad \mu_e = -\frac{1}{\omega^2 C'} \quad (3.23)$$

and

$$j\omega\epsilon_e = Y = \frac{1}{j\omega L'}, \quad \epsilon_e = -\frac{1}{\omega^2 L'}. \quad (3.24)$$

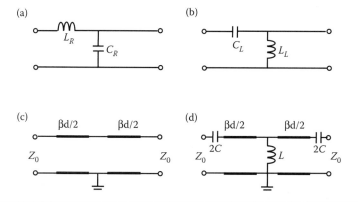

Figure 3.17 (a) The LC unit cell for right-handed materials (RHMs). (b) The LC unit cell for left-handed materials (LHMs). (c) Unloaded TL grid for RHMs. (d) Loaded TL grid for LHMs.

Clearly, the effective permittivity and permeability become negative. The corresponding propagation constant is chosen to be

$$\beta = -\sqrt{-ZY} = -\frac{1}{\omega\sqrt{L'C'}}. \quad (3.25)$$

And consequently the phase and group velocities are antiparallel as

$$v_\phi = \frac{\omega}{\beta} = -\omega^2\sqrt{L'C'}, \quad v_g = \left(\frac{\partial\beta}{\partial\omega}\right)^{-1} = +\omega^2\sqrt{L'C'}. \quad (3.26)$$

Note that a negative root is chosen in Equation 3.25 in order to ensure a positive group velocity as well as the Poynting vector which represents the flow of energy. And again the refractive index is calculated as

$$n = \frac{c}{v_\phi} = -\frac{1}{\omega^2\sqrt{L'C'}\sqrt{\mu_0\epsilon_0}} = \frac{\sqrt{\mu_e\epsilon_e}}{\sqrt{\mu_0\epsilon_0}}. \quad (3.27)$$

Because μ_e and ϵ_e are negative, a negative root is selected for Equation 3.27 and the refractive index is negative accordingly. In this way, an LHM is obtained. In addition, it has also been analyzed that lossless MNG (or ENG) media can be achieved by homogeneous TLs formed by cascading countless infinitesimally thin segments containing distributed series-shunt capacitance (or inductance). More discussions can be found in Chapter 4.

Ideally, the distributed LC networks should be infinite, including countless periodic unit cells shown in Figure 3.17a and b. For practical implementation, however, the model must be finite. A proved practical planar design is to use the LC loaded transmission line (TL) network. Figure 3.17c shows the unloaded TL grid for RHMs, whose equivalent circuit model is the unit cell in Figure 3.17a. In contrast, Figure 3.17d shows the loaded TL grid for LHMs, whose equivalent circuit model is the unit cell in Figure 3.17b. A relation has been proved for the loaded TL grid that

$$\beta d = -\sqrt{-ZY} = -\frac{1}{\omega\sqrt{LC}}. \tag{3.28}$$

The refractive index of the effective media composed of loaded TL grids is therefore calculated as

$$n = \frac{\beta d}{k_0 d} = \frac{-\sqrt{-ZY}}{\omega\sqrt{\mu_0 \epsilon_0} d} = -\frac{1}{\omega^2 \sqrt{LC}\sqrt{\mu_0 \epsilon_0} d}. \tag{3.29}$$

Note that the cell dimension d is designed to balance Equations 3.27 and 3.29.

Figure 3.18 presents an example of the LC loaded TL grid-based metamaterials. This is a planar left-handed lens that overcomes the diffraction limit. In the figure, the left-handed lens is a planar slab composed of printed metallic strips loaded with series capacitors and shunt inductors, while the right-handed surroundings are composed of printed metallic strips without any loading. The experiment

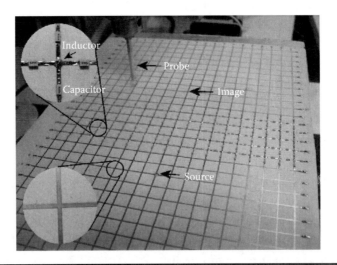

Figure 3.18 The left-handed planar transmission line lens. The unit cell of the left-handed (loaded) grid is shown in the top inset, while the unit cell of the positive-refractive-index (unloaded) grid is shown in the bottom inset. (Reprinted with permission from A. Grbic and G. V. Eleftheriades, *Phys. Rev. Lett.*, 92: 117403. Copyright 2004 by the American Physical Society.)

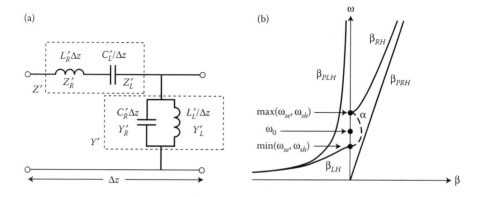

Figure 3.19 (a) Equivalent circuit model for the ideal CRLH TL. (b) Dispersion diagrams of the CRLH, PLH (β_{PLH}), and PRH (β_{PRH}) TLs. (C. Caloz and T. Itoh: *Electromagnetic Metamaterials: Transmission Line Theory and Microwave Applications.* 2004. Copyright Wiley-VCH Verlag GmbH & Co. KGaA. Reproduced with permission.)

has demonstrated a subdiffraction imaging with a half-power beamwidth of 0.21 effective wavelengths.

The LC loaded TL metamaterials have been developed by Itoh et al. around the year of 2004. Due to the fact that a pure left-handed (PLH) transmission line cannot exist physically because right-handed parasitic series inductance and shunt capacitance effects always exist for transmission lines, the CRLH transmission line model is proposed as a more general model to describe both the right-handed and left-handed properties for TL metamaterials [41,42]. A per-unit length CRLH TL consists of the impedance Z' and the admittance Y', as shown in Figure 3.19a. The impedance includes a right-handed (RH) component, the per-unit-length inductance L'_R, in series with a left-handed (LH) component, the times-unit-length capacitance C'_L. And the admittance also includes an RH component, the per-unit-length capacitance C'_R, in parallel with an LH component, the times-unit-length inductance L'_L. In this way, Z' and Y' are calculated as

$$Z'(\omega) = j\left(\omega L'_R - \frac{1}{\omega C'_L}\right), \quad Y'(\omega) = j\left(\omega C'_R - \frac{1}{\omega L'_L}\right). \tag{3.30}$$

We define the right-handed, left-handed, series, and shunt resonance frequencies respectively, as

$$\omega'_R = \frac{1}{\sqrt{L'_R C'_R}}, \quad \omega'_L = \frac{1}{\sqrt{L'_L C'_L}}, \quad \omega_{se} = \frac{1}{\sqrt{L'_R C'_L}}, \quad \omega_{sh} = \frac{1}{\sqrt{L'_L C'_R}}. \tag{3.31}$$

The explicit expression for the complex propagation constant is obtained [41]

$$\gamma = \alpha + j\beta = \sqrt{Z'Y'} = jS(\omega)\sqrt{\left(\frac{\omega}{\omega'_R}\right)^2 + \left(\frac{\omega'_L}{\omega}\right)^2 - \left(\frac{L'_R}{L'_L} + \frac{C'_R}{C'_L}\right)}, \quad (3.32)$$

where the sign function $S(\omega)$ is defined as

$$S(\omega) = \begin{cases} -1 & \text{if } \omega < \min(\omega_{se}, \omega_{sh}) \text{ LH range,} \\ 1 & \text{if } \omega > \max(\omega_{se}, \omega_{sh}) \text{ RH range.} \end{cases} \quad (3.33)$$

The corresponding dispersion diagram for CRLH is plotted in Figure 3.19b. This diagram, as well as Equations 3.32 and 3.33, indicate some important features of CRLH TL metamaterials. First of all, CRLH TL has an LH region when β is negative and the phase velocity $v_\phi = \omega/\beta$ is negative, and an RH region when β is positive and both the phase velocity v_ϕ and group velocity $v_g = \partial\omega/\partial\beta$ are positive. Accordingly, in Figure 3.19b, the CRLH dispersion curve tends to the PLH and PRH dispersion curves at lower and higher frequencies, respectively. Second, since the radicand in Equation 3.32 may be either positive or negative, the propagation constant γ can be purely real or imaginary, indicating a stop-band or a pass-band, respectively. Third, β may be zero at a nonzero frequency, which can be used to create a novel zeroth-order resonator. In addition, in Figure 3.19b, a CRLH gap is also observed between $\max(\omega_{se}, \omega_{sh})$ and $\min(\omega_{se}, \omega_{sh})$. This is due to the gap between the series resonance and the shunt one. This gap is filled when these two frequencies are equal, or, more specifically, under the condition

$$L'_R C'_L = L'_L C'_R. \quad (3.34)$$

In this case, the LH and RH contributions exactly balance each other at the frequency of ω_0, the CRLH gap closes up with nonzero group velocity, and the CRLH TL becomes a *balanced* one.

All the above-mentioned characteristics are unique for CRLH TL, and cannot be found in PLH or PRH TLs. Therefore, the CRLH metamaterials possess a variety of merits, such as reduced size, enhanced bandwidth, and composite LH/RH properties. A series of novel functional devices and antennas at microwave frequencies have been reported applying CRLH metamaterials. For example, the dual-band microwave components such as the phase shifters, matching networks, baluns, etc. [43], the broadband high-performance microstrip directional couplers [44,45], the backfire-to-endfire leaky-wave (LW) antenna [46], and the voltage-controlled LW antenna array [47]. More application examples can be found in References 41 and 42.

To implement CRLH TLs in practice, the LC network in Figure 3.19a must be realized with physical components that can generate the required capacitances

(C'_R and C'_L) and inductances (L'_R and L'_L). Two realization methods have been studied and investigated. The first one is to adopt surface-mount technology (SMT) chip components, and the second one is to use distributed components. Although the SMT chip components are readily available and easier to design, they are only available in discrete values and are limited to low frequencies. Therefore, distributed component-based CRLH TLs are also developed for many real applications. Different kinds of distributed components, for instance, the interdigital capacitors and stub inductors, have been implemented on microstrip, stripline, CPW, etc., to provide the LH and RH contributions [41].

Last, but not least, the 1D TL metamaterials can be extended to 2D models by constructing the network in one more direction. The readers are kindly referred to References 39 and 41 for details.

3.8 D.C. Particles

The above-discussed artificial particles respond to the time-varying EM field as they essentially are electric or magnetic resonators. In fact, the static fields also play an important role in various sectors. For example, they are involved in photocopy machines, electrostatic spraying systems, and electric impedance tomography (EIT). In 2007, metamaterials at zero frequency were proposed by Wood and Pendry using superconducting materials, together with the idea of the cloaking for static magnetic field [48]. Actually, Greenleaf et al. discovered the conductivity cloaking in EIT even before Pendry's design [49]. To date, the static invisibility cloaks have been investigated using conducting materials or superconductors, including two experiments on D.C. magnetic cloaks that have been delivered by two groups independently [50–54]. Recently, another approach has been proposed to realize a closed D.C. electric invisibility cloak with the aid of resistor network [55], and has been extended to achieve ultrathin closed D.C. cloak [59], D.C. ground-plane cloak [56], D.C. illusion device [57], and D. C. concentrator [58].

Figure 3.20 is the diagrammatic sketch of a D.C. cloak [55]. In Figure 3.20a, the inner sphere represents an object (could be either conducting or nonconducting) and the spherical shell represents the D.C. cloak. A bundle of electric currents flow in the background with homogeneous and isotropic conducting material in a straight way. When the inner sphere is embedded in the same background, due to the change of the conductivity profile, the same bundle of currents have significantly different distributions from their original paths. In this way, the inner sphere is detected. When the D.C. cloak is wrapped on the inner object, as shown in Figure 3.20a, it can smoothly guide the electric currents around the object. Outside the cloak, the current lines return to their original direction as if nothing happens, which makes the inner object "invisible." Figure 3.20b gives a cross section view of the current and potential distributions near the cloak. In this figure, a point source is used as the excitation.

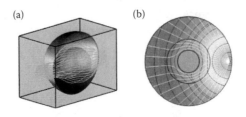

Figure 3.20 The principle of D.C. invisibility cloak. (a) The currents distribution when the object is covered by the cloak. The cloak smoothly guides the electric currents around the inner object and turns them back to the original direction. (b) The equipotential lines and current density vectors on a cross section when the cloaked object is illuminated by a point source, in which the white curves denote the current density vectors and the black curves represent the equipotential lines. (Reprinted with permission from F. Yang et al., *Phys. Rev. Lett.*, 109: 053902. Copyright 2012 by the American Physical Society.)

The EM parameters for the invisibility cloak have been investigated for the time-varying EM fields using the transformation optics (TO) theory. In the D.C. limit, it has been demonstrated in References 55 and 59 that a D.C. cloak could be realized using inhomogeneous conductivities. Detailed description on how to design a D.C. cloak will be presented in Chapter 9. Figure 3.21 plots the required radial and tangential components of the conductivity tensor for the D.C. cloak with inner and outer radii of 6 and 10 cm, respectively. It is observed that the designed D.C. cloak is made of continuous materials with anisotropic and inhomogeneous conductivities, which are difficult to obtain. However, it is possible to make an equivalence of the material to a resistor network and emulate the conductivities using the circuit theory.

Figure 3.22a illustrates a conducting plate which is discretized using the polar grids. Based on Ohm's law, each elementary cell in the grid can be implemented by two resistors

$$R_\rho = \frac{\Delta\rho}{\sigma_\rho \rho \Delta\varphi h}, \quad R_\varphi = \frac{\rho\Delta\varphi}{\sigma_\varphi \Delta\rho h}, \quad (3.35)$$

where $\Delta\rho$ and $\Delta\varphi$ are step lengths in the radial and tangential directions, respectively. Thus, the anisotropic conductivity tensor can be implemented easily using different resistors in different directions, as illustrated in Figure 3.22b. According to Figure 3.21, all required resistors have moderate values and can be easily obtained in electric stores.

In practice, the infinitely large material should be tailored to have a suitable size. Therefore, matching resistors are needed in the outer ring to emulate an infinite material, indicated as R_m in Figure 3.22b. To calculate the matching resistors, a schematic diagram is plotted in Figure 3.22d. The inner circle centering O represents a finite-sized conducting material, and the outer circle centering S is a reference

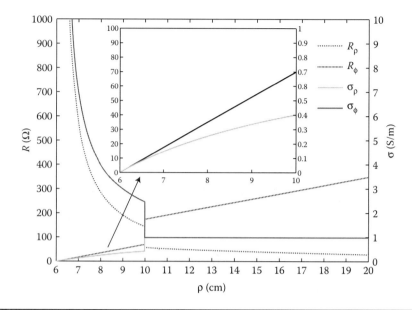

Figure 3.21 Components of anisotropic conductivity tensor required for the D.C. invisibility cloak and their corresponding resistors. The inset gives an enlarged view for details. (Reprinted with permission from F. Yang et al., *Phys. Rev. Lett.*, 109: 053902. Copyright 2012 by the American Physical Society.)

circle. A point source is located at S, and the potential can be expressed as

$$\Phi = k \ln \frac{r_0}{r}, \quad (3.36)$$

where k is a constant determined by the source strength, r_0 is the distance for the zero potential, and r is the distance to the source. After that, the potential on the periphery is obtained as

$$\Phi_p = k \ln \frac{r_0}{d}. \quad (3.37)$$

Hence the electric field is written as

$$\vec{E} = (-\nabla \varphi)_p = \frac{k}{d} \vec{e}_{\rho'}, \quad (3.38)$$

in which $\vec{e}_{\rho'} = \vec{SP}/|SP|$. Based on Figure 3.22d, we get

$$\vec{E}_\rho = |\vec{E}| \cos\beta \vec{e}_\rho = k \frac{\cos\beta}{d} \vec{e}_\rho, \quad (3.39)$$

60 ■ *Metamaterials*

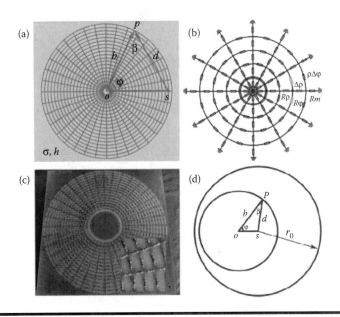

Figure 3.22 A conducting material plate, its equivalent resistor network, and the fabricated device. (a) A conducting material plate with thickness *h* and its polar grids. (b) The equivalent resistor network of the continuous material. (c) The fabricated resistor network with an enlarged view for details. (d) The diagram for derivation of the matching resistors. (Reprinted with permission from F. Yang et al., *Phys. Rev. Lett.*, 109: 053902. Copyright 2012 by the American Physical Society.)

where $\vec{e}_\rho = \vec{OP}/|OP|$ is the unit radial vector in the cylindrical system, and

$$\cos \beta = \frac{b - s \cos \varphi}{d}, \quad (3.40)$$

in which $s = |OS|$ is the distance between the source and the origin. The current flowing out of P in the radial direction is calculated as

$$I_\rho = \sigma_\rho E_\rho \cdot b \Delta \varphi \cdot h = k \frac{\sigma_\rho \cos \beta}{d} \cdot b \Delta \varphi \cdot h. \quad (3.41)$$

Hence the matching resistor is derived as

$$R_m = \frac{\phi_p}{I_\rho} = \frac{d \cdot (\ln r_0 - \ln d)}{\sigma b \Delta \varphi h \cos \beta}. \quad (3.42)$$

The resistor network can be extended to realize negative conductivity and resistance, which are necessary in some specific applications, for example, to design an

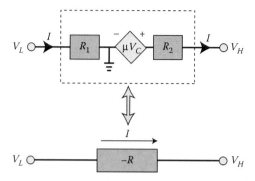

Figure 3.23 The negative-resistor model to realize negative conductivity and resistance. (F. Yang et al.: A negative conductivity material makes a dc invisibility cloak hide an object at a distance. Adv. Funct. Mater. 2013. 23(35). 4306. Copyright Wiley-VCH Verlag GmbH & Co. KGaA. Reproduced with permission.)

exterior D.C. cloak that hides an object at a distance, not inside it [60]. A negative resistor model is presented in Figure 3.23, which consists of two resistors and a controlled voltage source. Here, I represents the current running through the negative resistor, μV_C stands for the electromotive force of the controlled voltage source, μ is the controlling ratio, and V_C is the voltage of excitation in the network. Resistors R_1 and R_2 are used to match the current and voltage of the port, while V_H and V_L denote the high and low voltages, respectively, and $-R$ is the resistance of the ideal negative resistor. Since the current I running through $-R$ is $(V_H - V_L)/R$, and the proposed model is used to replace the ideal negative resistor, we can easily obtain the resistances of R_1 and R_2 as

$$R_1 = \frac{V_L}{I}, \quad R_2 = \frac{\mu V_C - V_H}{I}. \tag{3.43}$$

Because the network we fabricated is composed of linear components, the current and voltage at each node in the network change proportionally with the excitation signal, making R_1 and R_2 unchanged. Hence the model can be used to mimic an ideally negative resistor.

In the design process, the negative conductivity is first calculated using the TO theory, and then mapped to the negative resistors on a polar grid according to the relationship between the conductivity tensor and the resistor tensor, for example, the one presented in Equation 3.35. Next, a circuit simulation is performed on the resulted resistor network with ideal negative resistors, which gives the voltage and current distributions at each node, including V_H, V_L, and I mentioned above. Finally, Equation 3.43 is used to give the corresponding resistors in the circuit model.

In this chapter, we have presented an overview of the constructing unit cells of metamaterials, the electrically resonant particles and magnetically resonant particles. They serve as artificial composing atoms, and generate controllable permittivity and permeability, respectively. Designing methods, as well as fabrication techniques of resonant particles have also been discussed. Different arrangement of these particles will result in different EM characteristics, and categorize metamaterials into three groups: the "super crystal," the "super noncrystal," and the "super quasicrystal." In the following chapters, we will study these three categories of metamaterials individually.

References

1. J. B. Pendry, A. J. Holden, W. J. Stewart, and I. Youngs. Extremely low frequency plasmons in metallic mesostructures. *Phys. Rev. Lett.*, 76: 4773, 1996.
2. R. Shelby, D. R. Smith, and S. Schultz. Experimental verification of a negative index of refraction. *Science*, 292: 77, 2001.
3. W. X. Tang, H. Zhao, X. Zhou, J. Y. Chin, and T. J. Cui. Negative index material composed of meander line and SRRs. *Prog. Electromagn. Res. B*, 8: 103, 2008.
4. D. Schurig, J. J. Mock, and D. R. Smith. Electric-field-coupled resonators for negative permittivity metamaterials. *Appl. Phys. Lett.*, 88: 041109, 2006.
5. J. B. Pendry, A. J. Holden, D. J. Robbins, and W. J. Stewart. Magnetism from conductors and enhanced nonlinear phenomena. *IEEE Trans. Microw. Theory Techn.*, 47(11): 2075, 1999.
6. J. D. Baena, R. Marques, F. Medina, and J. Martel. Artificial magnetic metamaterial design by using spiral resonators. *Phys. Rev. B*, 69: 14402, 2004.
7. J. D. Baena, J. Bonache, F. Martin, R. M. Sillero, F. Falcone, T. Lopetegi, M. A. G. Laso et al. Equivalent-circuit models for split-ring resonators and complementary split-ring resonators coupled to planar transmission lines. *IEEE Trans. Microw. Theory Techn.*, 53(4): 1451, 2005.
8. H. Chen, L. Ran, J. Huangfu, X. Zhang, and K. Chen. Left-handed materials composed of only S-shaped resonators. *Phys. Rev. E*, 70: 057605, 2004.
9. R. Liu, C. Ji, J. J. Mock, J. Y. Chin, T. J. Cui, and D. R. Smith. Broadband ground-plane cloak. *Science*, 323: 366, 2009.
10. Q. Cheng, W. X. Jiang, and T. J. Cui. Radiation of planar electromagnetic waves by a line source in anisotropic metamaterials. *J. Phys. D: Appl. Phys.*, 43: 335406, 2010.
11. I. V. Shadrivov, S. K. Morrison, and Y. S. Kivshar. Tunable split-ring resonators for nonlinear negative-index metamaterials. *Opt. Express*, 14: 9344, 2006.
12. I. Gil, J. Bonache, J. Garcia-Garcia, and F. Martin. Tunable metamaterial transmission lines based on varactor-loaded split-ring resonators. *IEEE Trans. Microw. Theory Techn.*, 54(6): 2665, 2006.
13. D. Huang, E. Poutrina, and D. R. Smith. Analysis of the power dependent tuning of a varactor-loaded metamaterial at microwave frequencies. *Appl. Phys. Lett.*, 96: 104104, 2010.
14. N. Xiang, Q. Cheng, J. Zhao, T. J. Cui, H. F. Ma, and W. X. Jiang. Switchable zero-index metamaterials by loading positive-intrinsic-negative diodes. *Appl. Phys. Lett.*, 104: 053504, 2014.

15. F. Falcone, T. Lopetegi, M. A. G. Laso, J. D. Baena, J. Bonache, M. Beruete, R. Marques, F. Martin, and M. Sorolla. Babinet principle applied to the design of metasurfaces and metamaterials. *Phys. Rev. Lett.*, 93: 197401, 2004.
16. D. Schurig, J. J. Mock, B. J. Justice, S. A. Cummer, J. B. Pendry, A. F. Starr, and D. R. Smith. Metamaterial electromagnetic cloak at microwave frequencies. *Science*, 314: 977, 2006.
17. D. R. Smith, J. J. Mock, A. F. Starr, and D. Schurig. Gradient index metamaterials. *Phys. Rev. E*, 71: 036609, 2005.
18. F. Martin, F. Falcone, J. Bonache, R. Marques, and M. Sorolla. Miniaturized coplanar waveguide stop band filters based on multiple tuned split ring resonators. *IEEE Microw. Wireless Compon. Lett.*, 13: 511, 2003.
19. J. Garcia, F. Martin, F. Falcone, J. Bonache, I. Gil, T. Lopetegi, M. A. G. Laso, M. Sorolla, and R. Marques. Spurious passband suppression in microstrip coupled line band pass filters by means of split ring resonators. *IEEE Microw. Wireless Compon. Lett.*, 14(9): 416, 2004.
20. J. Naqui, A. Fernandez-Prieto, M. Duran-Sindreu, F. Mesa, J. Martel, F. Medina, and F. Martin. Common-mode suppression in microstrip differential lines by means of complementary split ring resonators: Theory and applications. *IEEE Trans. Microw. Theory Techn.*, 60(10): 3023, 2012.
21. R. Liu, Q. Cheng, T. Hand, J. J. Mock, T. J. Cui, S. A. Cummer, and D. R. Smith. Experimental demonstration of electromagnetic tunneling through an epsilon-near-zero metamaterial at microwave frequencies. *Phys. Rev. Lett.*, 100: 023903, 2008.
22. M. M. Bait-Suwailam, O. F. Siddiqui, and O. M. Ramahi. Mutual coupling reduction between microstrip patch antennas using slotted-complementary split-ring resonators. *IEEE Antenn. Wireless Propag. Lett.*, 9: 876, 2010.
23. L. Lewin. The electrical constants of a material loaded with spherical particles. *Inst. Elec. Eng.*, 94: 65, 1947.
24. C. L. Holloway, E. F. Kuester, J. Baker-Jarvis, and P. Kabos. A double negative (DNG) composite medium composed of magnetodielectric spherical particles embedded in a matrix. *IEEE Trans. Antenn. Propag.*, 51(10): 2596, 2003.
25. D. Bao, E. Kallos, W. Tang, C. Argyropoulos, Y. Hao, and T. J. Cui. A broadband simplified free space cloak realized by nonmagnetic dielectric cylinders. *Front. Phys. China*, 5(3): 319, 2010.
26. H. F. Ma, B. G. Cai, T. X. Zhang, Y. Yang, W. X. Jiang, and T. J. Cui. Three-dimensional gradient-index materials and their applications in microwave lens antennas. *IEEE Trans. Antenn. Propag.*, 61(5): 2561, 2013.
27. H. F. Ma and T. J. Cui. Three-dimensional broadband and broad-angle transformation-optics lens. *Nat. Commun.*, 1: 124, 2010.
28. V. G. Veselago. The electrodynamics of substances with simultaneously negative values of ϵ and μ. *Sov. Phys. Usp.*, 10: 509, 1968.
29. D. Smith, W. Padilla, D. Vier, S. Nemat-Nasser, and S. Schultz. Composite medium with simultaneously negative permeability and permittivity. *Phys. Rev. Lett.*, 84: 4184, 2000.
30. J. Pendry. Negative refraction makes a perfect lens. *Phys. Rev. Lett.*, 85: 3966, 2000.
31. R. W. Ziolkowski. Superluminal transmission of information through an electromagnetic metamaterial. *Phys. Rev. E*, 63: 046604, 2001.
32. D. R. Smith, J. B. Pendry, and M. C. K. Wiltshire. Metamaterials and negative refractive index. *Science*, 305: 788, 2004.

33. J. Pendry and S. Ramakrishna. Focusing light using negative refraction. *J. Phys. Cond. Matter*, 15: 6345, 2003.
34. A. Grbic and G. Eleftheriades. Overcoming the diffraction limit with a planar left-handed transmission-line lens. *Phys. Rev. Lett.*, 92: 117403, 2004.
35. J. Baena, L. Jelinek, R. Marques, and F. Medina. Near-perfect tunneling and amplification of evanescent electromagnetic waves in a waveguide filled by a metamaterial: Theory and experiments. *Phys. Rev. B*, 72: 075116, 2005.
36. R. Greegor, C. Parazzoli, K. Li, and M. Tanielian. Origin of dissipative losses in negative index of refraction materials. *Appl. Phys. Lett.*, 82: 2356, 2003.
37. D. R. Smith, J. J. Mock, A. F. Starr, and D. Schurig. Gradient index metamaterials. *Phys. Rev. E*, 71: 036609, 2005.
38. H. F. Ma and T. J. Cui. Three-dimensional broadband ground-plane cloak made of metamaterials. *Nat. Commun.*, 1: 21, 2010.
39. G. V. Eleftheriades, A. K. Iyer, and P. C. Kremer. Planar negative refractive index media using periodically L-C loaded transmission lines. *IEEE Trans. Microw. Theory Techn.*, 50(12): 2702, 2002.
40. A. Grbic and G. V. Eleftheriades. Overcoming the diffraction limit with a planar left-handed transmission-line lens. *Phys. Rev. Lett.*, 92: 117403, 2004.
41. C. Caloz and T. Itoh. TL theory of MTMs. *Electromagnetic Metamaterials: Transmission Line Theory and Microwave Applications*. John Wiley & Sons, Inc., Hoboken, New Jersey, USA, 2004.
42. A. Lai, C. Caloz, and T. Itoh. Composite right/left-handed transmission line metamaterials. *IEEE Microw. Mag.*, 5(3): 34, 2004.
43. I. Lin, M. DeVincentis, C. Caloz, and T. Itoh. Arbitrary dual-band components using composite right/left-handed transmission lines. *IEEE Trans. Microw. Theory Techn.*, 52(1): 1142, 2004.
44. C. Caloz and T. Itoh. A novel mixed conventional microstrip and composite right/left-handed backward-wave directional coupler with broadband and tight coupling characteristics. *IEEE Microw. Wireless Compon. Lett.*, 14: 31, 2004.
45. C. Caloz, A. Sanada, and T. Itoh. A novel composite right/lefthanded coupled-line directional coupler with arbitrary coupling level and broad bandwidth. *IEEE Trans. Microw. Theory Techn.*, 52: 980, 2004.
46. L. Liu, C. Caloz, and T. Itoh. Dominant mode (DM) leaky-wave antenna with backfire-to-endfire scanning capability. *Electron. Lett.*, 38(23): 1414, 2000.
47. S. Lim, C. Caloz, and T. Itoh. Metamaterial-based electronically controlled transmission line structure as a novel leaky-wave antenna with tunable radiation angle and beamwidth. *IEEE Trans. Microw. Theory Techn.*, 53(1): 161, 2005.
48. B. Wood and J. B. Pendry. Metamaterials at zero frequency. *J. Phys. Condens. Matter*, 19: 076208, 2007.
49. A. Greenleaf, M. Lassas, and G. Uhlmann. Anisotropic conductivities that cannot be detected by EIT. *Uhlmann, Physiol. Meas.*, 24: 413, 2003.
50. T. Chen, C. N. Weng, and J. S. Chen. Cloak for curvilinearly anisotropic media in conduction. *Appl. Phys. Lett.*, 93: 114103, 2008.
51. F. Magnus, B. Wood, J. Moore, K. Morrison, G. Perkins, J. Fyson, M. C. K. Wiltshire, D. Caplin, L. F. Cohen, and J. B. Pendry. A d.c. magnetic metamaterial. *Nat. Mater.*, 7: 295, 2008.
52. A. Sanchez, C. Navau, J. Prat-Camps, and D. Chen. Antimagnetic: Controlling magnetic fields with superconductor-metamaterial hybrids. *New. J. Phys.*, 13: 093034, 2011.

53. S. Narayana and Y. Sato. DC magnetic cloak. *Adv. Mater.*, 24: 71, 2012.
54. F. Gomory, M. Solovyov, J. Souc, C. Navau, J. Prat-Camps, and A. Sanchez. Experimental realization of a magnetic cloak. *Science*, 335: 1466, 2012.
55. F. Yang, Z. L. Mei, T. Y. Jin, and T. J. Cui. DC electric invisibility cloak. *Phys. Rev. Lett.*, 109: 053902, 2012.
56. Z. L. Mei, Y. S. Liu, F. Yang, and T. J. Cui. A dc carpet cloak based on resistor networks. *Opt. Express*, 20: 25758, 2012.
57. M. Liu, Z. L. Mei, X. Ma, and T. J. Cui. DC illusion and its experimental verification. *Appl. Phys. Lett.*, 101: 051905, 2012.
58. W. X. Jiang, C. Y. Luo, H. F. Ma, Z. L. Mei, and T. J. Cui. Enhancement of current density by dc electric concentrator. *Sci. Rep.*, 2: 956, 2012.
59. W. X. Jiang, C. Y. Luo, Z. L. Mei, and T. J. Cui. An ultrathin but nearly perfect direct current electric cloak. *Appl. Phys. Lett.*, 102: 014102, 2012.
60. F. Yang, Z. L. Mei, X. Y. Yang, T. Y. Jin, and T. J. Cui. A negative conductivity material makes a dc invisibility cloak hide an object at a distance. *Adv. Funct. Mater.*, 23(35): 4306, 2013.

Chapter 4
Homogeneous Metamaterials: Super Crystals

As addressed in Chapter 1, homogeneous metamaterials are constructed by periodically arranged identical artificial "atoms" with subwavelength scales. In the microwave region, these meta-atoms usually have macroscopic sizes and may take one of the following forms: 3D volumetric particles, planar complementary or waveguided (WG) elements, and LC-loaded transmission line (TL) grids or meshes (see Chapter 3). Metamaterials implemented from these three kinds of meta-atoms are referred to as bulk, WG, and TL metamaterials, respectively. By reasonably choosing the periodicity, the structure and the dimensions or other characteristic values of those meta-atoms, homogeneous metamaterials are able to achieve numerous kinds of effective medium properties, including those difficult or even impossible to realize by naturally occurring materials. Heretofore, homogeneous metamaterials have been mainly involved in realizing four categories of effective media: SNG media, DNG media, ZIM, and DPS media. An introduction on how to build homogeneous metamaterials with these effective media properties in the microwave band is given in Section 4.1. Afterward, some typical EM wave-controlling applications of those homogeneous metamaterials are presented in Sections 4.2 through 4.5.

4.1 Homogeneous Metamaterials: Periodic Arrangements of Particles

4.1.1 SNG Metamaterials

SNG media is defined as media in which only *one* of the two parameters ϵ and μ has the negative real part. A medium with negative real permittivity (and positive real permeability) is called ENG medium, while a medium with negative real permeability (and positive real permittivity) is called μ-negative (MNG) medium. Naturally occurring ENG media, also known as electric plasma, operate effectively in extremely high-frequency regimes. For example, noble metals (e.g., silver, gold) exhibit a low-loss ENG characteristic in the infrared (IR), visible, or even ultraviolet frequency domains due to their high plasma frequencies. As for MNG media, the generation of MNG property must rely on the magnetic response of the media. However, at frequencies up to the low end of microwave band, the magnetic response of most natural materials begins to tail off. Though individual materials, such as ferrites, remain moderately active and may exhibit MNG characteristic at a few gigahertz, they are often heavy, difficult to tune, and may not have desirable mechanical properties. Contrary to natural materials, metamaterials composed of artificial constructs are able to conquer the above constraints and have proved to be a reasonable means to obtain satisfactory SNG properties in microwave band. Strictly speaking, the definition of SNG medium here applies to isotropic materials, but most practical and realistic homogeneous metamaterials with SNG property (SNG metamaterials for short) to date are anisotropic. Typical anisotropic ENG (or MNG) metamaterials have diagonal permittivity and permeability tensors and only one or two elements of the diagonal permittivity (or permeability) tensor have negative real part. The following text of this section will focus on such SNG metamaterials.

The first MNG metamaterial capable of operating at microwave frequencies was proposed by Pendry and his coworkers in 1999 [1] and is illustrated in Figure 4.1a. The constituent particle of this metamaterial, as shown in the inset of Figure 4.1a, consists of two thin metallic sheets of thickness t_m and conductivity σ_m. These two sheets are in the shape of two flat concentric rings, each interrupted by a small gap, and the two similar split rings are coupled by means of a strong distributed capacitance in the region between the rings. Such planar particle, which is called edge-coupled split ring resonator (EC-SRR for short) [2], is usually printed on the surface of a dielectric substrate for which the thickness, relative permittivity, and loss tangent are denoted by t_s, Dk, and Df, respectively. In practice, the composite structure or metamaterial in Figure 4.1a is assembled by periodically stacking a certain number of identical solid boards along the z axis. Every solid board contains the same two-dimensionally periodic array of EC-SRR on one side and can be fabricated from metal-clad laminate using conventional photolithography. Note that in each of these EC-SRR arrays, every two neighboring EC-SRRs along the y axis are oriented "symmetrically," that is to say, each successive EC-SRR along the y axis is rotated in plane by 180° relative to its neighbor, as depicted in Figure 4.1a. These rotations symmetrize the EC-SRR array, adding an xz mirror plane and doubling the unit cell, and hence suppress the magnetoelectric coupling that could give rise to bianisotropic response for the composite structure. As discussed in Chapter 3, the EC-SRR particle exhibits a resonant magnetic response to external (or local) magnetic field applied along the z axis, resulting in a negative effective permeability along this direction in the band just above the resonance frequency. Since the gaps of split rings and the coupling between concentric rings strongly decrease the resonance frequency of the system, the electrical size of the EC-SRR in the MNG band may be designed to be adequately small. Typical response curves of effective relative permittivity ϵ_y and permeability μ_z of EC-SRR-based metamaterial (EC-SRR MTM) with certain configurations are obtained using the standard retrieval method introduced in Chapter 2 and demonstrated in Figure 4.1b. The scattering parameters used for retrieval are simulated by Ansoft HFSS (a finite element Maxwell's equations solver). In the simulation, a single EC-SRR particle is excited by plane waves propagating along the x axis and having electric field polarized along the y axis. Such polarization is realized by assigning perfect electric and magnetic conducting boundaries to the bounding surfaces normal to the y and z axis, respectively. Note that the lattice size along each axis is 3.333 mm in the instance here, which is apparently less than a tenth of the free space wavelength in the MNG band

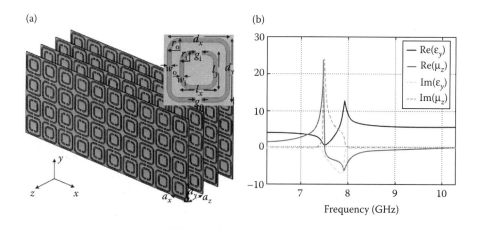

Figure 4.1 (a) Visualization of metamaterial formed by periodic and symmetric arrangement of EC-SRRs, where a_x, a_y, and a_z are the lattice constants. The geometric parameters of EC-SRR: d_x, d_y, l_x, l_y, g_i, g_o, r_i, r_o, w_i, and w_o are defined in the inset. (b) Frequency response of effective relative permittivity ϵ_y and permeability μ_z for EC-SRR-based metamaterial with the following specifications: $a_x = a_y = a_z = 3.333$ mm, $d_x = d_y = 3.2$ mm, $l_x = l_y = 2.2$ mm, $g_i = g_o = 0.2$ mm, $r_i = 0.4$ mm, $r_o = 0.6$ mm, $w_i = w_o = 0.3$ mm, $t_m = 0.035$ mm, $\sigma_m = 5.8 \times 10^7$ S/m, $t_s = 0.48$ mm, $Dk = 2.65$, and $Df = 0.001$.

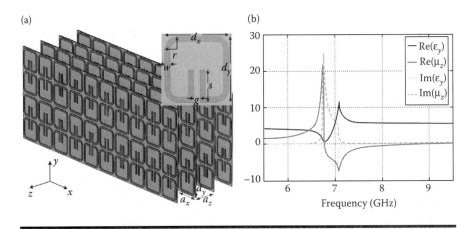

Figure 4.2 (a) Visualization of metamaterial formed by periodic and symmetric arrangement of SRRs, where a_x, a_y, and a_z are the lattice constants. The geometric parameters of SRR: d_x, d_y, r, s, w, and g are defined in the inset. (b) Frequency response of effective relative permittivity ϵ_y and permeability μ_z for SRR-based metamaterial with the following specifications: $a_x = a_y = a_z = 3.333$ mm, $d_x = d_y = 3.2$ mm, $r = 0.6$ mm, $s = 1.5$ mm, $w = 0.3$ mm, $g = 0.2$ mm, $t_m = 0.035$ mm, $\sigma_m = 5.8 \times 10^7$ S/m, $t_s = 0.48$ mm, $Dk = 2.65$, and $Df = 0.001$.

(about 8–9.6 GHz). Figure 4.1b also reveals an enhanced positive permittivity ϵ_y, indicating that the EC-SRR particle also couples to electric fields along the y axis. The EC-SRR MTM is well investigated in many publications [1–9].

Figure 4.2a depicts a modification of EC-SRR MTM, which is implemented by replacing all EC-SRRs in Figure 4.1a with single-ring SRRs [10]. The SRR-based metamaterial (SRR MTM) also exhibits MNG property when local magnetic field is directed along the axis of split rings. It is easier to synthesize than the EC-SRR MTM since the SRR particle is less complicated than the EC-SRR particle. Although the SRR particle does not own strong capacitance between concentric rings as the EC-SRR does, its local magnetic resonance can still occur at long wavelength as long as the split capacitance is tuned to be sufficiently large by increasing the length of split s. Typical response curves of retrieved relative permittivity ϵ_y and permeability μ_z for SRR MTM are demonstrated in Figure 4.2b.

The MNG metamaterial shown in Figure 4.3a is constructed by periodically arranging cylindrical disks with pure dielectric properties in three dimensions [11]. Each of these disks has very high permittivity (denoted by ϵ_{DR}) and is a high-quality resonator with most of EM fields localized inside the disk. When illuminated with a plane wave having magnetic field polarized along the z axis and propagating along the x axis, the dielectric disk provides magnetic dipole moment aligned with the external magnetic field at its lowest-order resonate mode since the electric field is oriented in a loop (i.e., in the yoz plane) around the disk for this mode. As a result,

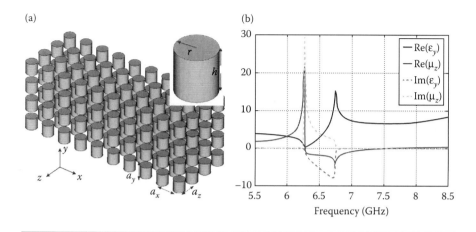

Figure 4.3 (a) Visualization of metamaterial formed by periodic arrangement of dielectric disk resonators, where a_x, a_y, and a_z are the lattice constants and r and h are the radius and height of disk, respectively. (b) Frequency response of effective relative permittivity ϵ_y and permeability μ_z for dielectric disk-based metamaterial with the following specifications: $a_x = a_y = a_z = 5$ mm, $\epsilon_{DR} = 100 + i0.1$, $h = 4$ mm, $r = 2$ mm.

the dielectric disk-based metamaterial exhibits strong magnetism at this resonant mode and its effective permeability μ_z turns negative just above the resonant frequency. An instance of retrieved frequency-dependent relative permittivity ϵ_y and permeability μ_z for dielectric disk-based metamaterial is shown in Figure 4.3b. In addition to cylindrical disk, other dielectric resonators such as dielectric sphere and rectangular block can also be utilized as the constituent particle of MNG metamaterials [11,12]. These dielectric resonator-based metamaterials (DR MTM) are free of conduction loss and offer higher-efficiency performance in comparison to the aforementioned metallic resonator-based metamaterials (e.g., SRR MTM).

In 1996, Pendry et al. realized an artificially electric plasma comprising regular lattices of thin metallic wires in which the effective plasma frequency can be depressed into the far IR or even gigahertz band [13] (see Chapter 3). Such manmade electric plasma, which is called wire media, is the first ENG metamaterial proposed for microwave application. A simple case of wire media, which characterizes a doubly periodic array of parallel conducting wires, is shown in Figure 4.4a. The simple wire media is uniaxially anisotropic and only its axial permittivity (along the z axis) behaves like electric plasma. Figure 4.4b illustrates a typical frequency response of the axial permittivity, which is retrieved from simulated scattering parameters with plane waves propagating along the x axis and incident electric field polarized along the wires. From Figure 4.4b, a fairly broad ENG band is observed below the effective plasma frequency (10.67 GHz in this instance). Generally, to design wire media with lattice periods which are much smaller than the operating

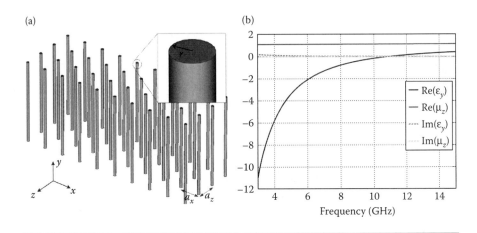

Figure 4.4 (a) Visualization of simple wire media formed by 2D periodic arrangement of parallel and continuous conducting wires, where a_x and a_z are the lattice constants and r is the radius of wire. (b) Frequency response of effective relative permittivity ϵ_y and permeability μ_z for simple wire media with array periodicity $a_x = a_y = 5$ mm and wire radius $r = 10$ μm. The wires are made from copper.

wavelength in microwave region requires extremely thin (and lossy) wires which are hard to fabricate. For example, when the array periods of the simple wire media shown in Figure 4.4a are $a_x = a_y = 5$ mm, the radii of the wires should be as small as 10 μm to achieve an effective plasma frequency of about 10.9 GHz. Another remarkable point about wire media is that they exhibit strong spatial dispersion regardless of the wavelength relative to the lattice spacing [14,15]. Moreover, ideal wire media requires infinitely long wires and breaks or terminations of the wires in critical areas (which is inevitable in realization) will cause significant property changes in these areas. Finally, to obtain an isotropic wire media, one should bring in three doubly periodic arrays of parallel metallic wires which are orthogonal to one another and the three sets of orthogonal wires must be electrically connected to each other. This is very tedious to implement and may be a barrier to mass production.

To overcome the drawbacks of wire media, Schurig et al. introduced a novel planar inductive-capacitive (LC) resonator as the building block of ENG metamaterials [16]. Such a resonator, referred to as an ELCR, is a local and self-contained oscillator. It consists of two identical single-ring SRRs put together on the split gap side, as shown in the inset of Figure 4.5a. The two split gaps are combined to form a capacitor-like structure, which couples strongly to electric field normal to the capacitor arms and is connected in parallel to two inductive loops. This allows the external electric field to drive the fundamental LC resonance for the resonator, making it possible to realize negative effective permittivity along the y axis above the resonate frequency. Besides, because the two inductive loops are equivalent but oppositely

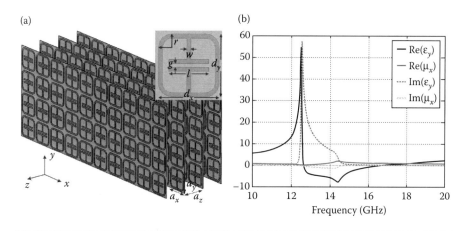

Figure 4.5 (a) Visualization of metamaterial formed by periodic arrangement of ELC resonators (ELCRs), where a_x, a_y, and a_z are the lattice constants. The geometric parameters of ELCR: d_x, d_y, r, l, w, and g are defined in the inset. (b) Frequency response of effective relative permittivity ϵ_y and permeability μ_z for ELCR-based metamaterial with the following specifications: $a_x = a_y = a_z = 3.333$ mm, $d_x = d_y = 3$ mm, $r = 0$ mm, $l = 1$ mm, $w = 0.25$ mm, $g = 0.25$ mm, $t_m = 0.017$ mm, $\sigma_m = 5.8 \times 10^7$ S/m, $t_s = 0.203$ mm, $Dk = 3.75$, and $Df = 0.02$.

wound with respect to the capacitor, a uniform magnetic field cannot drive the fundamental LC resonance and the electrically induced currents flow in such a manner as to cancel out the magnetoelectric coupling. An ENG metamaterial built from ELCRs is depicted in Figure 4.5a. It is constructed by periodically stacking several dielectric boards each printed with 2D periodic array of ELCRs. Such construction scheme is similar to that of SRR MTM presented above. Typical resonant responses of permittivity ϵ_y and permeability μ_x, obtained via standard retrieval process, are shown in Figure 4.5b, where the ENG band is found to appear at 15–16.6 GHz. Compared with wire media, the ELCR-based metamaterial (ELCR MTM) is more robust with regard to maintaining its bulk properties close to a boundary or interface. Moreover, it does not require intercell or interplane electrical connectivity to extend the anisotropic metamaterial shown in Figure 4.5a to those with negative-permittivity properties in two or three dimensions. The ELCR particle can also take another form, in which the two identical SRRs are placed back to back on the no-split side, as shown in the inset of Figure 4.6a [17]. Since now two capacitive gaps exist for each ELCR particle, the electric resonant frequency and hence the ENG band is pushed down to some extent, which can be seen by comparing the effective medium parameters shown in Figures 4.5b and 4.6b. Note that the two instances of ELCR-based metamaterials involved in the comparison hold the same laminate, lattice constants, and particle size. Besides, the metallic line width and gap width of particles are fixed as 0.25 mm for these two instances.

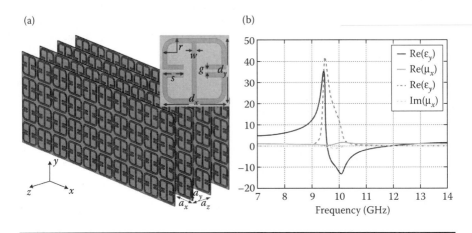

Figure 4.6 (a) Visualization of metamaterial formed by periodic arrangement of modified ELCRs, where a_x, a_y, and a_z are the lattice constants. The geometric parameters of modified ELCR: d_x, d_y, r, s, w, and g are defined in the inset. (b) Frequency response of effective relative permittivity ϵ_y and permeability μ_z for modified ELCR-based metamaterial with the following specifications: $a_x = a_y = a_z = 3.333$ mm, $d_x = d_y = 3$ mm, $r = 0$ mm, $s = 1$ mm, $w = 0.25$ mm, $g = 0.25$ mm, $t_m = 0.017$ mm, $\sigma_m = 5.8 \times 10^7$ S/m, $t_s = 0.203$ mm, $Dk = 3.75$, and $Df = 0.02$.

Figure 4.7a demonstrates another approach to implementing bulk ENG metamaterials with relatively small unit-to-wavelength ratio. The planar constituent particle involved in this scheme is called MLR [18], whose geometry is shown in the inset of Figure 4.7a. The MLR itself provides internal inductance and capacitance with respect to local electric field along the y axis, giving resonate electric response and hence negative permittivity component ϵ_y. In contrast to the self-contained resonance of ELCR, the cell-to-cell capacitive coupling between neighboring MLRs along the y axis also contributes to the meander line resonance. Such mutual coupling together with the large inductance offered by folded conducting lines drastically reduces the resonant frequency of MLR. Typical effective medium performance of MLR-based metamaterial (MLR MTM) is shown in Figure 4.7b.

The SNG metamaterials presented above are all made up of 3D volumetric particles and belong to the category of bulk metamaterial, which is inconvenient to incorporate with planar microwave circuits or devices. Alternatively, Figures 4.8a and 4.9a demonstrate another two instances of SNG metamaterials implemented from WG elements rather than volumetric particles. These two instances belong to the metamaterial category called WG-MTM [19,20], which is appropriate for metamaterial applications in planar waveguide environments. As shown, WG-MTM generally consists of two parallel metallic plates constituting a

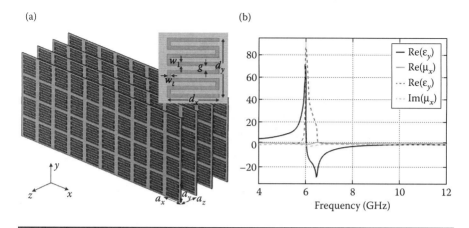

Figure 4.7 (a) Visualization of metamaterial formed by periodic arrangement of MLRs, where a_x, a_y, and a_z are the lattice constants. The geometric parameters of MLR: d_x, d_y, w_l, w_t, and g are defined in the inset. (b) Frequency response of effective relative permittivity ϵ_y and permeability μ_z for MLR-based metamaterial with the following specifications: $a_x = a_y = a_z = 4$ mm, $d_x = 2.7$ mm, $d_y = 3.77$ mm $w_l = w_t = 0.2$ mm, $g = 0.31$ mm, $t_m = 0.035$ mm, $\sigma_m = 5.8 \times 10^7$ S/m, $t_s = 0.5$ mm, $Dk = 2.65$, and $Df = 0.001$.

planar waveguide, and the lower plate is etched with planar pattern. This pattern includes many identically structured slits, which are arranged periodically along the two orthogonal directions (i.e., the x and y directions here). It is these structured slits that are referred to as WG elements in the context. The spacing between the two plates h is kept to be smaller than half wavelength to ensure that the second-order mode (TE mode) is cut off and only the dominant TEM mode is supported by the planar waveguide. When each element has electrically small size (usually less than one eighth of a wavelength) and behaves as a local electric or magnetic responser, the volume in between the two metallic plates and right above the etched pattern can be viewed as being occupied by a block of effective medium, namely, WG-MTM, with respect to the dominant TEM mode. By altering the geometry and dimensions of the WG elements, different effective medium properties can be achieved for WG-MTM. However, considering the field orientations of the dominant TEM mode, only the permittivity component ϵ_z and permeability components μ_x and μ_y are meaningful for WG-MTM.

The WG-MTM shown in Figure 4.8a employs WG element called CSRR. The CSRR structure was proposed by Falcone et al. [21], who showed by use of the Babinet principle that the CSRR couples to an external electric field directed along the normal of the CSRR surface, leading to an electric resonance. Hence it is possible for the CSRR-based WG-MTM to provide negative permittivity ϵ_z just above the resonance frequency [19,22]. In contrast, the WG-MTM shown in Figure 4.9a, which

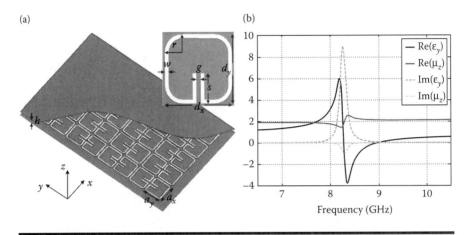

Figure 4.8 (a) Visualization of WG metamaterial formed by pattering periodic array of CSRR in the lower plate of a planar waveguide, where a_x and a_y are the periodicity of the array and h is the height of waveguide. The geometric parameters of CSRR: d_x, d_y, r, w, g, and s are defined in the inset. (b) Frequency response of effective relative permittivity ϵ_z and permeability μ_y for CSRR-based WG metamaterial with the following specifications: $a_x = a_y = 3.333$ mm, $d_x = d_y = 3$ mm, $r = 0$, $w = 0.3$ mm, $g = 0.3$ mm, $s = 1.967$ mm, $h = 1$ mm, $t_m = 0.018$ mm, $t_s = 0.2026$ mm, $Dk = 4.4$, and $Df = 0.02$.

characterizes WG element called complementary electric-LC resonator (CELCR), offers an MNG performance owing to the magnetic resonant behavior of CELCR driven by external magnetic field applied along the y axis [23]. Typical frequency response of effective relative permittivity ϵ_z and permeability μ_y for CSRR-based and CELCR-based WG-MTM are shown in Figures 4.8b and 4.9b, respectively. Note that for both examples of WG-MTM here, the bottom plate etched with periodic pattern is fabricated from single-sided copper-clad laminate (CCL)[*] and hence there is a dielectric substrate attached to and beneath the bottom plate. Besides, the CSRRs in Figure 4.8a are symmetrically arranged along the y axis as to suppress the magnetoelectric coupling for the CSRR-based WG-MTM.

The effective constitutive parameters in Figures 4.8b and 4.9b are retrieved using the same retrieval method as that for bulk metamaterials composed of 3D volumetric particles. The scattering parameters used for retrieval are obtained through simulations on a single WG element (i.e., CSRR or CELCR), using the commercial full-wave EM solver, Ansoft HFSS. In the simulations, the single-element WG-MTM is excited by the dominant guided TEM wave propagating along the

---

[*] The characteristic of the laminate is described by the following quantities: copper thickness t_m, thickness of dielectric substrate t_s, relative permittivity Dk, and loss tangent Df for the substrate.

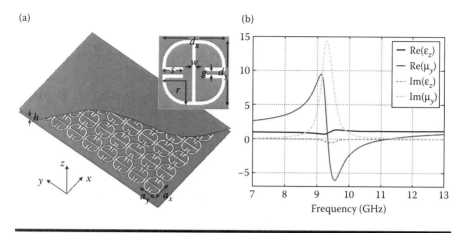

Figure 4.9 (a) Visualization of WG metamaterial formed by pattering periodic array of CELCR in the lower plate of a planar waveguide, where a_x and a_y are the periodicity of the array and h is the height of waveguide. The geometric parameters of CELCR: d_x, d_y, r, w, g, and s are defined in the inset. (b) Frequency response of effective relative permittivity ϵ_z and permeability μ_y for CELCR-based WG metamaterial with the following specifications: $a_x = a_y = 3.333$ mm, $d_x = d_y = 3$ mm, $r = 0$, $w = 0.3$ mm, $g = 0.3$ mm, $s = 1.05$ mm, $h = 1$ mm, $t_m = 0.018$ mm, $t_s = 0.2026$ mm, $Dk = 4.4$, and $Df = 0.02$.

x axis and the polarization of the TEM wave is constrained by perfect magnetic conducting (PMC) boundaries employed on two sides of the computational domain along the y axis.

Figure 4.10 presents typical TL implementation (or representation) of SNG media* in a lossless and one-dimensional (1D) sense. As demonstrated in

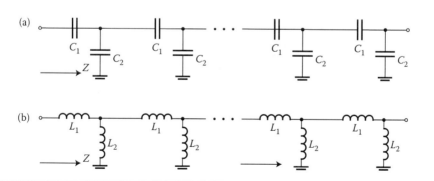

Figure 4.10 (a) The C–C cascaded circuit representing MNG media. (b) The L–L cascaded circuit representing ENG media.

---

* In the context, TL representation of SNG media is also referred to as TL SNG metamaterial.

Reference 24, lossless MNG (or ENG) media can be simulated by homogeneous TLs formed by cascading countless infinitesimally thin segments containing distributed series-shunt capacitance (or inductance). However, such homogeneous TLs do not appear to exist in nature. In practice, they are approximated by periodically cascaded networks with finite-length unit cells for the sake of physical realization. For MNG (or ENG) media, the cascading unit cells of these periodic networks take the form of lumped series-shunt capacitor (or inductor), resulting in the cascaded structure shown in Figure 4.10a or b.

Like SNG media, the series-shunt capacitor (C–C) and the series-shunt inductor (L–L) cascaded circuits depicted in Figure 4.10 support only evanescent waves. By applying periodic boundary conditions (PBCs) related to the Bloch–Floquet theorem to the unit cell, the evanescent eigenmodes of the C–C and L–L cascaded circuits can be solved. Then the effective permittivity and permeability of these two structures can be derived in closed forms from the attenuation constants and the purely imaginary Bloch impedances of the eigenmodes. For the C–C cascaded circuit, the per-unit cell attenuation index corresponding to the eigenmode decaying along the positive z axis reads

$$\theta_C = 2\ln\left(\sqrt{\frac{C_2}{4C_1} + 1} - \sqrt{\frac{C_2}{4C_1}}\right). \tag{4.1}$$

And the Bloch impedance of such a mode is expressed as

$$Z_B = \frac{\left(e^{\theta_C} - 1\right)}{i\omega C_2}. \tag{4.2}$$

Therefore, the effective permittivity and permeability of the C–C cascaded circuit can be extracted from Equations 4.1 and 4.2,

$$\epsilon_{CC} = \frac{C_2 \theta_C}{\left(e^{\theta_C} - 1\right)p}, \quad \mu_{CC} = -\frac{\theta_C\left(e^{\theta_C} - 1\right)}{\omega^2 C_2 p}, \tag{4.3}$$

where p is the unit cell length, which is much smaller than free space wavelength. Similarly, the effective permittivity and permeability for the L–L cascaded circuit can be derived as

$$\epsilon_{LL} = -\frac{\theta_L\left(e^{\theta_L} - 1\right)}{\omega^2 L_1 p}, \quad \mu_{LL} = \frac{L_1 \theta_L}{\left(e^{\theta_L} - 1\right)p}, \tag{4.4}$$

where

$$\theta_L = 2\ln\left(\sqrt{\frac{L_1}{4L_2} + 1} - \sqrt{\frac{L_1}{4L_2}}\right). \tag{4.5}$$

Apparently, effective permittivity is always positive and it is always negative for the C–C cascaded circuit, while the situation is opposite for the L–L cascaded circuit. In other words, the C–C (L–L) cascaded circuit is a kind of TL MNG (ENG) metamaterial.

4.1.2 DNG Metamaterials

DNG media, hypothesized by the Russian physicist Veselago in 1967 [25], are media in which the electric permittivity and the magnetic permeability are simultaneously negative. Such media are also termed LHM since the electric field, magnetic field, and wavevector of an EM wave form a left-handed triad in these media. By contrast, ordinary media whose permittivity and permeability are both positive are named as DPS media or RHM. It has been shown that LHM exhibit numerous unusual EM properties or phenomena not available in ordinary RHM, such as the reversal of Snell's law related to negative index of refraction, the antiparallel group and phase velocities of propagating waves, the reversal of Doppler effect, and the amplitude restoration of evanescent waves enabling super-resolution imaging [25–27]. To date, DNG media have only been demonstrated with artificially structured metamaterial. The existing DNG metamaterials can be classified into two categories in terms of implementation scheme. One is bulk DNG metamaterial composed of 3D volumetric particles and the other is TL DNG metamaterial constructed by periodically connecting TL sections or meshes with proper LC loading.

The most straightforward approach to constructing bulk DNG metamaterials is to combine two sets of constituent particles providing ENG and MNG behaviors respectively over the same frequency region. For example, the first DNG metamaterial demonstrated by Smith's group at microwave frequencies in 2000 consists of a two-dimensionally periodic array of copper EC-SRRs and continuous conducting wires [28]. Based on this DNG metamaterial, the famous experiment demonstrating the negative refraction phenomenon [29] was carried out.[*] However, as described in the last section, obtaining negative permittivity by use of continuous wires suffers from many significant disadvantages. In view of this, Liu et al. proposed an improved DNG metamaterial in which the ENG and MNG performances are provided by ELCRs and EC-SRRs, respectively [30]. Since both ELCR and EC-SRR are self-contained subwavelength resonators, the difficulties posed by the use of wires can be avoided with this improved DNG metamaterial. In practice, the

---

[*] Strictly speaking, the conducting wires used in the experiment were truncated and did not maintain electrical continuity. The resulting cut wire structures exhibit negative permittivity in a band between a lower resonance frequency and an upper cutoff frequency, rather than at all frequencies below an effective plasma frequency.

80 ■ *Metamaterials*

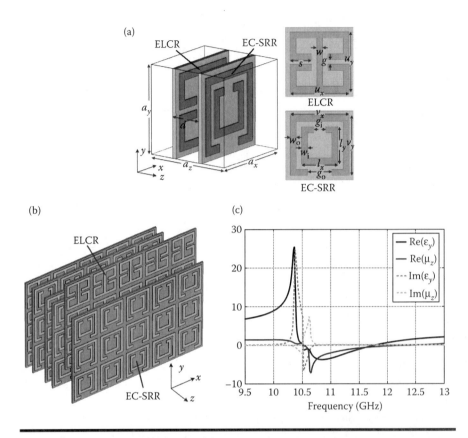

Figure 4.11 (a) Cubic lattice of DNG metamaterial formed by periodic arrangement of pairs of ELCR and EC-SRR, where a_x, a_y, and a_z are the lattice constants, d is the spacing between the ELCR and EC-SRR planes within a cell, u_x, u_y, s, w, and g are the geometric parameters of ELCR, and v_x, v_y, l_x, l_y, w_o, w_i, g_o, and g_i are the geometric parameters of EC-SRR. (b) Visualization of DNG metamaterial combining ELCRs and EC-SRRs. (c) Frequency response of effective relative permittivity ϵ_y and permeability μ_z for DNG metamaterial combining ELCRs and EC-SRRs with the following specifications: $a_x = a_y = a_z = 3.333\,\text{mm}$, $d = 1.67\,\text{mm}$, $u_x = u_z = 2.9\,\text{mm}$, $g = 0.2\,\text{mm}$, $s = 1.05\,\text{mm}$, $w = 0.3\,\text{mm}$, $v_x = v_z = 2.9\,\text{mm}$, $l_x = l_z = 1.8\,\text{mm}$, $g_i = 0.6\,\text{mm}$, $g_o = 1.2\,\text{mm}$, $w_o = w_i = 0.3\,\text{mm}$, $t_m = 0.035\,\text{mm}$, $\sigma_m = 5.8 \times 10^7\,\text{S/m}$, $t_s = 0.35\,\text{mm}$, $Dk = 2.33$, and $Df = 0.003$.

metallic ELCR array and EC-SRR array of such a DNG metamaterial are fabricated from metal-clad laminates separately and the assembly of the metamaterial is achieved by alternating the laminates patterned with ELCR array and the laminates patterned with EC-SRR array along the direction perpendicular to the ELCR or EC-SRR plane, as illustrated by Figure 4.11b. Actually, this DNG metamaterial can be regarded as being composed of compound particles combining EC-SRRs and

ELCRs, as indicated by its cubic unit cell shown in Figure 4.11a. It exhibits negative permittivity along the y axis and negative permeability along the z axis, which is consistent with the anisotropic SNG response offered by EC-SRR and ELCR. Typical retrieved resonate curves of relative permittivity ϵ_y and permeability μ_z for DNG metamaterial combining EC-SRRs and ELCRs are shown in Figure 4.11c, from which the low-loss DNG behavior is found to occur in the frequency range from 10.8 to 11.75 GHz. The compound particles used for implementing bulk DNG metamaterials can also be constructed by combining other appropriate electric and magnetic resonators. For example, the DNG metamaterials studied in References 31–33 have their building block unit cells constructed from two dielectric spheres with the same high permittivity but different radii. The sphere with large radius operates at its dominant electric resonant mode and provides negative permittivity, while the sphere with small radius operates at its dominant magnetic resonant mode and provides negative permeability.

For the bulk DNG metamaterials mentioned above, the ENG and MNG elements are separately included in each unit cell. These metamaterials require relatively high fabrication complexity and cost. In the literatures, efforts have also been made to investigate bulk DNG metamaterials incorporating both ENG and MNG features in a single geometry [34–39]. Such an incorporation scheme is expected to ease the fabrication task significantly. Figure 4.12 demonstrates an instance proposed by Xu et al., in which each unit cell contains a single planar metallic particle printed on single dielectric layer [39]. As shown in Figure 4.12a, the particle mainly characterizes a metallic fractal ring in the shape of a Sierpinski curve of the second iteration order. There are four subrings in the four corners of the fractal ring and four interdigital capacitors are loaded in the middle of four concaves formed between adjacent subrings. Besides, four side meandered arms each with length s are employed to electrically connect the neighboring particles in a coplanar fashion. The bulk DNG metamaterial based on the fractal-meandering particle is formed by periodically arranging the particles in the *xoy* plane and then layering the resulting structures along the z direction, as displayed in Figure 4.12b. Like the EC-SRR MTM introduced in the last section, this DNG metamaterial is realized completely in planar fully printed layout. When the incident magnetic field is normal to the particle plane, the interaction between the particle and EM wave produces inductive full-loop currents flowing along the interdigitals and the ring, yielding a resonant magnetic response accounting for negative effective permeability μ_z. On the other hand, the impinging of electric field (directed along the x or y axis) upon the particle creates an electric resonant effect contributing to negative effective permittivity ϵ_x or ϵ_y. The effect is related to the oscillating current flowing in parallel with the electric field through both the meandered arms and the fractal-interdigital structure. Note that the electric responses (or effective permittivity components) along the x and y axes are identical since the fractal-meandering particle possesses a fourfold rotational symmetry in the *xoy* plane. Moreover, moderate loss can be achieved in virtue of the tight coupling among unit cells. Figure 4.12c shows the retrieval

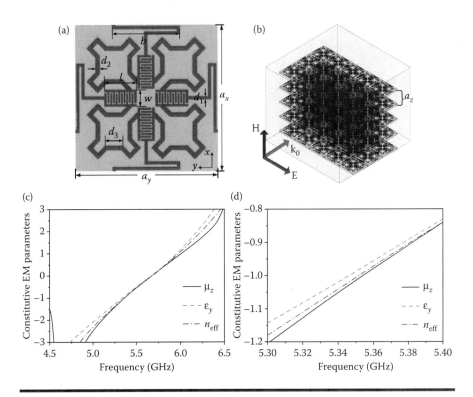

Figure 4.12 (a) Schematic of the planar fractal-meandering particle for DNG metamaterial development, where a_x and a_y are unit cell size in the xoy plane, b, l, w, d_1, d_2, and d_3 are the geometric parameters of the particle. (b) Visualization of bulk DNG metamaterial composed of the fractal-meandering particles, where a_z is the periodicity along the z axis. (c,d) Effective constitutive parameters μ_z and ϵ_y (real parts) for the fractal-based DNG metamaterial with the following specifications: $a_x = a_y = 10.6$ mm, $a_z = 12$ mm, $b = 5$ mm, $l = 2.4$ mm, $w = 1.2$ mm, $d_1 = 0.2$ mm, $d_2 = 0.3$ mm, $d_3 = 1.32$ mm, $s = 6.8$ mm, $t_m = 0.035$ mm, $\sigma_m = 5.8 \times 10^7$ S/m, $t_s = 3$ mm, $Dk = 4.2$, and $Df = 0.01$. The graph in (d) gives a microcosmic view around 5.35 GHz for the graph (c). (H.-X. Xu et al.: Three-dimensional super lens composed of fractal left-handed materials. Adv. Opt. Mater. 2013. 1(7). 495. Copyright Wiley-VCH Verlag GmbH & Co. KGaA. Reproduced with permission.)

results of constitutive parameters μ_z and ϵ_y (real parts) for the fractal-based DNG metamaterial with certain specifications. As is shown, this specific metamaterial exhibits good impedance matching to free space in the DNG band from 5.3 to 5.4 GHz.

As suggested by the TL representation of SNG media shown in Figure 4.10, when the ordinary series-shunt topology is used to realize equivalent media, negative permeability and negative permittivity can be obtained by introducing capacitance

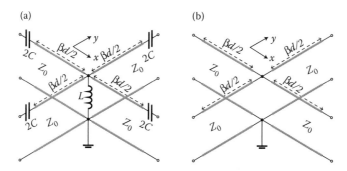

Figure 4.13 (a) Two-dimensional TL grid loaded with series capacitance and shunt inductance, acting as the unit cell of 2D NRI TL metamaterial. (b) Unloaded 2D TL grid, acting as the unit cell of 2D positive refractive index (PRI) TL metamaterial.

to the series branch and introducing inductance to the shunt branch, respectively. Therefore, the key point of constructing TL DNG metamaterial lies in incorporating both series capacitance and shunt inductance into the unit cell. Up to now, there have been mainly two types of TL DNG metamaterial. One of them was proposed by Itoh's group and is called as CRLH TL metamaterial [40,41]. The other was promoted by Eleftheriades's group, named as negative refractive index (NRI) TL metamaterial [42–46]. The unit cell of CRLH TL metamaterial consists of series capacitor-inductor circuit and shunt capacitor-inductor circuit, in its series and shunt branches, respectively. In general, CRLH TL metamaterial possesses two effective medium bands separated by series and shunt resonant frequencies. In the lower band it behaves like a DNG medium, while in the higher band it behaves like a DPS medium. The series inductance and shunt capacitance, which originate from the unavoidable parasitic effects in physical implementation, are responsible for the DPS band of CRLH TL metamaterial. As for NRI TL metamaterial, it is constructed by repeating TL grids loaded with series capacitance and shunt inductance. Figure 4.13a depicts the unit cell of 2D NRI TL metamaterial, where Z_0 is the characteristic impedance, β is the propagation constant, and d is the length of the interconnecting TL section directed along the x or y axis, and L and C are loaded inductance and capacitance [44,46]. Applying the Bloch boundary conditions, the dispersion relation for voltage/current plane waves in the 2D NRI TL metamaterial is written as

$$\sin^2\left(\frac{1}{2}k_x d\right) + \sin^2\left(\frac{1}{2}k_y d\right) = \left[\sin\left(\frac{1}{2}\beta d\right) - \frac{1}{2Z_0 \omega C}\cos\left(\frac{1}{2}\beta d\right)\right]$$
$$\cdot \left[2\sin\left(\frac{1}{2}\beta d\right) - \frac{Z_0}{2\omega L}\cos\left(\frac{1}{2}\beta d\right)\right], \qquad (4.6)$$

where k_x and k_y are the effective wave numbers in the x and y directions, respectively. And the Bloch impedances looking into the x and y directions are given by

$$\begin{cases} Z_x = \dfrac{Z_0}{\tan(1/2k_x d)} \cdot \left[\tan\left(\dfrac{1}{2}\beta d\right) - \dfrac{1}{2Z_0 \omega C}\right], \\ Z_y = \dfrac{Z_0}{\tan(1/2k_y d)} \cdot \left[\tan\left(\dfrac{1}{2}\beta d\right) - \dfrac{1}{2Z_0 \omega C}\right]. \end{cases} \quad (4.7)$$

For the case where the interconnecting TL sections are electrically short ($\beta d \ll 1$) and the phase delay per unit cell is small ($k_x d \ll 1$, $k_y d \ll 1$), the dispersion relation can be simplified using Taylor's expansions as

$$k_x^2 + k_y^2 \approx \left[\beta Z_0 - \dfrac{1}{\omega C d}\cos\left(\dfrac{1}{2}\beta d\right)\right] \cdot \left[\dfrac{2\beta}{Z_0} - \dfrac{1}{\omega L d}\cos\left(\dfrac{1}{2}\beta d\right)\right], \quad (4.8)$$

which indicates nearly isotropic propagation of plane wave in the xoy plane. Supposing that the effective relative permittivity and permeability of the TL metamaterial are μ_L and ϵ_L, respectively, we have

$$k_x^2 + k_y^2 = \omega^2 \mu_0 \epsilon_0 \mu_L \epsilon_L. \quad (4.9)$$

Substituting Equation 4.9 into Equation 4.8 and matching the Bloch impedances with the 2D radial impedances in the equivalent homogeneous medium yields the following expressions for μ_L and ϵ_L:

$$\mu_L = \dfrac{\beta Z_0}{k_0 \eta_0} - \dfrac{1}{\omega^2 \mu_0 C d}\cos\left(\dfrac{1}{2}\beta d\right), \quad \epsilon_L = \dfrac{2\beta \eta_0}{k_0 Z_0} - \dfrac{1}{\omega^2 \epsilon_0 L d}\cos\left(\dfrac{1}{2}\beta d\right), \quad (4.10)$$

where $k_0 = \omega\sqrt{\mu_0 \epsilon_0}$ and $\eta_0 = \sqrt{\mu_0/\epsilon_0}$ are the wave number and the wave impedance in free space, respectively. Equation 4.10 suggests that as long as the frequency is sufficiently low, the loaded reactances dominate and both μ_L and ϵ_L become negative, leading to DNG performance associated with the NRI property for the TL metamaterial. However, at high frequencies the effect of the loaded reactances diminishes and the effective medium parameters approach those of the unperturbed periodic TL structure shown in Figure 4.13b.

The above discussion reveals that the TL metamaterial approach produces DNG features through nonresonance mechanism. Hence it is more likely to achieve lower loss and wider bandwidth compared with the bulk metamaterial approach. Besides, 1D or 2D TL DNG metamaterials are well suited for applications in RF/microwave engineering since they inherently support waves guided along circuit components or confined within TL grids.

4.1.3 Zero-Index Metamaterials

ZIM is an interesting class of material in which the permittivity and/or permeability are equal to or near zero and hence the refractive index also reaches or approaches zero. There are three categories of ZIM: ϵ-near zero (ENZ) media, μ-near zero (MNZ) media, and nihility. Media in which only the electric permittivity or the magnetic permeability approaches zero is called ENZ or MNZ to, whereas nihility refers to media whose permittivity and permeability equal to zero simultaneously. Such media is not physically realizable, as is pointed out in References 47 and 48.

Natural ZIM may be found at IR and optical frequencies. For example, some low-loss noble metals, certain semiconductors, and plasmonic dielectrics such as silicon carbide have permittivity close to zero near their plasma frequency. In microwave region, however, a practical realization of ZIM requires in principle the use of metamaterials. Similar to SNG and DNG metamaterials, most practical ZIMs are essentially anisotropic, although the definitions above apply to isotropic media in a strict meaning. In general, the ENG (MNG) metamaterials composed of resonant constituents (in the form of 3D volumetric particles or WG elements) described in Section 4.1.1 also exhibit an ENZ (MNZ) property within a specific frequency band, where the permittivity (or permeability) as a function of frequency continuously passes through zero (giving a zero index) as it transitions from a negative region to a positive region. Hence, these ENG (MNG) metamaterials also turn out to be ENZ (MNZ) metamaterials, but in frequency regimes different from and close to the ENG (MNG) bands.

4.1.4 DPS Metamaterials

Most of the bulk and WG metamaterials introduced in the previous sections are composed of resonate unit cells. When we take a deep look into the effective medium curves of these metamaterials, it is found that although negative constitutive parameter or zero index characteristics are achieved in the resonant bands of the constituent unit cells, both permittivity and permeability remain positive in the frequency regions ahead of the resonant bands, giving rise to DPS metamaterials in these regions. Sometimes, DPS behaviors may even be obtained in the bands after resonances providing the homogenization condition (both the unit cell size and lattice periodicity are insignificant with respect to the wavelength) is satisfied in these bands. Like conventional DPS materials, DPS metamaterials can be classified as pure dielectric, pure magnetic, or magnetodielectric, depending on whether the unit cells offer only electric dipole moments, only magnetic dipole moments, or both electric and magnetic dipole moments. Among the bulk and WG metamaterials introduced above, ELCR MTM and MLR MTM are pure dielectric; CELCR-based WG-MTM is pure magnetic; EC-SRR MTM, SRR MTM, DR MTM, CSRR-based WG-MTM, and MTM combining EC-SRRs and ELCRs are magnetodielectric, in their respective DPS bands. DPS metamaterial has its unique advantages over conventional DPS material, although the latter exists widely in nature. First, DPS metamaterial provide more freedom to tailor the dielectric, magnetic, and anisotropic properties since artificial "atoms" are more flexible to control than natural "atoms." Second, due to the feasibility of designing constituent artificial "atoms" individually and arranging them spatially with certain rules, homogeneous DPS metamaterials can be easily extended to inhomogeneous DPS metamaterials with steerable complicated EM-wave-controlling capabilities. The second advantage will be explored vastly in Chapters 5 through 9.

In practice, DPS media with negligible dispersion or broadband performance are required in many circumstances. Figure 4.14a illustrates a typical bulk DPS metamaterial with consistent pure dielectric property over a wide bandwidth. This metamaterial is formed by arranging planar particles with I-shaped geometry in cubic lattice [49]. These constituent particles are metallic and deposited on thin dielectric substrates. Basically, the impinging of incident wave on the I-shape-based metamaterial will induce a resonance at some frequency if the applied electric field is directed along the y axis. Both the inductance of conducting line and the capacitance between neighboring particles along the y direction contribute to the electric resonance. Such a resonance can be tuned to occur at a sufficiently high frequency such that the effective medium parameters are nearly constants over a

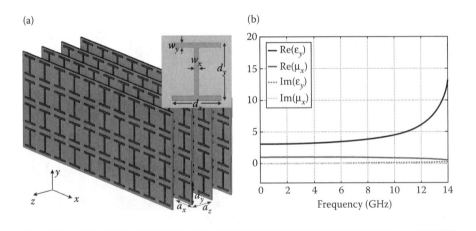

Figure 4.14 (a) Visualization of DPS metamaterial formed by periodic arrangement of I-shaped particles, where a_x, a_y, and a_z are the lattice constants. The geometric parameters of I-shaped particle: d_x, d_y, w_x, and w_y are defined in the inset. (b) Frequency response of effective relative permittivity ϵ_y and permeability μ_x for I-shape-based metamaterial with the following specifications: $a_x = a_y = a_z = 3\,\text{mm}$, $d_x = 2.1\,\text{mm}$, $d_y = 2.7\,\text{mm}$, $w_x = w_y = 0.2\,\text{mm}$, $t_m = 0.035\,\text{mm}$, $\sigma_m = 5.8 \times 10^7\,\text{S/m}$, $t_s = 0.2\,\text{mm}$, $Dk = 3.9$, and $Df = 0.02$.

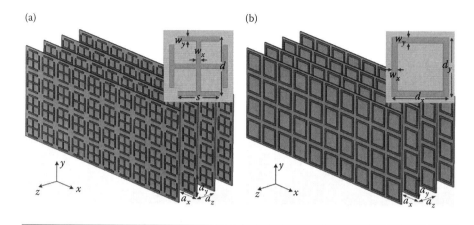

Figure 4.15 (a) Visualization of DPS metamaterial formed by periodic arrangement of Jerusalem cross (or crossed-I) particles. (b) Visualization of DPS metamaterial formed by periodic arrangement of closed ring particles.

large frequency region before resonance. Figure 4.14b depicts the retrieved response curves of effective relative permittivity ϵ_y and permeability μ_x for I-shape-based metamaterial with certain specifications. Apparently, both ϵ_y and μ_x are almost invariant in the DPS frequency band of roughly zero to 5 GHz. Note that in this band, ϵ_y and μ_x approximate to 3.2 and 1, respectively, validating the pure dielectric nature of the I-shape-based metamaterial. The value of ϵ_y can be easily tuned by changing the geometric parameters d_x and d_y. Figure 4.15 demonstrates another two examples of bulk DPS metamaterials with broadband dielectric performance. These two metamaterials are formed by periodic arrangements of Jerusalem cross (or crossed-I) particles[*] and closed ring particles, respectively. Since both particles possess a fourfold rotational symmetry in the *xoy* plane, both metamaterials exhibit isotropic permittivity responses with respect to electric fields lying in the *xoy* plane. The metamaterial composed of closed ring particles was investigated in detail in Reference 50.

Figure 4.16 presents a strategy to realize WG DPS metamaterial with negligible dispersion [20]. This strategy utilizes the WG element called embedded meander line (EML), which characterizes a meander line embedded in a square area defect in the lower metal plate (see inset of Figure 4.16a). When a dominant planar waveguide mode propagating along the *x* axis is applied, a resonant magnetic moment

---

[*] The Jerusalem cross is formed by orthogonally stacking two I-shaped structures with identical dimensions in the same plane. It is a well-known structure in the field of frequency selective surface (FSS), for which one usually cares about the resonant band of unit cell. In contrast, the DPS metamaterial application discussed here mainly employs the band far before resonance, in which the condition for homogenization is met.

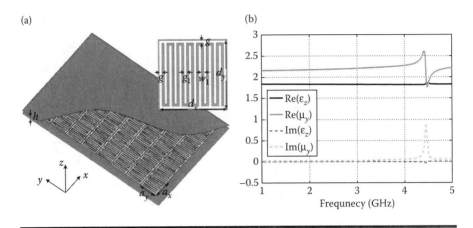

Figure 4.16 (a) Visualization of WG DPS metamaterial formed by patterning periodic array of EML in the lower plate of a planar waveguide, where a_x and a_y are the periodicity of the array and h is the height of waveguide. The geometric parameters of EML: d_x, d_y, g, w_1, and g_1 are defined in the inset. (b) Frequency response of effective relative permittivity ϵ_z and permeability μ_y for EML-based WG metamaterial with the following specifications: $a_x = a_y = 4.2$ mm, $d_x = d_y = 4.05$ mm, $g = 0.15$ mm, $w_1 = g_1 = 0.15$ mm, $t_m = 0.035$ mm, $\sigma_m = 5.8 \times 10^7$ S/m, $h = 1.43$ mm, $Dk = 2.65$, and $Df = 0.001$.

aligned with the y axis is induced by external magnetic fields for each EML element, giving rise to an enhanced effective permeability component μ_y. As the magnetic resonance is very weak, the effective medium parameters are almost nondispersive before resonance. The retrieved frequency-dependent relative permittivity ϵ_z and permeability μ_y for an instance of EML-based WG-MTM is shown in Figure 4.16b. Note that for this instance, both the upper and lower plates are copper layers with thickness denoted by t_m; they are sandwiched with a supporting dielectric substrate whose permittivity and loss tangent are represented by Dk and Df, respectively. Such a configuration makes the EML-based WG-MTM easy to fabricate with printed circuit technique. It can be observed from Figure 4.16b that μ_y is kept about 2.2 and ϵ_z is kept about 1.8 in the band below 4 GHz, which in fact guarantees a consistent magneto-dielectric performance in this band. The enhancement of permittivity component ϵ_z is due to the existence of the supporting dielectric substrate. However, ϵ_z is much lower than Dk because the defects in the lower metal plate decrease the capacitance between the upper and lower metal plates. Moreover, though such defects will cause some energy leakage, the effective medium loss of this instance (seen from the imaginary part of μ_y and ϵ_z in Figure 4.16b) is reasonably low in the aforementioned magneto-dielectric band.

As to the TL representation of DPS media, it is well known that conventional TLs are good imitators for DPS media. A typical 2D implementation of DPS media using conventional TLs are presented in Figure 4.13b and is called 2D PRI TL metamaterial [44,46]. It is actually a limiting case of the 2D NRI TL metamaterial shown in Figure 4.13a, where C is short-circuited ($C \to \infty$) and L is open-circuited ($L \to \infty$). Hence, the dispersion relation and the Bloch impedances for voltage/current plane waves in the 2D PRI TL metamaterial can be easily obtained through a modification on Equations 4.6 and 4.7 and are written as

$$\sin^2\left(\frac{1}{2}k_x d\right) + \sin^2\left(\frac{1}{2}k_y d\right) = 2\sin^2\left(\frac{1}{2}\beta d\right), \quad (4.11)$$

$$Z_x = \frac{Z_0}{\tan\left((1/2)k_x d\right)} \cdot \tan\left(\frac{1}{2}\beta d\right), \quad Z_y = \frac{Z_0}{\tan\left((1/2)k_y d\right)} \cdot \tan\left(\frac{1}{2}\beta d\right). \quad (4.12)$$

Similarly, when the interconnecting TL sections for each unit cell are electrically short ($\beta d \ll 1$), the effective relative permittivity ϵ_R and permeability μ_R for the 2D PRI TL metamaterial can be obtained from Equation 4.10 by letting $C \to \infty$ and $L \to \infty$. The resultant formulas are expressed as

$$\mu_R = \frac{\beta Z_0}{k_0 \eta_0}, \quad \epsilon_R = \frac{2\beta \eta_0}{k_0 Z_0}. \quad (4.13)$$

4.2 Single-Negative Metamaterials

4.2.1 Evanescent-Wave Amplification in MNG–ENG Bilayer Slabs

Evanescent-wave amplification (EWA), which plays an important role in high-resolution imaging, has aroused great interest in the metamaterials community. In 2003, Alú and Engheta analyzed the EM wave interaction with a pair of MNG and ENG slabs juxtaposed with each other [24]. They showed that if certain conditions are satisfied, such bilayered structure would achieve zero reflection, complete tunneling and transparency accompanied with EWA. In this section, two kinds of periodically lumped-element-loaded TLs, which are actually series-shunt capacitor cascaded circuit and series-shunt inductor cascaded circuit, are used to simulate the MNG–ENG bilayer structure. And we show that the evanescent waves which exist in two single layers independently can be amplified exponentially and the energy will "tunnel" through the bilayer structure if an antimatching condition for the effective permittivity and permeability of the bilayer structure is achieved [51].

Consider the problem geometry shown in Figure 4.17. It characterizes a bilayer slab residing in free space. The first layer with the relative permittivity ϵ_{r1} and relative permeability μ_{r1} denotes the MNG medium or magnetic plasma, and the

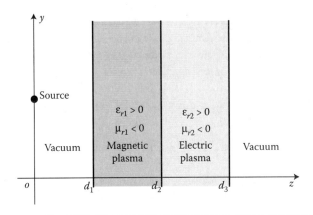

Figure 4.17 An MNG–ENG bilayer slab with a source located before the MNG layer.

second layer with the relative permittivity ϵ_{r2} and relative permeability μ_{r2} represents the ENG medium or electric plasma. The whole space is divided into four regions: 0, 1, 2, and 3. The incident wave from free space (region 0) to the bilayer slab is assumed to be TE-polarized and the electric fields in all regions have only x components under the Cartesian coordinate shown. The incident electric field is expressed as

$$\vec{E}^{\text{inc}} = E_0\, e^{ik_y y}\, e^{ik_{0z} z}\hat{x}. \tag{4.14}$$

Supposing that the reference plane is located at the entrance surface of the bilayer slab (i.e., $z = d_1$), the reflection and transmission coefficients of the slab can be derived as [52]

$$R = \frac{(a+b)e^{i2k_{0z}d_1}}{d+e}, \quad T = \frac{c}{d+e}, \tag{4.15}$$

where

$$\begin{cases} a = r_{01} + r_{01} r_{12} r_{23}\, e^{i2k_{2z}h_2}, \\ b = r_{12}\, e^{i2k_{1z}h_1} + r_{23}\, e^{i2(k_{1z}h_1 + k_{2z}h_2)}, \\ c = t_{01} t_{12} t_{23}\, e^{i(k_{1z}h_1 + k_{2z}h_2)}\, e^{-ik_{0z}(h_1 + h_2)}, \\ d = 1 + r_{12} r_{23}\, e^{i2k_{2z}h_2}, \\ e = r_{01} r_{12}\, e^{i2k_{1z}h_1} + r_{01} r_{23}\, e^{i2(k_{1z}h_1 + k_{2z}h_2)}. \end{cases} \tag{4.16}$$

In the above equations, $h_1 = d_2 - d_1$ and $h_2 = d_3 - d_2$ represent the thicknesses of the two layers; $k_{mz} = \sqrt{k_m^2 - k_y^2}$ is the longitudinal wave number in region m; and

r_{mn} and t_{mn} are the Fresnel reflection and transmission coefficients, respectively, with

$$r_{mn} = \frac{\mu_n k_{mz} - \mu_m k_{nz}}{\mu_n k_{mz} + \mu_m k_{nz}}, \quad t_{mn} = \frac{2\mu_n k_{mz}}{\mu_n k_{mz} + \mu_m k_{nz}}, \quad (4.17)$$

in which $m, n = 0, 1, 2,$ or 3.

When the two layers have equal thickness ($h_1 = h_2$) and satisfy the following antimatching condition

$$\epsilon_{r1} = -\epsilon_{r2} > 0, \quad \mu_{r1} = -\mu_{r2} < 0, \quad (4.18)$$

it can be deduced that the reflection coefficient R is always equal to zero for both propagating waves (k_{0z} is real) and evanescent waves (k_{0z} is imaginary). Under this condition, the bilayer slab is transparent to any incident waves, corresponding to the complete wave transmission effect [24].

It is well known that, whether it is propagating or evanescent, the incident wave turns into an evanescent wave when entering the SNG or plasma medium. This implies that no propagating waves can be supported by either single layer. The internal electric fields in the bilayer slab are given by

$$\vec{E}_i = E_0(E_i^+ e^{ik_{iz}z} + E_i^- e^{-ik_{iz}z})e^{ik_y y}\hat{x}, \quad (4.19)$$

where $i = 1, 2$, $k_{1z} = k_{2z} = i\alpha$, and E_i^+ and E_i^- are the forward and backward evanescent coefficients in region i. After simple manipulation, such coefficients are obtained as

$$E_1^\pm = \frac{1}{2}(1 \pm p_{10})e^{ik_{0z}d_1}e^{\pm \alpha d_1}, \quad E_2^\pm = \frac{1}{2}(1 \mp p_{10})e^{ik_{0z}d_1}e^{\pm \alpha d_3}, \quad (4.20)$$

in which $p_{10} = \mu_{r1}k_{0z}/(i\alpha)$. It is clear that the evanescent wave mainly grows exponentially as $e^{\alpha(z-d_1)}$ in region 1 and mainly decays exponentially as $e^{-\alpha(z-d_3)}$ in region 2 when transmitting through the slab. Hence, there exists a peak field value at the interface of the two plasma layers, which is in fact the surface plasmon excited by the interaction of source and the bilayer slab. It is the surface plasmon that helps to enhance the evanescent waves within the layers and to tunnel the incident waves through the slab region. The transmission coefficients through the antimatched bilayer slab can be written as

$$T = e^{-ik_{0z}(h_1+h_2)} \quad (4.21)$$

and

$$T = e^{\text{Im}(k_{0z})(h_1+h_2)} \quad (4.22)$$

for propagating and evanescent incident waves, respectively. It is evident that if a point source is located at the left boundary of the slab, the propagating components

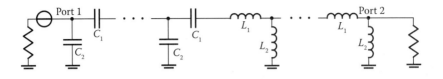

Figure 4.18 An ideal CC–LL bilayer structure, in which the voltage source represents an effective plane wave incident from free space to the CC circuit or effective MNG layer.

emitted from the source experience phase compensation, while the evanescent components have their amplitude to be recovered within the slab. As a result, both kinds of components at the exit surface of the slab are exactly the same as when they enter the slab.

To further study the aforementioned complete tunneling and EWA effect, an effective MNG–ENG bilayer slab is constructed by connecting two segments of cascaded circuits to each other, as shown in Figure 4.18. The left circuit segment is a series-shunt capacitor cascaded circuit (CC circuit for short) and the right segment is a series-shunt inductor cascaded circuit (LL circuit for short). The number of cells in each segment is $n = 4$. As discussed in Section 4.1, the CC circuit and LL circuit are equivalent to MNG and ENG medium, respectively. The CC–LL bilayer structure can be regarded as a two-port network and each port is terminated with a matching impedance to simulate the free space background. Besides, an ideal voltage source is placed in series at Port 1, representing a propagating plane wave normally incident from free space to the effective MNG layer.

Considering the formulas for calculating the effective permittivity and permeability of CC and LL circuit (Equations 4.3 and 4.4 in Section 4.1), the aforementioned antimatching condition can be translated into the following electric and magnetic plasma resonant conditions:

$$\omega = \omega_\epsilon = \sqrt{\frac{\theta_L(e^{\theta_L} - 1)(e^{\theta_C} - 1)}{\theta_C L_1 C_2}}, \quad (4.23)$$

$$\omega = \omega_\mu = \sqrt{\frac{\theta_C(e^{\theta_C} - 1)(e^{\theta_L} - 1)}{\theta_L L_1 C_2}}, \quad (4.24)$$

in which ω_ϵ and ω_μ are termed electric and magnetic plasma resonant frequencies, respectively. The antimatching of the effective MNG and ENG layers requires that those two resonant frequencies should be equal to each other. Hence, deriving from $\omega_\epsilon = \omega_\mu$, a final antimatching condition for the CC–LL bilayer structure is obtained as

$$L_1 C_1 = L_2 C_2, \quad (4.25)$$

and the corresponding resonant frequency is

$$\omega_0 = \frac{1}{\sqrt{L_1 C_1}} = \frac{1}{\sqrt{L_2 C_2}}. \tag{4.26}$$

As long as Equation 4.26 is satisfied, the effective MNG and ENG layers will be antimatched and a total transmission and EWA phenomenon will be expected at the frequency $\omega = \omega_0$. As an example, the *CC–LL* bilayer structure here is designed to be antimatched and the circuit elements are set as $C_1 = C_2 = 10 \text{ pF}$ and $L_1 = L_2 = 10 \text{ nH}$. Hence, the operating frequency for complete tunneling and EWA of the bilayer structure is predicted to be 502.3 MHz.

The transmission coefficient or scattering parameter S_{21} of the *CC–LL* bilayer structure or the effective MNG–ENG bilayer slab can be obtained by two means. One is performing a circuit simulation with the aid of Agilent's advanced design system (ADS) and the other is theoretical calculation through substituting the effective permittivity and permeability of the *CC* and *LL* circuits in Equations 4.3 and 4.4 into Equation 4.15. Figure 4.19 shows the comparison results of circuit simulation and theoretical calculation, which obviously have excellent agreement. From Figure 4.19, there is an obvious peak (nearly 0 dB) at the frequency of 502.3 MHz, indicating that a total transmission does occur as predicted.

Circuit simulation also gives the voltage distribution along the *CC–LL* bilayer structure and the theoretical calculation also gives the electric field distribution in

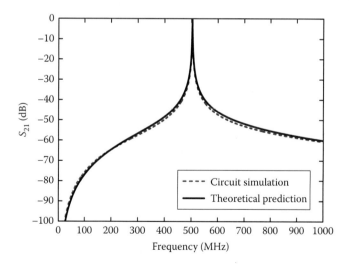

Figure 4.19 Comparison of transmission coefficients between circuit simulation and theoretical prediction for the ideal *CC–LL* bilayer structure. (Reprinted with permission from R. Liu et al., *Phys. Rev. B*, 75(12): 125118, 2007. Copyright 2007 by the American Physical Society.)

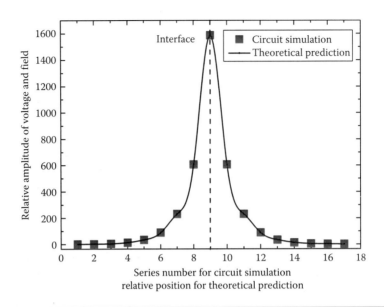

Figure 4.20 Comparison of field distribution between circuit simulation and theoretical prediction for the ideal CC–LL bilayer structure. (Reprinted with permission from R. Liu et al., *Phys. Rev. B,* 75(12): 125118, 2007. Copyright 2007 by the American Physical Society.)

the effective MNG–ENG bilayer slab. Both distributions, which agree well with each other, are shown in Figure 4.20. It is evident that the amplitude of evanescent waves increased exponentially in the first layer of the bilayer structure and then decreased exponentially in the second layer. Hence, the surface plasmon resonance and the EWA effect have also been observed within the *CC–LL* bilayer structure under microwave regime.

It should be remarked that when the unit cell number n of each circuit segment varies, the peak (resonant) frequency remains the same based on the above discussion. In other words, the resonant frequency is independent of the unit cell number or the length of the bilayer structure. This is also a well-known property for the plasma resonance. However, the variation of structure length (or cell number) will affect the bandwidth of the peak. The longer the structure, the narrower the bandwidth, and the higher the quality factor.

To verify the above analysis on EWA, a realistic *CC–LL* bilayer structure is designed using lumped capacitors and inductors. Here, the unit cell number for both *CC* and *LL* circuits is 4. All lumped elements are mounted on a PCB board along a microstrip and the distance between neighboring lumped elements is 5 mm, as illustrated in Figure 4.21. An F4B substrate made of polytetra-fluoroethylene and glass fiber is used, which has a thickness of $h = 0.8$ mm and a relative permittivity of

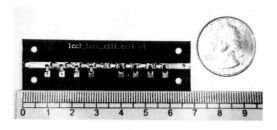

Figure 4.21 Fabricated sample of the realistic CC–LL bilayer structure. (Reprinted with permission from R. Liu et al., *Phys. Rev. B*, 75(12): 125118, 2007. Copyright 2007 by the American Physical Society.)

$\epsilon_r = 2.65$. The width of the microstrip is set to be 2.2 mm, ensuring a characteristic impedance of 50 Ω. The design value of lumped capacitors and inductors are $C_1 = 8.2\,\text{pF}$, $C_2 = 4\,\text{pF}$, $L_1 = 10\,\text{nH}$, and $L_2 = 15\,\text{nH}$, which are chosen to meet the antimatching condition and ensure a resonant frequency of about 500 MHz. We remark that the parasitical effect of the microstrip line has been considered in the above design.

The transmission coefficient (or scattering parameter S_{21}) of the CC–LL bilayer structure is measured using a vector network analyzer (VNA) and illustrated in Figure 4.22, in which the full-wave simulation result considering the substrate

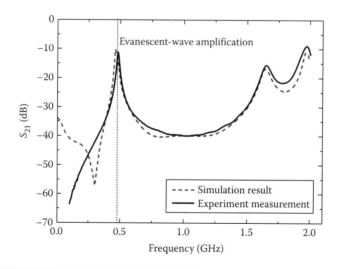

Figure 4.22 Comparison of transmission coefficients between measurement and CST simulation for the realistic CC–LL bilayer structure. (Reprinted with permission from R. Liu et al., *Phys. Rev. B*, 75(12): 125118, 2007. Copyright 2007 by the American Physical Society.)

effect (obtained from CST Microwave Studio) is also given for comparison. From Figure 4.22, the full-wave simulation and measurement results have good agreement and an obvious peak is observed near the frequency of 500 MHz. Hence the complete tunneling effect in the *CC–LL* bilayer structure has been experimentally verified. Note that the transmission peak in Figure 4.22 has been reduced to −10 dB from 0 dB in Figure 4.19. The reason for this reduction is that the simulation results in Figure 4.19 are obtained utilizing ideally lumped capacitors and inductors in ADS, while the realistic lumped capacitors and inductors have to be mounted on PCB. Hence, the loss in PCB substrate, the distributed capacitance and inductance of the microstrip and the insertion loss caused by sub-miniature-A (SMA) adapters in measurement are responsible for the drastic reduction of the transmission peak.

Moreover, in order to get an experimental observation of the EWA phenomenon, we place a vertical EM probe 1 mm away above the microstrip to detect the near electric field distribution along the *CC–LL* bilayer structure. Here, the probe is connected to the receiving port of a VNA and the transmitting port of the VNA, which serves as the excitation, is connected to Port 1 of the *CC–LL* bilayer structure. Port 2 of the bilayer structure is matched with a 50 Ω load. The measured electric field distribution along the *CC–LL* bilayer structure are compared with the CST simulation results, as illustrated in Figure 4.23. From Figure 4.23, a strong surface wave is observed at the interface between the *CC* and *LL* circuits. The evanescent waves in the *CC* circuit are amplified exponentially by the

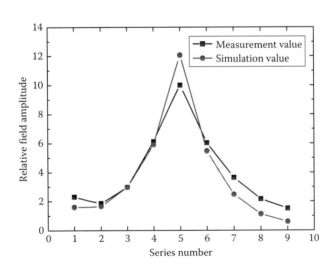

Figure 4.23 Comparison of field distribution between measurement and CST simulation for the realistic *CC–LL* bilayer structure. (Reprinted with permission from R. Liu et al., *Phys. Rev. B*, 75(12): 125118, 2007. Copyright 2007 by the American Physical Society.)

fierce plasma resonance. Considering the earlier analysis and equivalence between the circuit network and the medium model, this experiment confirms the EWA phenomenon.

4.2.2 Partial Focusing by Anisotropic MNG Metamaterials

Anisotropic SNG metamaterials can find many exotic characteristics and applications due to their complex dispersion relations. In 2004, Smith et al. showed that an anisotropic MNG slab, whose permeability is negative only along the longitudinal axis, will redirect *s*-polarized EM waves from a nearby source to a partial focus [53]. In this section, we present an experimental demonstration of this partial focusing phenomenon in a 2D sense, utilizing typical anisotropic MNG metamaterials [23].

The problem geometry under consideration is shown in Figure 4.24, where the space between two parallel perfect electric conducting (PEC) plates is divided into three regions. The left and the right regions are occupied by air and the middle region is covered by an anisotropic MNG slab, which has diagonal permittivity and permeability tensors. For this slab, only the permeability component μ_y along the propagation direction, or *y* axis, is negative, while all other principal elements of the permeability and permittivity tensors are positive. A current line source, whose ends are connected with the upper and lower PEC plates, respectively, is placed before the slab. Such a line source yields *s*-polarized waves (electric field perpendicular to the *xoy* plane) within the planar waveguide. Hence the EM fields in the problem is constrained to two dimensions and only three components of the permittivity and permeability tensors μ_x, μ_y, and ϵ_z are relevant to the problem. The dispersion

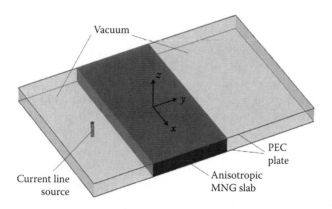

Figure 4.24 Theoretical model for partial focusing in a planar waveguide, where an anisotropic MNG slab is sandwiched in between two PEC plates and a current line source is placed in front of the slab.

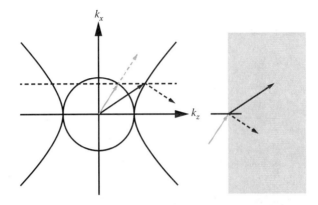

Figure 4.25 Isofrequency curves for free space (circle) and the anisotropic MNG slab (hyperbolic sheets). The incident phase propagation direction (solid gray arrow) is parallel to the incident group velocity (dashed gray arrow). The solid and dashed black arrows indicate the refracted wave vector and group velocity in the slab, respectively. Note that the group refraction is negative while the phase refraction is positive. (Reprinted with permission from D. R. Smith et al. Partial focusing of radiation by a slab of indefinite media. *Appl. Phys. Lett.*, 84(13): 2244, 2004. Copyright 2004, American Institute of Physics.)

relation for TE waves in the anisotropic slab can be written as [54]

$$\frac{k_x^2}{\mu_y} + \frac{k_y^2}{\mu_x} = \left(\frac{\omega}{c}\right)^2 \epsilon_z, \qquad (4.27)$$

where c is the light velocity in free space. It is obvious that the isofrequency curve for Equation 4.27 is a hyperbola, which is illustrated in Figure 4.25. It can be easily seen from Figure 4.25 that the phase velocity of the incident waves undergoes a positive refraction, while the group (or energy) velocity undergoes a negative refraction at the boundary between free space and the slab, providing some degree of refocusing inside or outside the slab. In Reference 53, ray-tracing diagram and full-wave simulation have been given to describe the propagation of the waves emitted from a line source in front of the slab, showing the existence of partial focusing for incident waves.

To confirm the above partial focusing effect by experiment, we implement the MNG slab using the CELCR-based waveguide metamaterial (see Section 4.1) and realize the geometry shown in Figure 4.24 in the parallel-plate waveguide chamber of a 2D near-field mapping apparatus (2D mapper, see appendix). As shown in Figure 4.26a, a cubic Styrofoam is fixed to the bottom plate of the 2D mapper and a piece of single-sided CCL is glued onto the top surface of the Styrofoam. Besides,

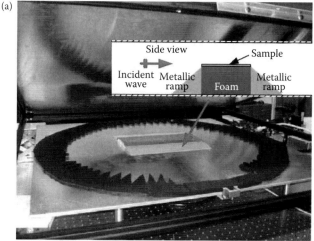

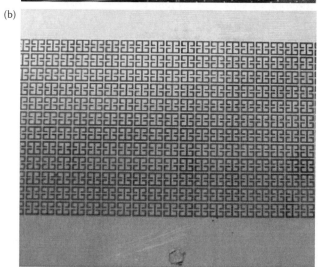

Figure 4.26 (a) Experimental setup for partial focusing utilizing 2D mapper. (Inset) Side view of the experimental setup. (b) Photograph of the fabricated sample etched with periodic CELCR array. (Reprinted with permission from Q. Cheng et al., *Phys. Rev. B*, 78(12): 121102(R), 2008. Copyright 2008 by the American Physical Society.)

the middle region of the copper layer of the CCL is etched with a periodic CELCR array, as illustrated by Figure 4.26b. The resultant WG metamaterial between the CELCR region and the top plate of the 2D mapper can be regarded as an anisotropic slab with different permeability components in the x and y directions since the shape of the CELCR element is not identical in these two directions. Hence, the top plate of the 2D mapper, the CCL with CELCR pattern, and the volume between them effectively constitute the three-region geometry described above. As to the line source, it is approximated by a probe antenna protruding above CCL after penetrating the styrofoam. The hole in front of the CELCR pattern, as shown in Figure 4.26b, indicates the position of the source antenna. Moreover, since the height of the 2D mapper is much larger than the spacing between the CELCR region and the top mapper plate, two metallic ramps are placed on each side of CCL in order to avoid the severe impedance mismatch due to the change of geometry (see the inset in Figure 4.26a).

In the experiment presented here, CCL is chosen to be FR4 circuit board with copper thickness of 0.018 mm, substrate thickness of 0.2026 mm, and dielectric permittivity of $4.4 + i0.088$. The gap between the top and bottom plates of the 2D mapper is 11 mm and the gap between CCL and the top plate is kept as 1 mm by properly setting the height of the Styrofoam. The CELCR array is fabricated using the standard photolithography, and there are altogether 12 units in the longitudinal direction and 60 units in the transverse direction. The dimensions for the CELCR unit shown in Figure 4.26 are selected as $a_x = a_y = 3.333$ mm, $d_x = d_y = 3$ mm, $s = 1.05$ mm, and $w = g = 0.3$ mm.

In order to get the effective medium parameters μ_x, μ_y and ϵ_z of the CELCR-based WG metamaterial, full-wave numerical simulations by commercial software HFSS have been made to obtain two sets of scattering parameters (S parameters) for CELCR-based metamaterial slabs with normally incident s-polarized waves. These two sets of S parameters correspond to metamaterial slabs with finite numbers of unit cells along the y and x directions, respectively. The first (or second) set is used to retrieve parameters μ_x and ϵ_z (or μ_y and ϵ_z). For the purpose of taking the strong mutual couplings between neighboring CELCR unit cells into account, both two-cell and three-cell metamaterial slabs are considered for each set of S parameters and an advanced retrieval method is applied for the subsequent calculation of the effective medium parameters from the S parameters. This advanced retrieval method, described in detail in Reference 55, is an updated version of the standard retrieval method introduced in Section 2.2. The effective permittivity and permeability curves retrieved from the two sets of S parameters are plotted in Figure 4.27a and b, respectively. By comparing Figure 4.27a with b, it can be observed that the curves of effective ϵ_z for the two cases differ a lot from each other in most of the frequency band. However, at the design frequency of 11.5 GHz, the values of both ϵ_z are quite close to each other ($1.085 - i0.1123$ in Figure 4.27a and $1.047 - i0.1338$ in Figure 4.27b), which is an evidence of consistent effective medium behaviors for waves incident from both the x and y directions at 11.5 GHz. Also it can be seen from Figure 4.27a and b that $\mu_x = 2.489 + i0.193$ and $\mu_y = -0.970 + i0.122$ at 11.5 GHz. Hence the CELCR-based WG metamaterial here exhibits the anisotropic MNG property mentioned above around 11.5 GHz, with respect to the s-polarized waves propagating in the xoy plane.

Using the experimental setup (as shown in Figure 4.26) based on the 2D mapper, the 2D electric field distribution inside and outside the CELCR-based

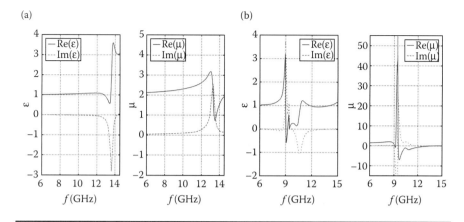

Figure 4.27 Retrieval results of effective medium parameters for the CELCR-based WG metamaterial. (a) Parameters ϵ_z and μ_x retrieved with s-polarized incident waves propagating along the y direction. (b) Parameters ϵ_z and μ_y retrieved with s-polarized incident waves propagating along the x direction. (Reprinted with permission from Q. Cheng et al., *Phys. Rev. B*, 78(12): 121102(R), 2008. Copyright 2008 by the American Physical Society.)

metamaterial slab is measured. The real part of the experimental electric field at 11.5 GHz is shown in Figure 4.28a, where the sign "X" stands for the location of the excitation antenna, and the region between the two dashed lines are occupied by the CELCR array. It is observed that there are several foci inside and outside the effective MNG slab and the waves continue to propagate radially behind the focus above the slab, just like cylindrical waves radiated from a 2D point source. Obviously, the experimental result is consistent with the theoretical prediction, which validates the partial focusing phenomenon.

We have also made numerical simulations at 11.5 GHz for the geometry shown in Figure 4.24 based on the extracted effective medium parameters mentioned above, using HFSS. The resulting electric field distribution is shown in Figure 4.28b and agrees well with the experimental result. Note that the field amplitudes in Figure 4.28a and b are not consistent since the excitation current in the simulation is not the same as that in the experiment.

4.3 Double-Negative Metamaterials

4.3.1 Strong Localization of EM Waves Using Four-Quadrant LHM–RHM Open Cavities

The localization of EM fields is very important in developing high-quality microwave/optical devices using the localized modes. In this section, a novel

102 ■ *Metamaterials*

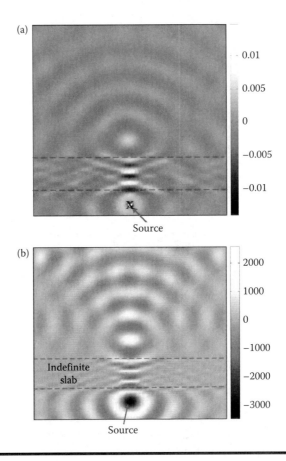

Figure 4.28 Comparison of 2D electric field distribution associated with the partial focusing between measurement and simulation. (a) Experimental result at 11.5 GHz obtained by 2D mapper. (b) Simulated result at 11.5 GHz obtained from HFSS. (Reprinted with permission from Q. Cheng et al., *Phys. Rev. B*, 78(12): 121102(R), 2008. Copyright 2008 by the American Physical Society.)

four-quadrant open cavity comprising DNG and DPS TL metamaterials is constructed and its physical properties are investigated. It is shown that large fields are concentrated at the center of the open cavity due to resonance, which is a strong localization of EM waves and energies [57].

As shown in Figure 4.29, the open cavity under study is a planar structure formed by DNG media (or LHM) in the first and third quadrants and DPS media (or RHM) in the second and fourth quadrants. It is assumed that the RHM and LHM here are antimatched (i.e., $\epsilon_L = -\epsilon_R$ and $\mu_L = -\mu_R$) and the cavity is excited by a 2D point source located in the second quadrant. Based on the ray-tracing method and the negative refraction of rays on the LHM–RHM interfaces,

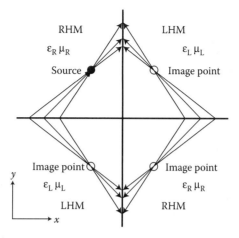

Figure 4.29 An open cavity formed by LHM in the first and third quadrants and RHM in the second and fourth quadrants. (Y. Qin, T. J. Cui, and Q. Cheng.: Strong localization of EM waves using open cavities made of left-handed transmision-line media. *Mic. Opt. Tech. Lett.* 2006. 48(8). 1662. Copyright Wiley-VCH Verlag GmbH & Co. KGaA. Reproduced with permission.)

infinite number of closed ray paths are formed around the center of the structure, emitting from the source and returning to the source, as depicted in Figure 4.29. Hence, the corresponding optical path is zero due to the positive index of refraction in RHM and the negative index of refraction in LHM, which results in an EM resonance. Note that there is no surrounding reflective walls in this cavity.

The open cavity is then implemented by TL metamaterials described in Section 4.1 for further investigation on the aforementioned physical properties. As shown in Figure 4.30, RHM in the second and fourth quadrants is realized by PRI TL metamaterials and LHM in the first and third quadrants is realized by NRI TL metamaterials. Each of the four blocks of TL-MTMs consists of 25 × 25 nodes (or meshes). Hence, the whole structure contains 50 × 50 nodes (or meshes). Each node is represented by a pair of natural numbers and the initial node (i.e., node numbered as (1,1)) is located at the bottom left corner. Moreover, the four edges of the TL-MTM-based open cavity are terminated with appropriate impedances with the purpose of simulating an infinitely periodic structure.

The PRI TL-MTM here is designed to be an equivalent pure dielectric with effective relative permittivity ϵ_R as n_e^2 and effective relative permeability μ_R as 1, where n_e is the index of refraction. From Equation 4.13, one can easily obtain the propagation constant β and the characteristic impedance Z_0 for the interconnecting TL sections of the PRI TL-MTM

$$\beta = \frac{n_e}{\sqrt{2}} k_0, \quad Z_0 = \frac{\sqrt{2}}{n_e} \eta_0, \tag{4.28}$$

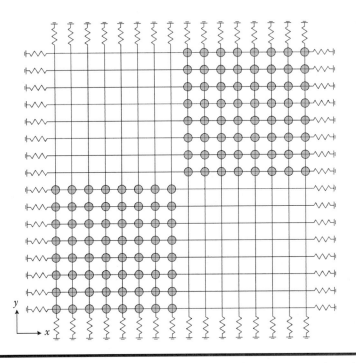

Figure 4.30 Realization of the four-quadrant LHM–RHM open cavity using NRI and PRI TL metamaterials. The four edges are terminated by Bloch impedances.

in which k_0 is the wave number and η_0 is the wave impedance in free space. For the NRI TL-MTM, which should be antimatched with the PRI TL-MTM, the effective relative permittivity ϵ_L and effective relative permeability μ_L have to be chosen as $-n_e^2$ and -1, respectively. Since the interconnecting TL sections of the NRI TL-MTM hold the same characteristic impedance and propagation constant as those of the PRI TL-MTM do, the expressions of the loaded series capacitance C and shunt inductance L can be deduced from Equation 4.10 as

$$C = \frac{\cos\left((1/2)\beta d\right)}{2\omega^2 \mu_0 d}, \quad L = \frac{\cos\left((1/2)\beta d\right)}{2n_e^2 \omega^2 \epsilon_0 d}, \quad (4.29)$$

where d is the length of mesh along x and y directions for both PRI and NRI TL-MTMs. Equations 4.28 and 4.29 are the general formulations to design antimatched PRI and NRI TL-MTMs with an arbitrary refractive index n_e. Once the antimatching condition is met, an impedance match between the PRI and NRI TL-MTMs is automatically achieved.

In order to make the characteristic impedance Z_0 small so that the TL sections can be conveniently realized by microstrip, the refractive index n_e of the open cavity is set as 2. Besides, the size of mesh is set as $d = 1$ mm and the operating frequency is chosen to be 1 GHz. The corresponding TL parameters and the loaded capacitance and inductance are calculated from Equations 4.28 and 4.29, which are $Z_0 = 266.573\,\Omega$, $\beta = 39.619$ rad/m, $L = 358.0593$ nH, and $C = 10.0775$ pF. Obviously, the ultra subwavelength condition ($k_x d \ll 1$, $k_y d \ll 1$ and $\beta d \ll 1$) is well satisfied with these values.

As to the terminal impedances truncating the open cavity, they must be matched with the Bloch impedances of the PRI or NRI TL-MTMs in order to eliminating the reflections at the edges [46,58]. Since the PRI and NRI TL-MTMs are antimatched as designed above, their Bloch impedances are the same and are written as

$$Z_x = Z_0 \frac{\beta}{k_x}, \quad Z_y = Z_0 \frac{\beta}{k_y}, \tag{4.30}$$

which is an approximation of Equation 4.12. Substituting Equation 4.28 into Equation 4.30, one gets

$$Z_x = \eta_0 \frac{k_0}{k_x}, \quad Z_y = \eta_0 \frac{k_0}{k_y}. \tag{4.31}$$

If the observation point (x, y) is far away from the source point (x', y'), the ray approximation can be used to determine k_x and k_y [46]:

$$k_x = k \frac{x - x'}{\rho}, \quad k_y = k \frac{y - y'}{\rho}, \tag{4.32}$$

where $k = n_e k_0$ is the effective wave number in both the PRI and NRI TL-MTMs, and ρ is the distance between the observation and source points. In such a case, the final expressions for determining the Bloch impedances are

$$Z_x = \frac{\eta_0}{n_e} \cdot \frac{\rho}{x - x'}, \quad Z_y = \frac{\eta_0}{n_e} \cdot \frac{\rho}{y - y'}. \tag{4.33}$$

It can be easily shown that Equation 4.33 is the same as the radial TL impedances under the far-field approximation. In the TL-MTM-based open cavity shown in Figure 4.30, the terminal loads for the PRI TL-MTM in the second quadrant are computed using Equation 4.33 with respect to the real voltage source, while the terminal loads for any of the other three blocks of TL-MTMs are computed using Equation 4.33 with respect to the image point in the relevant quadrant.

Circuit simulations using the Agilent's advanced design system (ADS) have been performed to get the voltage distributions of the designed open cavity for different locations of the driving source. When the voltage source is located at the node numbered as (24, 27), which is close to the center of the open cavity (25, 25), the voltage distributions are illustrated in Figure 4.31. In this case, the fields are

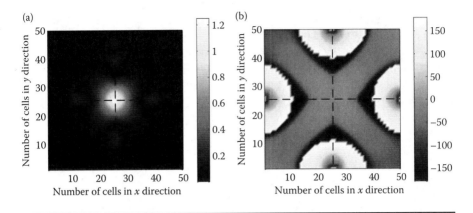

Figure 4.31 Simulated voltage (a) amplitude and (b) phase distributions in the open cavity, for which the voltage source resides at the node numbered as (24, 27). (Y. Qin, T. J. Cui, and Q. Cheng: Strong localization of EM waves using open cavities made of left-handed transmision-line media. *Mic. Opt. Tech. Lett.* 2006. 48(8). 1662. Copyright Wiley-VCH Verlag GmbH & Co. KGaA. Reproduced with permission.)

concentrated in a small region around the center due to the small region of closed-ray paths. It is also noticed that the field intensity in the center and surface waves on the four interfaces between the PRI and NRI TL-MTMs are relatively weak. Figure 4.32 demonstrates the voltage distributions when the source is located at the

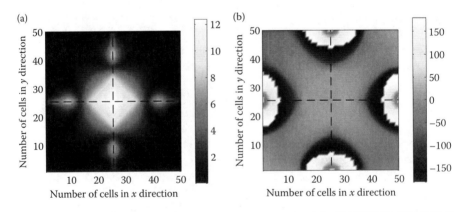

Figure 4.32 Simulated voltage (a) amplitude and (b) phase distributions in the open cavity, for which the voltage source resides at the node numbered as (13, 38). (Y. Qin, T. J. Cui, and Q. Cheng: Strong localization of EM waves using open cavities made of left-handed transmision-line media. *Mic. Opt. Tech. Lett.* 2006. 48(8). 1662. Copyright Wiley-VCH Verlag GmbH & Co. KGaA. Reproduced with permission.)

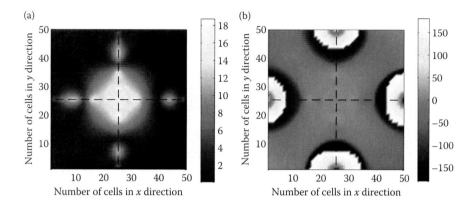

Figure 4.33 Simulated voltage (a) amplitude and (b) phase distributions in the open cavity, for which the voltage source resides at the node numbered as (10, 41). (Y. Qin, T. J. Cui, and Q. Cheng.: Strong localization of EM waves using open cavities made of left-handed transmission-line media. *Mic. Opt. Tech. Lett.* 2006. 48(8). 1662. Copyright Wiley-VCH Verlag GmbH & Co. KGaA. Reproduced with permission.)

node numbered as (13, 38), center of the second quadrant. Here, large fields are produced in the central region of the structure and strong surface waves are observed on the RHM–LHM interfaces. The third location of voltage source is the node numbered as (10, 41), which is far away from the center of open cavity. The computed voltage distributions for the third source location are shown in Figure 4.33. It is noticed that the field pattern and surface waves are quite similar to those in Figure 4.32, where larger fields are concentrated in nearly the same central region. The field patterns shown in Figures 4.31 through 4.33 are almost symmetrical in the four quadrants, although the driving source is only placed in the second quadrant. The strong localization of EM waves in the central region of the open cavity is due to the interactions of propagating and surface waves near the center, which is quite different from the optical behaviors in the photonic-crystal open cavity [59,60].

4.3.2 Free-Space LHM Super Lens Based on Fractal-Inspired DNG Metamaterials

This section presents an application of the DNG metamaterial adopting the fractal and meandering strategy (introduced in Section 4.1.2) in implementation of a 3D free-space LHM lens [39]. The free-space LHM lens was proposed by Veselago and Pendry in their seminal work [25,61] and has intrigued an enormous and long-holding interest in science and engineering communities. Such a lens is actually a planar LHM slab with both the relative permittivity and the relative permeability equal to −1. When the distance between a source and the slab is appropriately chosen and the transverse dimensions of the slab are sufficiently large, both the propagating and evanescent waves emitted from the source could be recovered completely at the exterior image point of the slab [58], making the LHM slab as a perfect lens. Although the lossless perfect lens has been proven unphysical, a slightly lossy LHM slab is still possible to be made as a super lens which is capable of achieving subwavelength-resolution imaging, overcoming the conventional diffraction limit associated with PRI material (i.e., DPS material).

As described in Section 4.1.2, the DNG metamaterial composed of fractal-meandering particles exhibits negative permeability response along the z axis and identical negative permittivity response in the x and y directions. Moreover, if this type of DNG metamaterial is designed as an instance with the same specifications (i.e., the periodicity and dimensions of the fractal-meandering particle and the characteristic parameters of the metal-clad laminates) as those listed in the caption of Figure 4.12, both the effective permittivity ϵ_y (or ϵ_x) and the effective permeability μ_y approach -1 in the vicinity of about 5.35 GHz. Hence, such an instance or specific DNG metamaterial can be directly utilized to implement an anisotropic LHM super lens, which operates effectively at about 5.35 GHz with respect to incident TM waves propagating or evanescent in the *xoy* plane.

To examine the subwavelength imaging functionality of the aforementioned specific DNG metamaterial, 2D full-wave simulations have been conducted by the finite-element-method-based Comsol Multiphysics software package. In the simulations, a hypothetical LHM slab which is 40 mm thick along the x axis, 200 mm wide along the y axis, and infinitely long along the z axis is placed in a free-space background. The constitutive parameters of the LHM slab are set to be $\mu_y = -1.006 + i0.132$ and $\epsilon_x = \epsilon_y = -0.995 - i0.049$, which coincides exactly with the retrieved medium parameters at 5.35 GHz for that specific metamaterial. Two cases of TM current line source (axial direction along the z axis) are considered. For the first case, a single current line source is employed in front of the slab, whereas for the second case, two identical current line sources in close proximity (interval of 34 mm) are placed in front of the slab. Both the sources and the slab are enclosed in a sufficiently large computation domain terminated with perfectly matched layer (PML) boundaries.

Figure 4.34 portrays the simulated magnetic field distribution at 5.35 GHz in the entire domain for the LHM slab. From this figure, very obvious focusing behavior is observed inside (interior focusing) and at the rear side (exterior focusing) of the slab in either single-source or dual-source case. The large field concentration on interfaces is attributed to the plasmonic surface waves at the interface of the LHM slab and free space. The slight discontinuity of the wave front and the slightly lower imaging intensity compared with the source intensity are due to the loss indicated by the imaginary part of constitutive parameters and the impedance mismatch between the LHM slab and free space. Moreover, the converged spots of the slab can be easily discerned for the dual-source case and the size of each converged spot measured by the -3 dB contour (half-power beamwidth, HPBW) are on the order

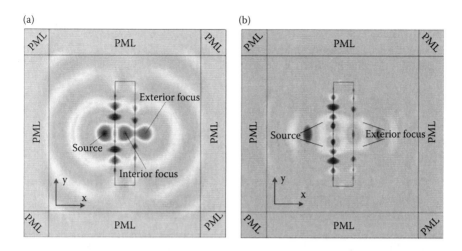

Figure 4.34 Numerically simulated magnetic field distributions at 5.35 GHz for a super lens-like LHM slab with $\mu_y = -1.006 + i0.132$ and $\epsilon_x = \epsilon_y = -0.995 - i0.049$. (a) Single-source case. (b) Dual-source case. (H.-X. Xu et al.: Three-dimensional super lens composed of fractal left-handed materials. *Adv. Opt. Mater.* 2013. 1(7). 495. Copyright Wiley-VCH Verlag GmbH & Co. KGaA. Reproduced with permission.)

of 19 mm, corresponding to $0.34\lambda_0$ at 5.35 GHz. Therefore, the subwavelength imaging beyond the diffraction limit is unambiguously demonstrated and hence the super lens identity of the slab is confirmed. A further inspection also indicates that in the dual-source case, the spot size characterized by HPBW is slightly narrower than the single-source case. This is attributable to the wave interference of the two sources.

A prototype of the free-space LHM super lens based on the aforementioned specific DNG metamaterial has been fabricated. First, 20 solid pieces each containing a 10 × 3 array of the fractal-meandering particle are fabricated from 3-mm-thick FR4 circuit boards using standard PCB technology. Next, to produce the air gap between every two successive pieces, each solid piece is attached to a plastic layer (foam with $\epsilon_{foam} = 1.2$) which has a thickness of 9 mm and identical footprint with the piece. Finally, the 20 solid pieces together with the 20 plastic layers are tightly stacked by adhesives and then reinforced by a hot press to obtain a bulk lens or metamaterial. The final LHM super lens occupies a volume of 31.8 × 106 × 240 mm^3. As a consequence, the transverse dimension along the y axis is approximately $1.94\lambda_0$ at 5.35 GHz which guarantees negligible diffraction effects around the edges of the lens. Figure 4.35 presents the photograph of the realistic free-space LHM lens.

To experimentally verify the focusing and subdiffraction imaging capability of the fabricated super lens, a free-space focusing and near-field measurement system

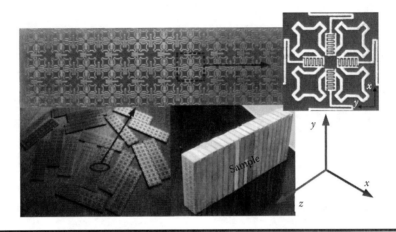

Figure 4.35 Photograph of a fabricated 3D free-space LHM super lens composed of fractal-meandering particles. (H.-X. Xu et al.: Three-dimensional super lens composed of fractal left-handed materials. *Adv. Opt. Mater.* 2013. 1(7). 495. Copyright Wiley-VCH Verlag GmbH & Co. KGaA. Reproduced with permission.)

is set up for the lens. Figure 4.36 schematically illustrates the system, where the fabricated prototype is placed in between two small loop antennas each with an inner diameter of 4.6 mm. These antennas are fabricated from 1.19-mm-diameter semirigid 50 Ω coaxial cable and are implemented in shielded-loop configuration

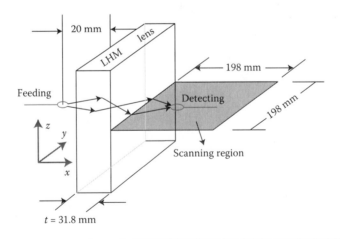

Figure 4.36 Schematic illustration of free-space focusing and near-field measurement system for the fabricated free-space LHM super lens. (H.-X. Xu et al.: Three-dimensional super lens composed of fractal left-handed materials. *Adv. Opt. Mater.* 2013. 1(7). 495. Copyright Wiley-VCH Verlag GmbH & Co. KGaA. Reproduced with permission.)

to mitigate the unbalanced currents. One of the antennas is employed to provide an approximative axial magnetic excitation for the lens and the other antenna is used to detect the magnetic components of transmitted field in the exterior focal region. They are connected to two ports of a VNA (Agilent PNA-LN5230C) via two coaxial cables and are affixed to foam platforms with an identical height to provide the mechanical robustness. The excitation or transmitting antenna is stationary relative to the lens while the detecting antenna covers a scanning region of 198 × 198 mm^2 at the rear side of the lens. To measure the near fields continuously, the transmitting antenna and the prototype lens are fixed on an aluminium plate, which moves automatically in two dimensions driven by a pair of orthogonally oriented computer-controlled step motors. The whole experimental configuration is consistent with the single-source simulation presented above.

Figure 4.37 plots the measured magnetic field intensity and phase distribution in increments of 1 mm at the rear side (focal region) of the prototype lens at two different frequencies. Note that the field intensity for each frequency has been normalized to its individual maximum amplitude. As suggested by the localized field

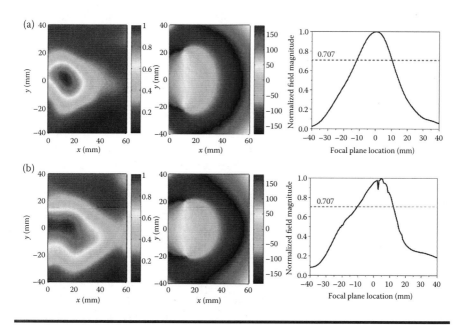

Figure 4.37 Measured magnetic field distributions at the rear side of the fabricated LHM slab lens at (a) 5.35 and (b) 5.4 GHz. The left column is the normalized magnetic field magnitude, the middle column is the phase distribution, and the right column is the −3 dB contour at the exterior focal plane. (H.-X. Xu et al.: Three-dimensional super lens composed of fractal left-handed materials. *Adv. Opt. Mater.* 2013. 1(7). 495. Copyright Wiley-VCH Verlag GmbH & Co. KGaA. Reproduced with permission.)

magnitude and the reversal of concavity of phase fronts shown in Figure 4.37, a tightly confined exterior focal region is evident at both 5.35 and 5.4 GHz. In fact, the prototype lens maintains a well-resolved imaging capability over a bandwidth of about 5.3–5.4 GHz except for some perturbed ripples at the focal plane. This is also implied by the mitigative slope of effective medium parameters shown in Figure 4.12. Although the amplification of evanescent wave cannot be directly visualized, the strong power concentration further confirms that the prototype lens is capable of focusing the propagating-wave components of a source. The measured exterior spot is in reasonable agreement with that obtained in the simulation. The HPBW indicated by the −3 dB intensity profile (along the y axis) is on the order of 21.9 mm ($0.39\lambda_0$), which goes beyond the conventional diffraction limits and enables an improved imaging resolution and discrimination of converged spots. The slight discrepancy between the simulated and measured results with respect to the spot size may be attributable to neglect of permittivity and loss tangent of the adhesives (i.e., the plastic layers) during the retrieval process and the subsequent focusing simulations.

4.4 Zero-Index Metamaterials

4.4.1 Electromagnetic Tunneling through a Thin Waveguide Channel Filled with ENZ Metamaterials

ZIM can offer extraordinary potential EM-wave-controlling applications. It has been suggested that ZIM can be used to shape curved wave fronts into planar ones, to narrow the main lobe of an antenna embedded in the media, and to design efficient coupler for some waveguide devices. In particular, Silveirinha and Engheta [62] proposed that ENZ media can couple guided EM waves through thin subwavelength waveguide channels, and thus reduce the reflection coefficient at a waveguide junction or bend. In this section, we design an effective ENZ medium using the CSRR-based WG metamaterial introduced in Section 4.1 and present an experimental verification of microwave tunneling between two planar waveguides separated by a thin channel filled with such ENZ metamaterial [22].

Consider the planar waveguide-based structure shown in Figure 4.38a. It is composed of three distinct planar waveguide sections, which are distinguished by the differing gap heights between the upper and lower metal plates. The gap heights between the upper and lower metal plates are denoted by h_p for both the input and output waveguides and h for the narrowed waveguide channel in the middle. The channel length is represented by l and the width of the whole structure is represented by w_c. The two sides of the structure along y direction are covered with PMC boundaries so that the structure supports TEM waves propagating along x direction. If all three waveguide sections are empty (i.e., filled with air), the wave impedances

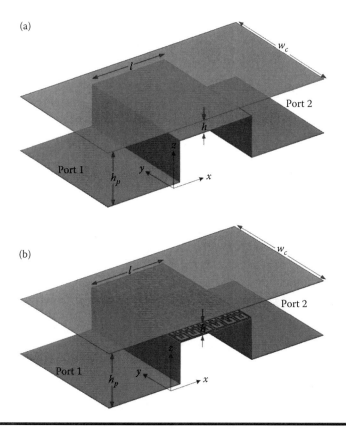

Figure 4.38 Two planar waveguides separated by (a) an empty thin channel and (b) a thin channel loaded with CSRR-based WG metamaterial.

corresponding to TEM waves in these sections are equivalent, since they are independent of the shape of the waveguide cross section. However, the characteristic impedances, which are in inverse proportion to the gap heights, are different for the input/output waveguide and the waveguide channel. When $h \ll h_p$, a severe characteristic impedance mismatch occurs at the ends of the thin channel and inhibits the transmission of TEM waves from the input waveguide to the output. Nevertheless, if the channel is loaded with ENZ medium, the characteristic impedance of the channel can be raised to the point where the three waveguide sections are matched and thus an enhanced transmission or tunneling of TEM waves would be expected.

The mechanism of such tunneling effect can be further explained based on a simplified two-port network model shown in Figure 4.39 [63]. In this model, regions inside and outside the channel are all represented by TLs, for which Z_p and Z_c refer to the characteristic impedances of the input/output waveguide and the waveguide channel, respectively. Note that the additional shunt admittance on account of the discontinuity between the channel and the input/output waveguide

114 ◼ *Metamaterials*

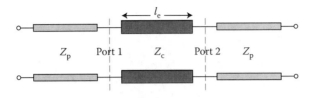

Figure 4.39 Simplified two-port network model for the thin waveguide channel.

is omitted for simplicity. The reflectance R (seen at Port 1) and the transmittance T (from Port 1 to Port 2) of the waveguide channel can be easily deduced as

$$\begin{cases} R = \dfrac{R_0(1 - e^{i2k_c l_e})}{1 - R_0^2 \, e^{i2k_c l_e}}, \\ \\ T = \dfrac{(1 - R_0^2)e^{ik_c l_e}}{1 - R_0^2 \, e^{i2k_c l_e}}, \end{cases} \qquad (4.34)$$

in which l_e is the effective length of the channel, k_c is the wave number inside the channel, and R_0 is the reflection coefficient between the input/output waveguide and the channel and is given by

$$R_0 = \frac{Z_c - Z_p}{Z_c + Z_p}. \qquad (4.35)$$

In the absence of ENZ medium, it is apparent that $Z_c/Z_p = h/h_p$ and R_0 approximates to -1 when $h \ll h_p$. In this case, the coupling between the input and output waveguides is weak, except possibly when a resonance condition is met and a Fabry–Perot oscillation occurs. If ENZ medium is now used to fill in the channel, Z_c/Z_p equals to $\sqrt{\mu_y/\epsilon_z} \cdot h/h_p$, in which μ_y and ϵ_z are the relative permeability and permittivity of the ENZ medium, respectively. Since ϵ_z is a small finite value approaching zero, Z_c/Z_p can be made close to unity, resulting in a fairly small magnitude of R_0 and hence a strong tunneling of EM waves through the waveguide channel.

To verify the tunneling effect provided by ENZ medium in microwave region, the three-section planar waveguide structure is implemented inside the parallel-plate waveguide chamber of the 2D mapper described in Section 4.3. The top and bottom aluminum plates of the 2D mapper are used as the upper and lower plates of the input/output waveguide, respectively. The distance between the top and bottom aluminum plates (which is actually the aforementioned gap heights h_p now) is controlled to be 11 mm. The waveguide channel is constructed by fixing a cubic Styrofoam support to the bottom aluminum plate and then gluing a single-sided CCL onto the top surface of the Styrofoam support. The dimensions of both CCL and Styrofoam in the plane of aluminum plates (i.e., the *xoy* plane) are 18.6 mm × 200 mm. Specifically, CCL is chosen as FR4 circuit board with copper thickness of 0.018 mm, substrate thickness of 0.2026 mm, and dielectric permittivity of $4.4 + i0.088$, and the height of the Styrofoam is chosen to be 9.78 mm. Hence, the gap heights of the waveguide channel is set to be $h = 1$ mm. Besides, the remaining surfaces of the Styrofoam are smoothly covered with copper tape so that the copper layer of the FR4 laminate would make good electrical contact with the bottom aluminum plate. As to the loading of ENZ medium, artificial ENZ medium or ENZ metamaterial is employed and the configuration of ENZ metamaterial loading in the waveguide channel is illustrated in Figure 4.38b. In this configuration, the ENZ medium is realized by the CSRR-based WG metamaterial presented in Section 4.1 and the loading of such metamaterial is implemented by simply patterning a CSRR array in the copper layer of the CCL using standard photolithography. The periodicity and unit cell dimensions of the CSRR array chosen for verification are $a_x = a_y = 3.333$ mm, $d_x = d_y = 3$ mm, $r = 0$ mm, $g = 0.3$ mm, $w = 0.3$ mm, and $s = 1.967$ mm. With such configuration, the CSRR-based WG metamaterial exhibits a near-zero permittivity around 8.8 GHz, as shown in Figure 4.40, with respect to a local electric field polarized along the normal of the

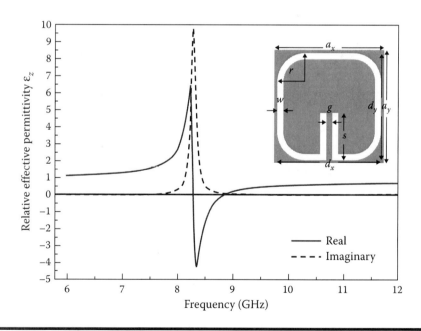

Figure 4.40 Retrieved permittivity ϵ_z of the CSRR-based WG metamaterial used for verification of the EM tunneling effect. (Reprinted with permission from R. Liu et al., *Phys. Rev. Lett.*, 100(2): 023903, 2008. Copyright 2008 by the American Physical Society.)

CSRR surface. The frequency response of the effective permittivity is obtained using the retrieval method introduced in Section 4.1. A total of 200 CSRR unit cells (5 in the propagation direction and 40 in the transverse direction) are used to form the ENZ metamaterial. No matter whether the waveguide channel is loaded with ENZ metamaterial or not, the whole waveguide structure is bounded on either side by layers of absorbing material, which approximate the PMC boundaries and also reduce reflections at the periphery.

The waveguide channel loaded with the CSRR-based WG metamaterial (CSRR channel for short) and the empty waveguide channel (control channel for short), both residing in the planar waveguide environment, are simulated using the finite element Maxwell's equations solver, Ansoft HFSS. The simulated transmission coefficients as a function of frequency for both channels are shown in Figure 4.41. It is noticed that at approximately the frequency 8.8 GHz where the effective permittivity of the CSRR-based metamaterial approaches zero, a transmission peak is

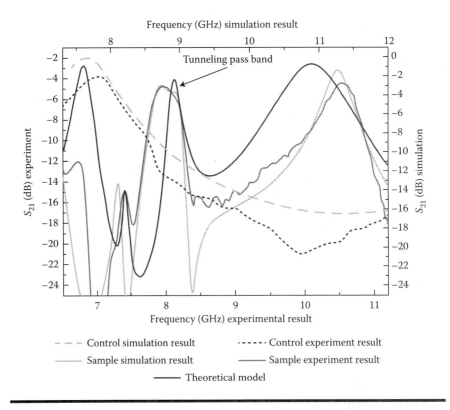

Figure 4.41 Simulated, theoretical, and experimental transmittances for both CSRR and control channels. (Reprinted with permission from R. Liu et al., *Phys. Rev. Lett.*, 100(2): 023903, 2008. Copyright 2008 by the American Physical Society.)

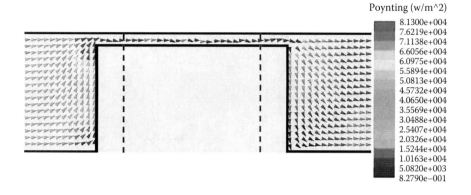

Figure 4.42 Simulated Poynting vector distribution at 8.8 GHz for thin waveguide channel loaded with the CSRR-based WG metamaterial. (Reprinted with permission from R. Liu et al., *Phys. Rev. Lett.*, 100(2): 023903, 2008. Copyright 2008 by the American Physical Society.)

found for the CSRR channel. This peak is due to the tunneling effect as expected and is absent when the control channel is present. Actually, the transmission peak for the control channel appears around 7.6 GHz and is due to the Fabry–Perot oscillation effect instead. A theoretical result of the transmission coefficient for the CSRR channel is also shown in Figure 4.41. This result is calculated using the simplified model presented by Equation 4.34, where the effective length l_e is set as 13 mm and the effective medium parameters (i.e., μ_y and ϵ_z) are obtained from the retrieval process for the CSRR-based metamaterial.[*] The theoretical result is seen to be in qualitative agreement with its simulated counterpart, supporting the interpretation of the transmission peak at about 8.8 GHz as an indication of tunneling effect. In addition, Figure 4.42 shows the simulated Poynting vector distribution at 8.8 GHz for the CSRR channel, revealing the squeezing of EM wave through the channel.

The experimental verification of the tunneling effect is carried out by making transmission and field mapping measurements for both the control channel and the CSRR channel, using aforementioned 2D mapper. For both kinds of measurements, the TEM wave for excitation is approximated by the collimated beam provided by the 2D mapper.

The transmission measurements are conducted by detecting the transmitted field in the output waveguide for both the control channel and the CSRR channel. The corresponding transmission coefficients are shown in Figure 4.41 and compared with the simulated results from Ansoft HFSS. The measured transmission

---

[*] The shunt admittance at the interface between the channel and the input/output waveguide is also included in this calculation.

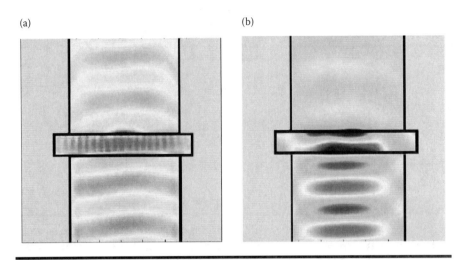

Figure 4.43 Two-dimensional electric field distributions at 8.04 GHz measured in the 2D mapper for (a) CSRR and (b) control channels. (Reprinted with permission from R. Liu et al., *Phys. Rev. Lett.*, 100(2): 023903, 2008. Copyright 2008 by the American Physical Society.)

peaks of the CSRR channel and the control channel are found to occur at about 7.9 and 7 GHz, respectively, which are lower than those found in the simulated results. The frequency deviation between measurements and simulations probably originates from the differences between the experimental environment and the simulation model (e.g., finite channel width and fluctuation of channel height in the experiment). By uniformly shifting the frequency scale, the measured and simulated curves are almost identical (note the two scales indicated on the top and bottom axes). Hence the measurements and simulations are in excellent qualitative agreement.

Through the field mapping measurements, we get phase-sensitive maps of 2D electric field distributions for both channels in the ENZ band of the CSRR-based metamaterial (or the tunneling band of the CSRR channel, around 7.9 GHz). Figure 4.43 shows the spatial electric field maps for both channels at 8.04 GHz. The field is normalized by the average field strength. It is clear that the CSRR channel allows a significant transmission of incident EM energy to the output waveguide, while most of incident EM energy is reflected back to the input waveguide when the control channel is present. Figure 4.43a also reveals that the phase variation across the CSRR channel is slow, which is predictable since the CSRR-based metamaterial has refractive index near zero at 8.04 GHz. Figure 4.44 further compares the pass band behaviors of the control and CSRR channels by plotting the field phases versus the longitudinal positions for both channels, illustrating the difference between two transmission mechanisms (i.e., the Fabry–Perot resonance and the EM tunneling

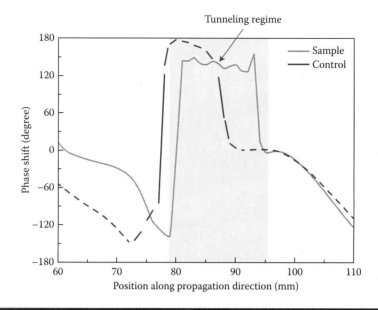

Figure 4.44 Measured phase distributions along the wave-propagation direction (the *x* direction) for CSRR channel at 8.04 GHz and control channel at 7 GHz. The shadow region indicates the location of thin waveguide channel. (Reprinted with permission from R. Liu et al., *Phys. Rev. Lett.*, 100(2): 023903, 2008. Copyright 2008 by the American Physical Society.)

presented here). As shown in Figure 4.44, a strong phase variation exists across the control channel at 7 GHz (where the transmission peak of the control channel occurs), indicating that the propagation constant is nonzero and the large transmittance results from a resonance condition related to the length of the channel. By contrast, the phase advance across the CSRR channel is negligible at 8.04 GHz (where the CSRR channel experiences its pass band), which is consistent with the behavior of the ENZ medium. Hence, it is confirmed through experiments that "squeezed waves" will tunnel without phase shift through extremely narrow channels filled with ENZ medium.

4.4.2 Highly Directive Radiation by a Line Source in Anisotropic Zero-Index Metamaterials

The theory of high emission efficiency and directivity based on anisotropic zero-index metamaterial (AZIM) has been proposed in Reference 64. AZIM discussed here is a kind of anisotropic material with a single constitutive component approaching zero and can be designed to match the surrounding media with proper field polarization.

120 ■ *Metamaterials*

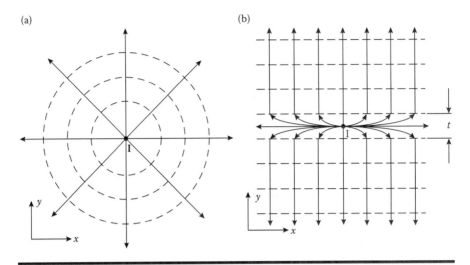

Figure 4.45 Sketch of emission patterns of (a) a line source in free space and (b) a line source embedded in an AZIM slab. (The source of the material Q. Cheng, W. X. Jiang, and T. J. Cui. Radiation of planar electromagnetic waves by a line source in anisotropic metamaterials. *J. Phys. D: Appl. Phys.*, 43(33): 335406. Copyright 2010 IOP Publishing. Reproduced by permission of IOP Publishing. All rights reserved is acknowledged.)

When a *z* axis-oriented current-line source is located in free space, cylindrical waves are generated, as illustrated in Figure 4.45a. When the source is embedded in an AZIM slab with thickness *t*, TE modes are emitted from the source. In such a case, relative medium parameters ε_z, μ_x, and μ_y of the AZIM slab are relevant to the propagation and scattering of the emitted wave. For the AZIM slab, as shown in Figure 4.45b, the emitted wave in free space will be a plane wave, in which the straight wave front will be parallel to the interface. To understand this phenomenon, we simplify the Maxwell's equation restricting the electric field inside the slab into

$$\frac{\partial E_z}{\partial x} = -i\omega\mu_y\mu_0 H_y, \qquad (4.36)$$

where ω is the angular frequency. In the limit of $\mu_y \to 0$, the electric field inside the slab will be constant along the *x* axis and only have variation or wave vector along the *y* direction. Because the tangential fields are continuous at the boundaries, the radiation will be of plane-wave-like character. In other words, the radiation wave is emitted to the far-field region as a highly directive beam. In particular, when $\varepsilon_z = \mu_x = 1$, the wave impedance of the propagating mode in the AZIM slab will be matched to that in free space and the plane wave will be emitted without any reflections. Furthermore, when the line source is embedded in a bounded AZIM

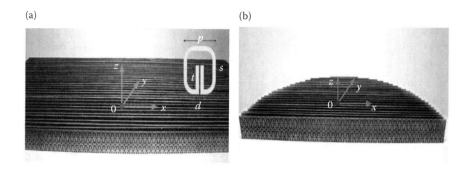

Figure 4.46 Fabricated AZIM lenses with (a) a slab shape and (b) a semicircular shape. (The source of the material Q. Cheng, W. X. Jiang, and T. J. Cui. Radiation of planar electromagnetic waves by a line source in anisotropic metamaterials. *J. Phys. D: Appl. Phys.*, 43(33): 335406. Copyright 2010 IOP Publishing. Reproduced by permission of IOP Publishing. All rights reserved is acknowledged.)

with arbitrary shape, the plane-wave feature will be conserved in the near-field region outside the AZIM, as long as the impedance-matching condition is satisfied. The directive emission feature is independent of shape variation of AZIM, which makes AZIM a good candidate for conformal antennas.

Two AZIM samples have been designed and fabricated in microwave region. One sample takes a rectangular shape and the other takes a semicircular shape [65], as illustrated in Figure 4.46a and b, respectively. Each AZIM sample is composed of a number of PCBs, and the gap between each two adjacent PCBs is 3.33 mm. The rectangular sample covers an area of 260×103 mm^2. In the semicircular sample, the radius is selected as 77 mm, and there are 23 PCBs in the longitudinal direction. To meet the requirement of near-zero permeability in the y direction, SRRs (see Section 4.1.1) were adopted as the unit cells. For each piece of PCB, array of SRRs was etched in the copper layer of F4B laminate. The configuration of these SRRs is the same as that listed in the caption of Figure 4.2. Following the standard retrieval procedure, the frequency response of relative permittivity component ε_z and relative permeability components μ_x and μ_y were obtained for the SRR unit cell. There exists a zero point for the μ_y curve at the frequency of 8.1 GHz, which is the design frequency of the AZIM samples.

The generation of plane waves and highly directive emission feature of AZIM were experimentally verified with the two samples. In the experiment, the 2D near-field scanning apparatus introduced in previous section was utilized to map the near-field distributions of the fabricated samples. A monopole antenna was placed inside the samples to mimic the line source.

The measured electric field distribution around the rectangular AZIM at 8.1 GHz is illustrated in Figure 4.47a, in which the feeding monopole antenna is

122 ■ *Metamaterials*

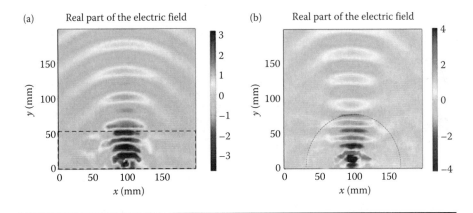

Figure 4.47 Scanned near electric fields of (a) the slab AZIM lens and (b) the semicircular AZIM lens. (The source of the material Q. Cheng, W. X. Jiang, and T. J. Cui. Radiation of planar electromagnetic waves by a line source in anisotropic metamaterials. *J. Phys. D: Appl. Phys.*, 43(33): 335406. Copyright 2010 IOP Publishing. Reproduced by permission of IOP Publishing. All rights reserved is acknowledged.)

located at $x = 100$ mm and $y = 15$ mm. It should be noted that the region surrounded by the dashed line corresponds to a half of the rectangular AZIM. The internal field inside the AZIM, as shown in Figure 4.47a, is like plane wave due to the anisotropic zero-index property. At the interface between the AZIM lens and free space, most of the EM energy is restricted along the y direction, which significantly enhances the directivity of the monopole antenna or line source.

In some engineering applications, antennas should be designed to have conformal shapes along the surface of the mechanical carrier. It is expected that the properties of conformal antennas are similar to those of planar antennas. Hence, the AZIM sample with a semicircular shape was also considered in Reference 65. The measured near-field distribution of the semicircular AZIM lens is demonstrated in Figure 4.47b, where the monopole antenna is placed at $x = 100$ mm and $y = 10$ mm. From Figure 4.47b, it is observed that plane wave was generated inside the semicircular AZIM and emitted at the curved interface. Moreover, the plane wave is still directed to the y direction. Hence, it is proved that the near-field distribution and the directive radiation property is insensitive to shape variation of AZIM.

4.4.3 Spatial Power Combination for Omnidirectional Radiation via Radial AZIM

A novel feature of ZIMs, to realize spatial power combination for omnidirectional radiation, was proposed in Reference 66. Consider a circular-ring region filled with radial anisotropic zero-index metamaterial (RAZIM). Under the cylindrical coordinate system, RAZIM is described by permittivity and permeability tensors as

$$\bar{\bar{\varepsilon}} = \hat{\rho}\hat{\rho}\varepsilon_\rho + \hat{\varphi}\hat{\varphi}\varepsilon_\varphi + \hat{z}\hat{z}\varepsilon_z, \qquad (4.37)$$

$$\bar{\bar{\mu}} = \hat{\rho}\hat{\rho}\mu_\rho + \hat{\varphi}\hat{\varphi}\mu_\varphi + \hat{z}\hat{z}\mu_z, \qquad (4.38)$$

in which the component μ_ρ approaches zero. If we consider the case of TE polarization, only z component of the electric field, E_z, and φ and ρ components of the magnetic field, H_φ and H_ρ are relevant to the problem. In such a case, the differential equation of E_z in the RAZIM ring is written as

$$\frac{1}{\mu_\varphi}\frac{\partial}{\partial\rho}\left(\rho\frac{\partial E}{\partial\rho}\right) + \frac{1}{\rho\mu_\rho}\frac{\partial^2 E_z}{\partial\varphi^2} + \omega^2\varepsilon_z\rho E_z = 0. \qquad (4.39)$$

Let $E_z = \Psi(\rho)\Theta(\varphi)$, it is derived that

$$\frac{1}{\rho^2}\frac{d^2\Psi}{d\rho^2} + \frac{1}{\rho}\frac{d\Psi}{d\rho} + \left(\omega^2\mu_\varphi\varepsilon_z\rho^2 - \frac{n^2}{\mu_\rho/\mu_\varphi}\right)\Psi = 0, \qquad (4.40)$$

and $\Theta(\varphi)$ has the wave form of $e^{in\varphi}$. After the boundary conditions on the air–RAZIM interface is considered with $\varepsilon_z = \varepsilon_0$ and $\mu_\varphi = \mu_0$, we get the solution of Equation 4.40 as

$$E_z = \Sigma_{n=-\infty}^{\infty} i^n (A_n H_n^{(1)}(k_s\rho))e^{in\varphi}, \qquad (4.41)$$

in which $k_s = \omega\sqrt{\mu_\varphi\varepsilon_z} = k_0$, and A_n is the amplitude. Since $\mu_\rho \to 0$, we have $n \to 0$, and

$$E_z = A_0 H_0^{(1)}(k_0\rho). \qquad (4.42)$$

Hence, the waves in the RAZIM ring only propagate along the radial direction and the fields have no variation in terms of φ. In other words, purely cylindrical waves are supported in the RAZIM ring regardless of the location of the source.

The power combination condition has been derived for the RAZIM ring [66]. When only source 1 exists, the electric field in the air region inside the ring is written as

$$E_{z1}^{(1)}(\vec{\rho}) = \frac{i}{4}\omega\mu_0 I_1 H_0^{(1)}(k_0|\vec{\rho} - \vec{\rho}_1|), \qquad (4.43)$$

in which I_1 and $\vec{\rho}_1$ are the electric current and the position of Source 1. From boundary conditions, the radiated electric field of Source 1 in the air region outside the ring can be written as

$$E_{z1}^{(3)}(\vec{\rho}) = \frac{i}{4}\omega\mu_0 I_1 J_0(k_0\rho_1) H_0^{(1)}(k_0\rho). \qquad (4.44)$$

When Sources 1 and 2 exist simultaneously, the total radiated electric field in the air region outside the ring is given by

$$E_z^{(3)}(\vec{\rho}) = \frac{i}{4}\omega\mu_0[I_1 J_0(k_0\rho_1) + I_2 J_0(k_0\rho_2)]H_0^{(1)}(k_0\rho), \quad (4.45)$$

in which I_1 and I_2 are the electric currents and $\vec{\rho}_1$ and $\vec{\rho}_2$ are the positions of Sources 1 and 2, and J_0 is the zero-order Bessel's function. From the property of the Bessel function, the total electric field radiated by both sources will be stronger than the electric field generated by any single source when $I_1 J_0(k_0\rho_1)$ and $I_2 J_0(k_0\rho_2)$ have the same sign, and the combined power will be larger than each individual power. In particular, when the two sources are close to the axis of the circular ring so that $k_0\rho_1$ and $k_0\rho_2$ are much less than the first zero point of Bessel function, both $J_0(k_0\rho_1)$ and $J_0(k_0\rho_2)$ are close to 1, leading to a significant power combination. Hence, high-efficiency power combination for perfect omnidirectional radiations can be achieved through the circular RAZIM ring.

To verify the predicted peculiar phenomenon experimentally, a circular RAZIM ring was realized using SRRs, which have strong magnetic-resonance response. The sample is composed of six concentric layers, and each layer is a thin PCB etched with a number of SRRs, as shown in Figure 4.48a. Such an arrangement ensures that the ring possesses radial anisotropy. Applying the parameter retrieval method, we got to obtained the effective permeability and permittivity components μ_ρ, μ_φ, and ε_z of the RAZIM. The radial component of permeability μ_ρ is nearly zero with tiny loss at the design frequency (10.4 GHz).

In the experiments, the sample was placed in the 2D Mapper. Two feeding probes, which were connected to two output ports of a power divider, acted as Sources 1 and 2, respectively. Three separate experiments were conducted to verify the omnidirectional radiation and power-combination properties. In the first two experiments, Sources 1 and 2 took effect sequentially. When the two sources were applied simultaneously in the third experiment, a significant enhancement of omnidirectional radiation was obtained, as verified by the measured near-field distribution shown in Figure 4.48b. The field enhancement directly indicates the achievement of high-efficiency power combination. Figure 4.48c demonstrates the measured power densities for the three experiments. It is shown that the combined power is nearly 4 times larger than the power of a single source, validating the phenomenon of highly efficient spatial-power combination. The strong combined power is due to the interactions between the two sources. Assuming that the current of Source 1 is I and its self-radiation impedance is R_{11}, the radiation power of Source 1 is $P_1 = I^2 R_{11}$. Similarly, for Source 2, we have $P_2 = I^2 R_{22}$. When Sources 1 and 2 are applied simultaneously, the combined radiation power will be $P = I^2(R_{11} + R_{12} + R_{21} + R_{22})$, in which R_{12} and R_{21} are mutual impedances of the two sources. The existence of RAZIM will result in mutual impedances

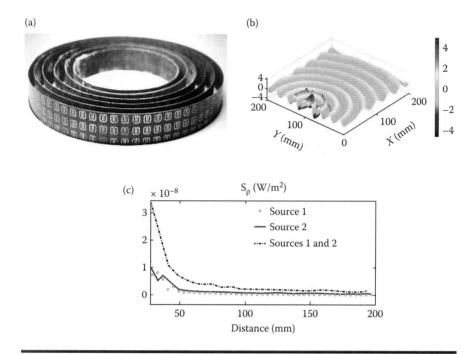

Figure 4.48 (a) Photograph of a fabricated RAZIM ring. (b) Real part of measured near electric field under the simultaneous excitations of Sources 1 and 2. (c) Measured power densities along the line $y = 100$ mm, for three cases of excitations. (Reprinted with permission from Q. Cheng, W. X. Jiang, and T. J. Cui, *Phys. Rev. Lett.*, 108(21): 213903, 2012. Copyright 2012 by the American Physical Society.)

nearly the same as the self-impedance, making the combined radiation power nearly 4 times larger than each individual power.

4.4.4 Directivity Enhancement to Vivaldi Antennas Using Compact AZIMs

A high-directivity Vivaldi antenna was developed in Reference 67, by combining the features of AZIM and traditional Vivaldi antenna together. Vivaldi antenna is a type of planar microstrip antenna with exponentially tapered slot. It works in a broad frequency band but suffers from low directivity since it radiates cylindrical-like waves. According to the principle introduced in Section 4.4.2, when proper AZIM is incorporated in the exit area of the tapered slot, the resulting modified Vivaldi antenna will generate plane-like waves instead of cylindrical-like waves. Hence, a significant enhancement of directivity can be obtained with AZIM loading.

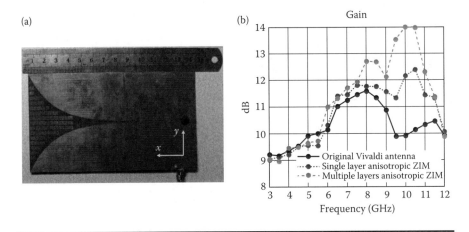

Figure 4.49 (a) Photograph of a Vivaldi antenna loaded with AZIM. (b) Measured gains of Vivaldi antennas with and without AZIM loading. (B. Zhou and T. J. Cui, Directivity enhancement to vivaldi antennas using compactly anisotropic zero-index metamaterials. *IEEE Ant. Wirel. Prop. Lett.*, 10: 326 © 2011 IEEE.)

In Reference 67, the MLR described in Section 4.1.1 was chosen as the constituent unit cell of AZIM. For each MLR unit cell, the longitudinal permittivity component ε_x is designed to be equal or close to zero, while the permeability component μ_z and the permittivity component ε_y are designed to match with the impedance in free space, both in the operating band of the Vivaldi antenna. Such a configuration ensures that the designed AZIM can enhance the antenna directivity and gain significantly. Moreover, the MLR-based AZIM can be embedded into the original Vivaldi antenna smoothly and efficiently. Hence, this kind of AZIM has great advantage in designing high-gain Vivaldi antennas.

Two different prototypes of modified Vivaldi antennas with AZIM loading were constructed in Reference 67. They employ a single AZIM layer and multiple AZIM layers in the vertical direction (i.e., the z direction), respectively. The fabricated single-AZIM-layer Vivaldi antenna is shown in Figure 4.49a. It can be observed that the antenna and the AZIM layer share the same substrate. Hence, the fabrication of the single-AZIM-layer Vivaldi antenna makes a smooth mixture of the metamaterial with the antenna. As to the Vivaldi antenna loaded with multilayer AZIM, the antenna performance can be further improved significantly, while the overall antenna structure is still compact.

The radiation property of both prototypes were measured and the measured gains are demonstrated in Figure 4.49b. For comparison, the performance of traditional Vivaldi antenna is also presented in the figure. As is shown, both the single-layer and the multilayer AZIM strategies are effective in greatly enhancing the antenna gain in the frequency band from 9.5 to 10.5 GHz. And the multilayer AZIM strategy proves to offer even more significant improvement of gain.

4.5 Double-Positive Metamaterials

4.5.1 Transmission Polarizer Based on Anisotropic DPS Metamaterials

This section presents the implementation of transmission polarizers with anisotropic DPS metamaterials [68]. We first examine the idea of transmission polarizer based on an anisotropic slab, in which the function and the transmission efficiency of the polarizer can be controlled freely. Then two examples of such transmission polarizers are presented. One of the polarizers converts linearly polarized waves to circularly polarized waves, while the other polarizer converts linearly polarized waves from one polarization to its cross polarization. These two polarizers are realized by DPS metamaterials composed of ELCRs introduced in Section 4.1, which have strong coupling to local electric field and show designable effective electric permittivity [16, 17]. Measurement results are shown for these two polarizers, which are in excellent agreement with the theoretical predictions.

The anisotropic slab used for the transmission polarizer has diagonal relative permittivity and permeability tensors whose components are denoted by ϵ_x, ϵ_y, ϵ_z, μ_x, μ_y, and μ_z. Assume that the incoming wave of the polarizer is normally incident on the slab (i.e., propagating along the minus z axis). As pointed out in Reference 69, the waves propagating in the anisotropic slab along the z axis can be decomposed into two orthogonal modes with the electric fields directed along the x axis and y axis, respectively. The transmission coefficients T^x and T^y of these two modes through the slab or polarizer are related to the constitutive properties of the slab and written as

$$T^{x,y} = \frac{(1 - r_{x,y}^2) e^{ik_{x,y}d}}{1 - r_{x,y}^2 e^{i2k_{x,y}d}}, \tag{4.46}$$

in which d is the thickness of the anisotropic slab, $k_{x,y}$ is the wave number for the two modes, and $r_{x,y}$ is the reflection coefficient on the air–slab interface for the two modes. The expressions for $k_{x,y}$ and $r_{x,y}$ are given by

$$k_{x,y} = \omega\sqrt{\mu_{y,x}\epsilon_{x,y}}, \quad r_{x,y} = \frac{\eta_0 - \eta_{x,y}}{\eta_0 + \eta_{x,y}}, \tag{4.47}$$

where η_0 is the wave impedance of air and $\eta_{x,y} = \sqrt{\mu_{y,x}/\epsilon_{x,y}}$ is the wave impedance of the anisotropic slab for the two wave modes.

Suppose that the electric field of the incident TEM wave is written as $\vec{E} = \hat{x}E_x + \hat{y}E_y$. The conditions for linearly polarized outgoing waves are derived as

$$|E_x| \cdot |T^x| = p |E_y| \cdot |T^y|, \tag{4.48}$$

$$\left[\arg(E_x) + \arg(T^x)\right] - \left[\arg(E_y) + \arg(T^y)\right] = m\pi, \tag{4.49}$$

where m is an integer and p is a proportional factor indicating the orientation of the transmitted electric field. The conditions for elliptically polarized outgoing waves are

$$|E_x| \cdot |T^x| = q |E_y| \cdot |T^y|, \tag{4.50}$$

$$[\arg(E_x) + \arg(T^x)] - [\arg(E_y) + \arg(T^y)] = \frac{n\pi}{2}, \tag{4.51}$$

where n is an odd number and q is the length ratio between the major and minor axes of the polarization ellipse. We remark that such conditions are put forward on the assumption that the ellipse foci lie on the coordinate axis (i.e., the x or y axis). The elliptical polarization turns to circular polarization when $q = 1$. The values of T^x and T^y can be controlled by designing the constitutive parameters of the anisotropic slab. And the values of $|E_x|/|E_y|$ and $\arg(E_x) - \arg(E_y)$ can be adjusted by rotating the slab, together with the coordinate system, around its central normal. By properly setting these values, various polarization transformations can be achieved. Moreover, if the transmission coefficient T^x and T^y are tuned as close to 0 dB as possible, the polarizer will have a high transmission efficiency. Compared with the reflector polarizer [70], the incoming and outgoing waves of the transmission polarizer here are naturally well separated, and therefore do not interfere with each other.

The idea of the transmission polarizer is further demonstrated by presenting two instances of converting linear polarization to circular polarization and to its cross linear polarization. For both instances, the incident waves are linearly polarized with $|E_x| = |E_y|$ and $\arg(E_x) = \arg(E_y)$. In the first instance, T^x and T^y of the anisotropic slab is designed to meet the following relations:

$$|T^x| = |T^y|, \quad \arg(T^x) - \arg(T^y) = \frac{3\pi}{2}. \tag{4.52}$$

Hence Equations 4.50 and 4.51 are satisfied with $q = 1$, $n = 3$ and the outgoing wave is circularly polarized. Besides, $|T^x|$ and $|T^y|$ are controlled to be close to 0 dB so that nearly all energy is transmitted to the targeted polarization. The anisotropic slab in the second instance has the same constitutive parameters as that in the first instance but has twice thickness. Therefore, for the second instance, Equations 4.48 and 4.49 are satisfied with $p = 1$, $m = 3$ and the incident linear polarization is converted to its cross polarization.

The two instances of the transmission polarizer are implemented at 9.5 GHz with the ELCR-based metamaterials, which act as the anisotropic slabs for polarization conversion. As described in Section 4.1.1, the ELCR-based metamaterial is constructed by successively stacking a certain number of single-sided CCL with a fixed spacing. The copper layer of each of these laminates is etched to form an identical 2D ELCR array. Among the differently shaped ELCRs investigated in Reference 17, ELCR with two pairs of arms (shown in the inset of Figure 4.6a) is

Homogeneous Metamaterials ■ 129

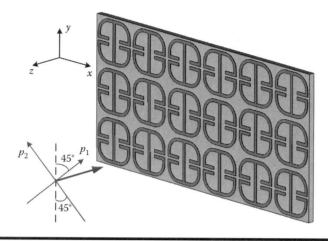

Figure 4.50 A transmission polarizer made from ELCRs and illuminated with normally incident plane waves.

adopted as the constituent particle here for its merit of being electrically small. The lattice constants of the metamaterials are chosen to be $a_x = a_y = a_z = 4$ mm. The CCL are selected to be 20 cm × 20 cm F4B circuit boards with copper thickness of 0.035 mm, substrate thickness of 0.75 mm, and dielectric permittivity of $3.0 + i0.01$. There are 48 × 48 ELCR unit cells on each laminate and the normal of each laminate is aligned with the z axis or the illumination direction of incident waves (see Figure 4.50). Two pieces of laminate patterned with periodic ELCR array are employed for the linear-to-circular polarizer, while four pieces of laminate are employed for the linear-to-linear polarizer. Hence the slab thicknesses of these two metamaterial polarizers are 8 and 16 mm, respectively.

The transmission coefficients T^x and T^y of those two metamaterial polarizers can be calculated from the effective medium parameters of the ELCR-based metamaterial (ϵ_x, ϵ_y, μ_y, and μ_x) by Equation 4.46. And these effective medium parameters can be retrieved from simulated scattering parameters using the standard retrieval technique introduced in Chapter 2. The simulations have been made on a single ELCR particle using the commercial software Ansoft HFSS. In the simulations, the ELCR particle is excited by a plane wave propagating along the minus z axis and combinations of perfect electric and magnetic boundary conditions are employed along the remaining two axes to model an infinite structure and confine the polarization of the plane wave. Two sets of scattering parameters are obtained from simulations. The first set of scattering parameters, for which the perfect electric and magnetic boundary conditions are assigned to the bounding surfaces normal to the x and y axis, respectively, are used to compute the effective medium parameters ϵ_x and μ_y. While ϵ_y and μ_x are calculated from the second set of scattering parameters with perfect electric and magnetic boundary conditions applied in the y

130 ■ *Metamaterials*

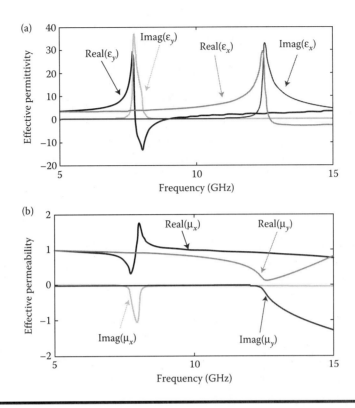

Figure 4.51 Retrieved anisotropic medium parameters of the optimized ELCR particle from 5 to 15 GHz. (a) Relative permittivity ϵ_x and ϵ_y. (b) Relative permeability μ_y and μ_x. (Reprinted with permission from J. Y. Chin, M. Lu, and T. J. Cui. Metamaterial polarizers by electric-field-coupled resonators. *Appl. Phys. Lett.*, 93(25): 251903. Copyright 2008, American Institute of Physics.)

and x axis, respectively. The geometric parameters of ELCR particle are optimized for the purpose of matching the values of T^x and T^y for the first instance with the relations of Equation 4.52 at 9.5 GHz to the utmost extent. The final geometry design of ELCR particle are $d_x = d_y = 3.6$ mm, $g = w = 0.2$ mm, $r = 1.38$ mm, and $s = 1.19$ mm, with which T^x and T^y for the first instance are kept to be greater than -1 dB and the phase difference between them is equal to $-270°$. The corresponding retrieved medium parameters are plotted in Figure 4.51.

The two metamaterial polarizers are fabricated and their performances are measured using a pair of X-band lens antennas connected to a VNA (Rohde and Schwarz ZVA40). The electric field of the transmitting antenna is polarized to the direction p_1 illustrated in Figure 4.50, which has a 45° angle with respect to the y axis so that the relation between the x and y components of the electric field is characterized by $|E_x| = |E_y|$ and $\arg(E_x) = \arg(E_y)$. Before either of the two fabricated polarizers are placed between the antenna pairs, a calibration is conducted by having the receiving

Homogeneous Metamaterials ■ 131

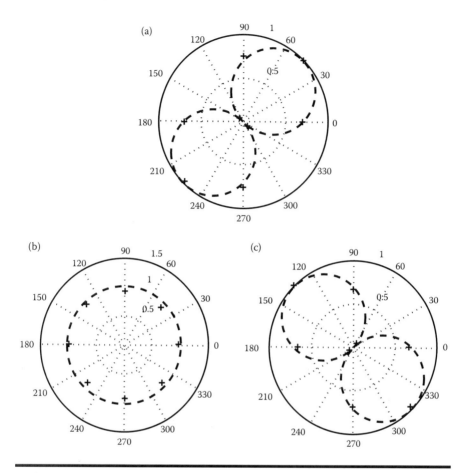

Figure 4.52 Theoretical polarization patterns (dashed lines) and measured polarization patterns (plus signs) of the (a) incident waves, (b) transmitted waves through the linear-to-circular polarizer, and (c) transmitted waves through the linear-to-linear polarizer. (Reprinted with permission from J. Y. Chin, M. Lu, and T. J. Cui. Metamaterial polarizers by electric-field-coupled resonators. *Appl. Phys. Lett.*, 93(25): 251903. Copyright 2008, American Institute of Physics.)

antenna aligned at the same direction with the transmitting antenna, which is marked as 45°. Then the receiving antenna are rotated from 0° to 315° by a step of 45° to achieve the polarization pattern of the incident waves. The comparison of the polarization patterns from the measurement and the theoretical prediction for the incident wave at 10 GHz is shown in Figure 4.52a. Afterward, the sample of linear-to-circular polarizer is put in between the antenna pairs and the polarization patterns of the transmitted waves are measured likewise. The comparison of this measured polarization pattern and its theoretical counterpart at 10 GHz is shown in Figure 4.52b. For circular polarization, the amplitude of electric field is supposed to be invariant with rotation (see the dashed line) and the measured variation is less than 1 dB. A similar measurement is conducted for the sample of linear-to-linear polarizer. The polarization pattern of the transmitted waves at 10 GHz is shown in Figure 4.52c. Comparing Figure 4.52a with Figure 4.52c, it is seen that the polarization pattern is rotated by 90°, which marks the conversion of incident polarization p_1 to its cross polarization p_2 (also defined in Figure 4.50). Measured polarization isolation is better than -20 dB.

For both polarizers, comparison has also been made between the simulated and measured transmissions from the incident polarization p_1 to the same polarization p_1 and to its cross polarization p_2. The comparisons for the linear-to-circular and linear-to-linear polarizers are shown in Figure 4.53a and b, respectively. The measured results agree well with the simulated results except that the central operating frequencies for both fabricated polarizers are about 10 GHz, which are higher than the design frequency 9.5 GHz. This frequency deviation mainly originates from the fabrication error and the random error in measurement. From Figure 4.53a, it is observed that the measured field intensity for the polarizations p_1 and p_2 around 10 GHz are roughly equal to each other at about -3 dB, indicating that the linear-to-circular polarization conversion is achieved with little loss of energy. From Figure 4.53b, it is observed that over the frequency range around 10 GHz, the measured field intensity for the cross polarization p_2 is close to 0 dB and that for the original polarization p_1 is less than -20 dB, which is a direct evidence of highly efficient conversion from a linear polarization to its cross linear polarization.

4.5.2 Increasing Bandwidth of Microstrip Antennas by Magneto-Dielectric Metamaterials Loading

This section demonstrates how to utilize the EML-based WG DPS metamaterial (see Section 4.1.4 to enhance the bandwidth of conventional microstrip patch antenna [20]). As shown in Figure 4.16b, the EML-based WG metamaterial is actually artificial magneto-dielectrics with both permittivity and permeability greater than one. It was first proposed by Hansen et al. that by loading magneto-dielectrics in the substrate volume between patch and ground plane, it is possible to design patch antennas with both broadened bandwidth and acceptable patch size [71]. The basic principle of this broadbanding technique is that the enhanced magnetic response of magneto-dielectrics lowers the quality factor of patch antenna while its refractive index lowers the resonant frequency as pure dielectric material does. Consider a microstrip rectangular patch antenna with a block of magneto-dielectrics loaded in the substrate volume right under the patch. The quality factor of the antenna at its resonance reads

$$Q = \frac{\omega W}{P}, \qquad (4.53)$$

where ω is the angular frequency, W is the time-averaged energy residing in the near fields of the antenna, and P is the time-averaged power dissipated during one

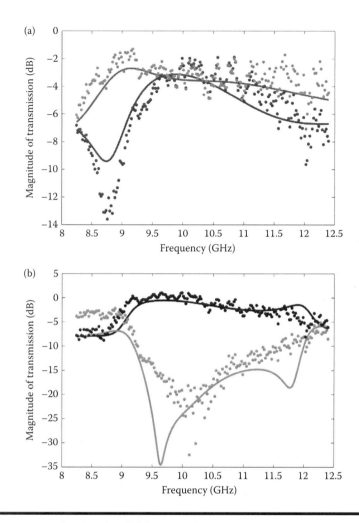

Figure 4.53 (a) Simulated (solid lines) and measured (dots) transmission of electric field from p1 to p1 (light color) and from p1 to p2 (deep color) for the linear-to-circular polarizer. (b) Simulated (solid lines) and measured (dots) transmission of electric field from p1 to p1 (gray) and from p1 to p2 (black) for the linear-to-linear polarizer. (Reprinted with permission from J. Y. Chin, M. Lu, and T. J. Cui. Metamaterial polarizers by electric-field-coupled resonators. *Appl. Phys. Lett.*, 93(25): 251903. Copyright 2008, American Institute of Physics.)

cycle. Known from the cavity model of patch antenna, W mainly owes to the energy stored right under the patch [72]:

$$W = \int \left(\frac{\epsilon_0 \epsilon_c}{4} |E|^2 + \frac{\mu_0 \mu_c}{4} |H|^2 \right) dV, \qquad (4.54)$$

where ϵ_c and μ_c are relative permittivity and relative permeability of the MDM, respectively. By applying the fundamental TM$_{10}$ mode resonance condition and making the integration, Equation 4.54 becomes

$$W = \frac{\pi \cdot w_p \cdot h \cdot E_0^2}{4\omega\eta_0 \sqrt{\mu_c/\epsilon_c}}, \quad (4.55)$$

where h is substrate thickness, E_0 is the electric field amplitude at the radiation edge (with length w_p) of the patch, and η_0 is the wave impedance in free space. As to P, it is a good approximation to take only the radiation power into account

$$P = \frac{G_r E_0^2 h^2}{2}, \quad (4.56)$$

where G_r is the radiation conductance. Inserting Equations 4.55 and 4.56 into Equation 4.53, we get

$$Q = \frac{\pi \cdot w_p \cdot h \cdot \eta_0}{2G_r} \sqrt{\frac{\mu_c}{\epsilon_c}}. \quad (4.57)$$

Equation 4.57 clearly reveals that the quality factor decreases as the ratio μ_c/ϵ_c increases. Reference 71 further gave the zero-order VSWR = 2 bandwidth of square patch antenna with magneto-dielectric substrate:

$$BW \approx \frac{96\sqrt{\mu_c/\epsilon_c}h/\lambda_0}{\sqrt{2}[4 + 17\sqrt{\mu_c\epsilon_c}]}, \quad (4.58)$$

where λ_0 is the free space wavelength at antenna resonance. Therefore one can enhance the bandwidth of patch antennas by increasing $\sqrt{\mu_c/\epsilon_c}$ for a given miniaturization factor of the patch (i.e., constant $\sqrt{\mu_c\epsilon_c}$). Theoretically, the magneto-dielectric loading technique has little influence upon the radiation characteristics of patch antennas.

According to the above theory, an improved bandwidth for a patch antenna will be expected if the aforementioned EML-based WG-MDM is properly incorporated with the antenna, acting as the antenna substrate. A proposed scheme of loading the patch antenna with the EML-based WG-MDM is illustrated in Figure 4.55a. The relevant novel patch antenna, referred to as the WG-MDM antenna, mainly comprises three layers: a normal supporting dielectric layer and two metallic layers attached to the two sides of the dielectric layer, respectively. The radiating rectangular patch is arranged in the top metallic layer and is fed by 50 Ω microstrip through a quarter-wavelength matching line. A 2D periodic array of EMLs is etched in the bottom ground right under the patch. In this way, a block of EML-based WG-MDM is successfully loaded into the substrate volume between the patch and the EML array and the top patch and the bottom ground naturally act as the upper and lower metallization for the WG-MDM, respectively. Considering that the field distribution under the patch can be approximated by a standing TEM wave for the fundamental TM_{10} mode, only the permittivity component ϵ_z and the permeability component μ_y of the WG-MDM are relevant to the problem. The whole structure is quite compact and requires only routine PCB process for fabrication as conventional patch antenna does. However, since etching EML array in the ground would inevitably lead to energy leakage and increase of back radiation, an additional metal shield plate for leakage inhibition is appended beneath and parallel to the ground, as shown in Figure 4.54b.

A design example of the WG-MDM antenna, operating at about 3.5 GHz, is presented below. In this design, the periodicity and dimensions of the EML element

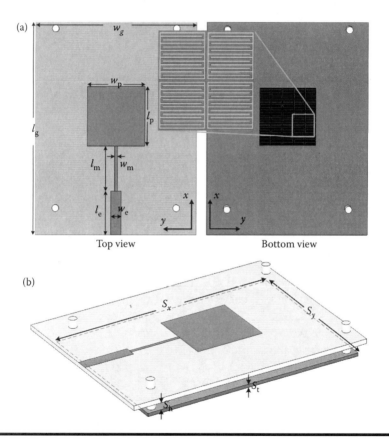

Figure 4.54 (a) Illustration of microstrip patch antenna loaded with EML-based WG magneto-dielectric material (WG-MDM). The deep color area indicates metallization and the light color area indicates supporting dielectric board. (b) Appearance of the WG-MDM-loaded patch antenna after an additional metal shield plate is appended to it.

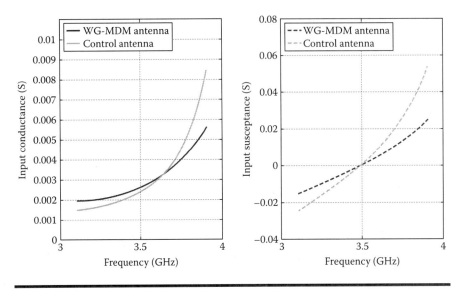

Figure 4.55 Simulated input conductance (left) and susceptance (right) of both the WG-MDM antenna (the deep color line) and the control antenna (the light color line).

and the characteristic parameters of the dielectric layer are the same as those listed in the caption of Figure 4.16. With this configuration, the effective medium parameters of the WG-MDM are $\epsilon_z \approx 1.822 - i0.0001$ and $\mu_y \approx 2.227 + i0.0364$ at 3.5 GHz, as revealed by Figure 4.16b. Hence the miniaturization factor for the rectangular patch is about $\sqrt{\mu_y \epsilon_z} \approx 2.014$. The scale of the EML array is chosen as 5 × 5, corresponding to a 21 mm×21 mm area in the ground plane. The array is located completely under the patch, given that the patch width is $w_p = 21$ mm and the patch length is $l_p = 22$ mm. Note that l_p is about half wavelength inside the WG-MDM and is slightly longer than the EML array size along the x direction so that the EML array has little influence on the radiation conductance. The geometries of the feeding network are determined as $w_m = 0.42$ mm, $l_m = 15.34$ mm, $w_e = 3.86$ mm, and $l_e = 16$ mm, where w_m and l_m are the width and length of matching line, respectively, and w_e and l_e are the width and length of 50 Ω microstrip, respectively. The dimensions of the shield plate are $s_x = 70$ mm, $s_y = 50$ mm, and $s_t = 1$ mm and the distance between the shield and antenna ground is $s_h = 5$ mm. The overall sizes of the WG-MDM antenna are $w_g = 60$ mm and $l_g = 76.34$ mm.

For comparison, a conventional rectangular patch antenna with a pure dielectric substrate (i.e., control antenna) is also designed. The permittivity of the substrate is tuned to be $\epsilon_d = 3.51$ (with dielectric loss tangent 0.001) so that the WG-MDM antenna and the control antenna have the same supporting dielectric height, the same patch size, and the same overall size. Hence, as implied by Equation 4.58, the

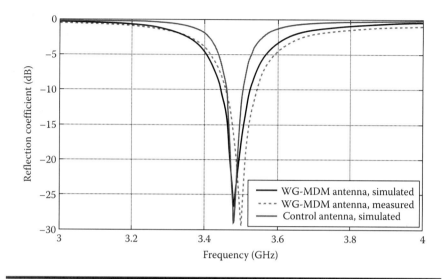

Figure 4.56 Simulated and measured reflection coefficients of the WG-MDM antenna (solid black line and dashed gray line, respectively) and simulated reflection coefficient of the control antenna (solid gray line).

bandwidth improvement factor of the WG-MDM antenna over the control antenna can be estimated by calculating $\sqrt{\epsilon_d \mu_y / \epsilon_z}$. When we neglect the imaginary part of μ_y and ϵ_z, the improvement factor is predicted to be 2.07.

Both the WG-MDM antenna and the control antenna are simulated using CST Microwave Studio. Figure 4.55 shows the simulated input admittances for both antennas. The phase reference plane for these admittances is located at the junction between the patch and the feeding network. Obviously, the susceptance curve of the WG-MDM antenna varies more slowly around its resonant frequency than that of the control antenna, indicating an improved impedance matching for the WG-MDM antenna. Figure 4.56 depicts the simulated reflection coefficients of both antennas. As is shown, the WG-MDM antenna and the control antenna operate around the frequency 3.486 GHz and their −10 dB impedance bandwidths are 77.7 and 42 MHz, respectively. Hence the simulated results announce a bandwidth improvement factor of about 1.85. The discrepancy between the theoretical and simulated improvement factors can be attributed to two reasons. First, the effective medium parameters retrieved for the WG-MDM are inaccurate since the retrieval process assumes an infinitely wide EML array and neglects the influence of shield plate on the effective medium behavior of the WG-MDM. Second, the analytical model used to evaluate impedance bandwidth of patch antenna (Equation 4.58) is approximative. Figure 4.57 demonstrates the simulated resonate magnetic field intensity under the patch for both antennas. As is shown, the magnetic field intensity of the WG-MDM antenna is much smaller than that of the control antenna,

Table 4.1 Comparison of Radiation Performance between the WG-MDM Antenna and the Control Antenna

	Radiation Efficiency	Gain Copolarization
WG-MDM antenna	0.903	7.001 dB
Control antenna	0.921	6.832 dB

validating that the impedance bandwidth improvement is a result of decreased stored energy (or quality factor) and hence the enhanced $\sqrt{\mu_y/\epsilon_z}$ value of the WG-MDM substrate. The periodic distribution of the magnetic intensity in Figure 4.57a is due to the periodicity of the EML array. Simulations also show that the radiation pattern of the WG-MDM antenna is close to that of the control antenna. They have almost the same radiation efficiency and forward copolarization gain (see Table 4.1) at the central working frequency.

A sample of the WG-MDM antenna is fabricated and its reflection and radiation performances are measured. Seen from the measured reflection coefficient (also shown in Figure 4.56), the sample antenna has a central frequency of about 3.494 GHz and a −10 dB bandwidth of 76 MHz, which are close to the simulation results. Figure 4.58 shows the radiation patterns of the sample at 3.494 GHz, using a spherical coordinate consistent with the Cartesian coordinate shown in Figure 4.54.

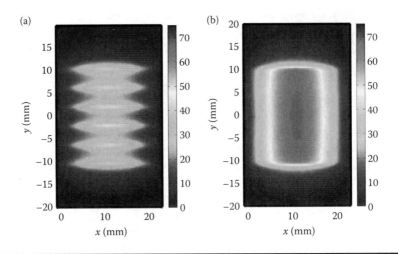

Figure 4.57 Simulated magnitude distribution of magnetic field intensity under the patch at 3.49 GHz for (a) the WG-MDM antenna and (b) the control antenna. (X. M. Yang et al., Increasing the bandwidth of microstrip patch antenna by loading compact artificial magneto-dielectrics. *IEEE Trans. Ant. Propag.*, 59(2): 373 © 2011 IEEE.)

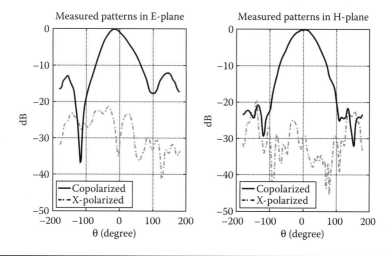

Figure 4.58 Measured radiation patterns of the WG-MDM antenna in E-plane (left) and H-plane (right) at 3.494 GHz. The positive/negative θ value for the E-plane patterns corresponds to the $\phi = 0°/\phi = 180°$ half plane, and the positive/negative θ value for the H-plane patterns corresponds to the $\phi = 90°/\phi = 270°$ half plane. (X. M. Yang et al., Increasing the bandwidth of microstrip patch antenna by loading compact artificial magneto-dielectrics. *IEEE Trans. Ant. Propag.*, 59(2): 373 © 2011 IEEE.)

It reveals that the measured radiation patterns of the WG-MDM antenna could compare favorably with conventional patch antenna. The front-to-back ratios of the sample are 11.98 and 20.76 dB for E- and H-planes, respectively, and the cross-polarizations for E- and H-planes are about −21.07 and −19.58 dB, respectively. Besides, measurement also confirms that the loading of EML-based WG-MDM has little influence on the forward copolarization gain of patch antenna.

In this chapter, we have presented an overview of homogeneous metamaterials developed in the past ten or more years. Four kinds of homogeneous metamaterials are involved in the overview: single-negative, double-negative, zero-index, and double-positive metamaterials. The constituents of these homogeneous metamaterials may take one of the three forms: 3D volumetric particles, planar complementary or WG elements, and *LC*-loaded TL grids or meshes. We have also introduced some potential applications of homogeneous metamaterials reported in recent years, including evanescent-wave amplification, partial focusing, EM-wave localization, subwavelength imaging, EM tunneling, spatial power combination, directivity enhancement, polarization conversion, and antenna broadbanding. We believe that, with the advancing of metamaterial technology, more and more applications will be exploited, which is expected to bring drastic changes to EM engineering.

Appendix: 2D Near-Field Mapping Apparatus

The 2D near-field mapping apparatus (2D mapper) is proposed and developed by Smith's group [56]. It mainly characterizes a planar waveguide chamber which, is composed of two parallel aluminum plates. It is designed to measure the 2D near electric field distribution internal and external to samples of artificial microstructures (e.g., metamaterials) placed in the chamber, at microwave frequencies. Usually, the distance between these two aluminum plates is set to be much smaller than the wavelength concerned so that only the dominant TEM waves with electric field perpendicular to the plates can propagate in the chamber. In such a case, the microwave scattering between the conducting plates is constrained to two dimensions. A photograph of the planar waveguide chamber built by Smith's group is shown in Figure 4A.1.

Samples in the chamber can be excited by either an approximate line source or a collimated beam approximating to a plane wave. When a line source is needed, a coaxial probe is fixed to the bottom plate, with its center pin protruding into the chamber through a hole drilled in the bottom plate. Cylindrical wave is launched around the pin when the probe is excited by an external source. To introduce an approximate plane wave into the chamber, a coaxial waveguide adapter is fixed to the edge of the bottom plate and the upper portion of the adapter has been milled away, allowing the top plate to replace the top wall of the adapter so that a tiny gap is

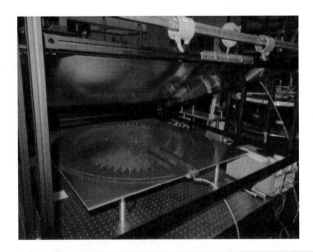

Figure 4A.1 Photograph of the 2D field mapping planar waveguide chamber in its open position. The upper plate (including the detector antenna) should be lowered parallel to the lower plate prior to the field mapping and after the sample is positioned relative to the source on the lower plate. (B. J. Justice et al. Spatial mapping of the internal and external EM fields of negative index metamaterials. *Opt. Express*, 14(19): 8694, 2006. With permission of Optical Society of America.)

kept between the plate and the adapter. The source wave generated from the adapter is shaped as a collimated finite-width beam by a long guiding channel formed using microwave absorbing material. A VNA provides the source microwave signal for either the coaxial probe or the waveguide adapter, which is connected to the output port of the VNA through a flexible coaxial cable. The input port of the same VNA is connected to another coaxial probe to detect the amplitude and phase of the electric field within the chamber. This coaxial probe is also mounted onto the upper plate such that the center pin and the dielectric sheath is inserted into, but not protruding below, the upper plate.

To achieve automated field scanning, the bottom plate is attached to a pair of orthogonally oriented motorized linear translation stages controlled by a computer. During measurements, samples under excitation are fixed on the upper surface of the bottom plate and avoid touching the lower surface of the top plate. Meanwhile, driven by the motorized stages, the bottom plate together with samples and source feed is translated step by step relative to the stationary top plate, in the plane of the aluminum plates. In this way, point-by-point phase-sensitive field detections are achieved and 2D spatial field maps internal and external to samples are obtained. In practice, a custom program can be used to coordinate the motion of the stages and the scan with the data acquisition of the VNA. Note that the bottom plate is completely covered by the top plate throughout the field mapping process since the top plate is sufficiently larger than the bottom plate. Besides, saw-toothed microwave absorber is arranged in a circular pattern around the mapping area and fixed to the lower plate. The absorber serves to minimize reflections from the plate edges back into the mapping region.

References

1. J. B. Pendry, A. J. Holden, D. J. Robbins, and W. J. Stewart. Magnetism from conductors and enhanced nonlinear phenomena. *IEEE Trans. Microw. Theory Tech.*, 47(11): 2075, 1999.

2. R. Marqués, F. Mesa, J. Martel, and F. Medina. Comparative analysis of edge- and broadside-coupled split ring resonators for metamaterial design—Theory and experiments. *IEEE Trans. Antenn. Propag.*, 51(10): 2572, 2003.

3. P. Gay-Balmaz and O. J. F. Martin. Electromagnetic resonances in individual and coupled split-ring resonators. *J. Appl. Phys.*, 92(5): 2929, 2002.

4. R. Marqués, F. Medina, and R. Rafii-El-Idrissi. Role of bianisotropy in negative permeability and left-handed metamaterials. *Phys. Rev. B*, 65(14): 144440, 2002.

5. N. Katsarakis, T. Koschny, M. Kafesaki, E. N. Economou, and C. M. Soukoulis. Electric coupling to the magnetic resonance of split ring resonators. *Appl. Phys. Lett.*, 84(15): 2943, 2004.

6. B. Sauviac, C. R. Simovski, and S. A. Tretyakov. Double split-ring resonators: Analytical modeling and numerical simulations. *Electromagnetics*, 24(5): 317, 2004.

7. D. R. Smith, J. Gollub, J. J. Mock, W. J. Padilla, and D. Schurig. Calculation and measurement of bianisotropy in a split ring resonator metamaterial. *J. Appl. Phys.*, 100(2): 024507, 2006.
8. L. Zhou and S. T. Chui. Magnetic resonances in metallic double split rings: Lower frequency limit and bianisotropy. *Appl. Phys. Lett.*, 90(4): 041903, 2007.
9. T. Driscoll, D. N. Basov, W. J. Padilla, J. J. Mock, and D. R. Smith. Electromagnetic characterization of planar metamaterials by oblique angle spectroscopic measurements. *Phys. Rev. B*, 75(11): 115114, 2007.
10. L. Zhou and S. T. Chui. Eigenmodes of metallic ring systems: A rigorous approach. *Phys. Rev. B*, 74(3): 035419, 2006.
11. A. Ahmadi and H. Mosallaei. Physical configuration and performance modeling of all-dielectric metamaterials. *Phys. Rev. B*, 77(4): 045104, 2008.
12. B.-I. Popa and S. A. Cummer. Compact dielectric particles as a building block for low-loss magnetic metamaterials. *Phys. Rev. Lett.*, 100(20): 207401, 2008.
13. J. B. Pendry, A. J. Holden, W. J. Stewart, and I. Youngs. Extremely low frequency plasmons in metallic mesostructures. *Phys. Rev. Lett.*, 76(25): 4773, 1996.
14. P. A. Belov, R. Marqués, S. I. Maslovski, I. S. Nefedov, M. Silveirinha, C. R. Simovski, and S. A. Tretyakov. Strong spatial dispersion in wire media in the very large wavelength limit. *Phys. Rev. B*, 67(11): 113103, 2003.
15. C. R. Simovski and P. A. Belov. Low-frequency spatial dispersion in wire media. *Phys. Rev. E*, 70(4): 046616, 2004.
16. D. Schurig, J. J. Mock, and D. R. Smith. Electric-field-coupled resonators for negative permittivity metamaterials. *Appl. Phys. Lett.*, 88(4): 041109, 2006.
17. W. J. Padilla, M. T. Aronsson, C. Highstrete, M. Lee, A. J. Taylor, and R. D. Averitt. Electrically resonant terahertz metamaterials: Theoretical and experimental investigations. *Phys. Rev. B*, 75(4): 041102(R), 2007.
18. W. X. Tang, H. Zhao, X. Y. Zhou, J. Y. Chin, and T. J. Cui. Negative index material composed of meander lines and SRRs. *Progr. Electromag. Res. B*, 8: 103, 2008.
19. R. Liu, X. M. Yang, J. G. Gollub, J. J. Mock, T. J. Cui, and D. R. Smith. Gradient index circuit by waveguided metamaterials. *Appl. Phys. Lett.*, 94(7): 073506, 2009.
20. X. M. Yang, Q. H. Sun, Y. Jing, Q. Cheng, X. Y. Zhou, H. W. Kong, and T. J. Cui. Increasing the bandwidth of microstrip patch antenna by loading compact artificial magneto-dielectrics. *IEEE Trans. Antenn. Propag.*, 59(2): 373, 2011.
21. F. Falcone, T. Lopetegi, M. A. G. Laso, J. D. Baena, J. Bonache, and M. Beruete. Babinet principle applied to the design of metasurfaces and metamaterials. *Phys. Rev. Lett.*, 93(19): 197401, 2004.
22. R. Liu, Q. Cheng, T. Hand, J. J. Mock, T. J. Cui, S. A. Cummer, and D. R. Smith. Experimental demonstration of electromagnetic tunneling through an epsilon-near-zero metamaterial at microwave frequencies. *Phys. Rev. Lett.*, 100(2): 023903, 2008.
23. Q. Cheng, R. Liu, J. J. Mock, T. J. Cui, and D. R. Smith. Partial focusing by indefinite complementary metamaterials. *Phys. Rev. B*, 78(12): 121102(R), 2008.
24. A. Alù and N. Engheta. Pairing an epsilon-negative slab with a mu-negative slab: Resonance, tunneling and transparency. *IEEE Trans. Antenn. Propag.*, 51(10): 2558, 2003.
25. V. G. Veselago. The electrodynamics of substances with simultaneously negative values of ϵ and μ. *Sov. Phys. Usp.*, 10(4): 509, 1968.
26. D. R. Smith, J. B. Pendry, and M. C. K. Wiltshire. Metamaterials and negative refractive index. *Science*, 305: 788, 2004.

27. W. J. Padilla, D. N. Basov, and D. R. Smith. Negative refractive index metamaterials. *Mater. Today*, 9: 28, 2006.
28. D. R. Smith, W. J. Padilla, D. C. Vier, S. C. Nemat-Nasser, and S. Schultz. Composite medium with simultaneously negative permeability and permittivity. *Phys. Rev. Lett.*, 84(18): 4184, 2000.
29. R. A. Shelby, D. R. Smith, and S. Schultz. Experimental verification of a negative index of refraction. *Science*, 292: 77, 2001.
30. R. Liu, A. Degiron, J. J. Mock, and D. R. Smith. Negative index material composed of electric and magnetic resonators. *Appl. Phys. Lett.*, 90(26): 263504, 2007.
31. O. G. Vendik and M. S. Gashinova. Artificial double negative (DNG) media composed by two different dielectric sphere lattices embedded in a dielectric matrix. In *Proceedings of the 34th European Microwave Conference*, Amsterdam, Netherlands, 3: 1209, 2004.
32. I. Vendik, O. Vendik, and M. Odit. Isotropoic artificial media with simutaneously negative permittivity and permeability. *Microw. Opt. Tech. Lett.*, 48(12): 2553, 2006.
33. L. Jylh, I. Kolmakov, S. Maslovski, and S. Tretyakova. Modeling of isotropic backward-wave materials composed of resonant spheres. *J. Appl. Phys.*, 99(4): 043102, 2006.
34. H. Chen, L. Ran, J. Huangfu, X. Zhang, K. Chen, T. M. Grzegorczyk, and J. A. Kong. Left-handed materials composed of only S-shaped resonators. *Phys. Rev. E*, 70(5): 057605, 2004.
35. D. Wang, L. Ran, H. Chen, M. Mu, J. A. Kong, and B.-I. Wu. Experimental validation of negative refraction of metamaterial composed of single side paired S-ring resonators. *Appl. Phys. Lett.*, 90(25): 254103, 2007.
36. L. Peng, L. Ran, H. Chen, H. Zhang, J. A. Kong, and T. M. Grzegorczyk. Experimental observation of left-handed behavior in an array of standard dielectric resonators. *Phys. Rev. Lett.*, 98(15): 157403, 2007.
37. J. Huangfu, L. Ran, H. Chen, X.-M. Zhang, K. Chen, T. M. Grzegorczyk, and J. A. Kong. Experimental confirmation of negative refractive index of a metamaterial composed of Ω-like metallic patterns. *Appl. Phys. Lett.*, 84(9): 1537, 2004.
38. M. Kafesaki, I. Tsiapa, N. Katsarakis, Th. Koschny, C. M. Soukoulis, and E. N. Economou. Left-handed metamaterials: The fishnet structure and its variations. *Phys. Rev. B*, 75(23): 235114, 2007.
39. H.-X. Xu, G.-M. Wang, M. Q. Qi, L.-M. Li, and T. J. Cui. Three-dimensional super lens composed of fractal left-handed materials. *Adv. Opt. Mater.*, 1(7): 495, 2013.
40. C. Caloz and T. Itoh. *Electromagnetic Metamaterials: Transmission Line Theory and Microwave Applications*. John Wiley & Sons, Inc., Hoboken, New Jersey; Florida, 2004.
41. A. Lai, C. Caloz, and T. Itoh. Composite right/left-handed transmission line metamaterials. *IEEE Microw. Mag.*, 5(3): 34, 2004.
42. G. V. Eleftheriades, A. K. Iyer, and P. C. Kremer. Planar negative refractive index media using periodically L-C loaded transmission lines. *IEEE Trans. Microw. Theory Tech.*, 50(12): 2702, 2002.
43. G. V. Eleftheriades, O. Siddiqui, and A. K. Iyer. Transmission line models for negative refractive index media and associated implementations without excess resonators. *IEEE Microw. Wireless Compon. Lett.*, 13(2): 51, 2003.
44. A. Grbic and G. V. Eleftheriades. Periodic analysis of a 2-D negative refractive index transmission line structure. *IEEE Trans. Antenn. Propag.*, 51(10): 2604, 2003.

45. A. Grbic and G. V. Elefheriades. Dispersion analysis of a microstrip-based negative refractive index periodic structure. *IEEE Microw. Wireless Compon. Lett.*, 13(4): 155, 2003.
46. T. J. Cui, Q. Cheng, Z. Z. Huang, and Y. J. Feng. Electromagnetic wave localization using a left-handed transmission-line superlens. *Phys. Rev. B*, 72(3): 035112, 2005.
47. A. Lakhtakia. On perfect lenses and nihility. *Int. J. Millimeter Infrared Waves*, 23(3): 339, 2002.
48. C. Qiu, H. Yao, L. Li, S. Zouhdi, and T. Yeo. Routes to left-handed materials by magnetoelectric couplings. *Phys. Rev. B*, 75(24): 245214, 2007.
49. R. Liu, C. Ji, J. J. Mock, J. Y. Chin, T. J. Cui, and D. R. Smith. Broadband ground-plane cloak. *Science*, 323: 366, 2009.
50. R. Liu, Q. Cheng, J. Y. Chin, J. J. Mock, T. J. Cui, and D. R. Smith. Broadband gradient index microwave quasioptical elements based on non-resonant metamaterials. *Opt. Express*, 17(23): 21030, 2009.
51. R. Liu, B. Zhao, X. Q. Lin, Q. Cheng, and T. J. Cui. Evanescent-wave amplification studied using a bilayer periodic circuit structure and its effective medium model. *Phys. Rev. B*, 75(12): 125118, 2007.
52. W. C. Chew. *Waves and Fields in Inhomogeneous Media*. Van Nostrand Reinhold, New York, 1990.
53. D. R. Smith, D. Schurig, J. J. Mock, P. Kolinko, and P. Rye. Partial focusing of radiation by a slab of indefinite media. *Appl. Phys. Lett.*, 84(13): 2244, 2004.
54. D. R. Smith, P. Kolinko, and D. Schurig. Negative refraction in indefinite media. *J. Opt. Soc. Am. B*, 21(5): 1032, 2004.
55. L. L. Hou, J. Y. Chin, X. M. Yang, X. Q. Lin, R. Liu, F. Y. Xu, and T. J. Cui. Advanced parameter retrievals for metamaterial slabs using an inhomogeneous model. *J. Appl. Phys.*, 103(6): 064904, 2008.
56. B. J. Justice, J. J. Mock, L. Guo, A. Degiron, D. Schurig, and D. R. Smith. Spatial mapping of the internal and external electromagnetic fields of negative index metamaterials. *Opt. Express*, 14(19): 8694, 2006.
57. Y. Qin, T. J. Cui, and Q. Cheng. Stong localization of EM waves using open cavities made of left-handed transmision-line media. *Microw. Opt. Tech. Lett.*, 48(8): 1662, 2006.
58. A. Grbic and G. V. Eleftheriades. Growing evanescent waves in negative-refractive-index transmission-line media. *Appl. Phys. Lett.*, 82(12): 1815, 2003.
59. S. L. He and Z. C. Ruan. A completely open cavity realized with photonic crystal wedges. *J. Zhejiang Univ. Science*, 6A(5): 355, 2005.
60. M. Notomi. Theory of light propagation in strongly modulated photonic crystals: Refractionlike behavior in the vicinity of the photonic band gap. *Phys. Rev. B*, 62(16): 10696, 2000.
61. J. B. Pendry. Negative refraction makes a perfect lens. *Phys. Rev. Lett.*, 85(18): 3966, 2000.
62. M. Silveirinha and N. Engheta. Tunneling of electromagnetic energy through subwavelength channels and bends using epsilon-near-zero materials. *Phys. Rev. Lett.*, 97(15): 157403, 2006.
63. R. E. Collin. *Field Theory of Guided Waves*. IEEE Press, New York, USA, 1991.
64. Y. G. Ma, P. Wang, X. Chen, and C. K. Ong. Near-field plane-wave-like beam emitting antenna fabricated by anisotropic metamaterial. *Appl. Phys. Lett.*, 94(4): 044107, 2009.
65. Q. Cheng, W. X. Jiang, and T. J. Cui. Radiation of planar electromagnetic waves by a line source in anisotropic metamaterials. *J. Phys. D: Appl. Phys.*, 43(33): 335406, 2010.

66. Q. Cheng, W. X. Jiang, and T. J. Cui. Spatial power combination for omnidirectional radiation via anisotropic metamaterials. *Phys. Rev. Lett.*, 108(21): 213903, 2012.
67. B. Zhou and T. J. Cui. Directivity enhancement to vivaldi antennas using compactly anisotropic zero-index metamaterials. *IEEE Antenn. Wireless Propag. Lett.*, 10: 326, 2011.
68. J. Y. Chin, M. Lu, and T. J. Cui. Metamaterial polarizers by electric-field-coupled resonators. *Appl. Phys. Lett.*, 93(25): 251903, 2008.
69. J. Y. Chin, M. Lu, and T. J. Cui. A transmission polarizer by anisotropic metamaterials. In *Proceedings of the 2008 IEEE Antennas and Propagation Society International Symposium and USNC/URSI National Radio Science Meeting*, San Diego, CA, USA, pp. 1–4, July 5–12, 2008.
70. J. Hao, Y. Yuan, L. Ran, T. Jiang, J. A. Kong, C. T. Chan, and L. Zhou. Manipulating electromagnetic wave polarizations by anisotropic metamaterials. *Phys. Rev. Lett.*, 99(6): 063908, 2007.
71. R. C. Hansen and M. Burke. Antenna with magneto-dielectrics. *Microw. Opt. Tech. Lett.*, 26(2): 75, 2000.
72. P. Ikonen and S. A. Tretyakov. On the advantages of magnetic materials in microstrip antenna miniaturization. *Microw. Opt. Tech. Lett.*, 50(12): 3131, 2008.

Chapter 5
Random Metamaterials: Super Noncrystals

Similar to homogeneous metamaterials, random metamaterials are constructed by arranging subwavelength artificial "atoms" in certain lattices. However, for random metamaterials, one or more aspects (e.g., lattice size, geometry, orientation, and structural dimensions) related to their constituent "atoms" should vary instead of keeping constant in space, for the sake of achieving spatially random distribution of one or several effective EM parameters (e.g., permittivity, permeability, refractive index or wave impedance). Random metamaterials possess peculiar EM-wave-controlling features. Intuitively, they would guide or scatter EM waves irregularly due to their randomness and may find applications in diffuse reflection, radar stealth, etc. This chapter mainly focuses on two categories of random metamaterials capable of diffusing EM waves. These two kinds of random metamaterials possess randomly distributed gradients of refractive index and randomly distributed reflection phase, respectively. Typical implementation schemes for these two types of metamaterials are presented in Section 5.1, and Sections 5.2 and 5.3 will focus on their diffuse scattering behaviors with respect to impinging EM waves.

5.1 Random Metamaterials: Random Arrangements of Particles

5.1.1 Randomly Gradient Index Metamaterial

Randomly gradient index metamaterial is the artificial realization of medium with randomly distributed gradients of refractive index [1]. The whole spatial region

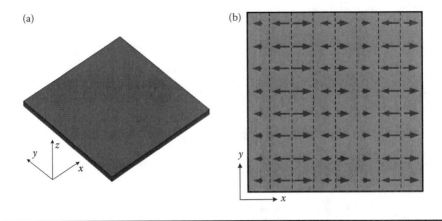

Figure 5.1 Schematic plot of a planar slab of 1D randomly gradient index medium. (a) Oblique view. (b) Elevation view, where the arrows represent the index gradients for subregions along the x axis.

occupied by the medium is divided into numerous subregions and each subregion characterizes a constant spatial linear gradient in refractive index; the index gradients of these subregions are randomly selected from certain distribution. Figure 5.1 gives a schematic plot of a planar slab of 1D randomly gradient index medium. As is shown, the subregion division of this slab is carried out only along the x direction, which means that the index gradients are randomly distributed only along this direction. The refractive index of the slab is invariant along both the y and z axis and the index gradients in each subregion is directed to either the positive or the negative x axis.

Generally speaking, 3D bulk metamaterials with randomly gradient index properties can be realized by arranging artificial volumetric particles of the same type in periodic lattices and assigning appropriate structural dimensions to each of these particles. The local effective refractive index related to each particle is independently engineered by tuning the dimensions of the corresponding particle. The constituent particles can be either local electric or local magnetic responsers and usually the DPS band before the first-order resonance is utilized for gradient index development. However, when broadband performance is required, it is more appropriate to implement randomly gradient index metamaterials with electric particles, since it is easier to achieve negligible dispersions by electric particles than by magnetic particles.

A sample of 1D randomly gradient index metamaterial composed of electric particles is shown in Figure 5.2a. It is actually an array of I-shaped particles with cubic lattices. The lattice constants are $a_x = a_y = a_z = 3$ mm. All the I-shaped particles are fabricated from single-sided copper-clad laminates with copper thickness of $t_m = 0.035$ mm and dielectric substrate thickness of $t_s = 0.93$ mm. The relative

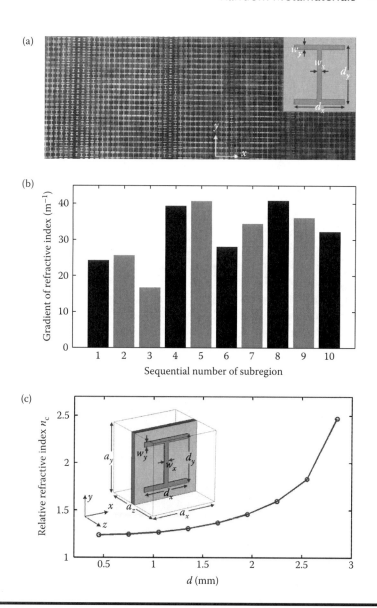

Figure 5.2 (a) Surface view of a fabricated sample of 1D randomly gradient index metamaterial composed of I-shaped particles. (b) Distribution of index gradients at 9.3 GHz for the randomly gradient index sample. The sequential number of subregion increases toward the positive x direction. (c) A design curve relating refractive index n_c at 9.3 GHz with the dimension $d = d_x = d_y$ for I-shaped particles with certain specifications, where n_c refers to $(n_{yx} + n_{yz})/2$.

permittivity and loss tangent of the substrate are $Dk = 2.65$ and $Df = 0.002$, respectively. The metallic line width of all particles are fixed ($w_x = w_y = 0.2$ mm) and the particle sizes along the x and y axis are the same ($d_x = d_y$) and denoted as d. The dimension d varies along the x axis in such a way that random distribution of index gradient is achieved along this direction. The relevant refractive index is $n_{yx} = \sqrt{\epsilon_y \mu_x}$ or $n_{yz} = \sqrt{\epsilon_y \mu_z}$. As described in Section 4.1, the I-shaped particle resonantly responses to external electric field polarized along the y axis, resulting in enhanced local effective permittivity ϵ_y. The degree of response and the value of ϵ_y strongly depend on the dimension d, whereas the local effective permeabilities μ_x and μ_z almost do not change with d, both approaching one. Hence, the index n_{yx} or n_{yz} can be easily tuned by adjusting the dimension d. The sample is divided into 10 subregions and the index gradients of each subregion at 9.3 GHz are shown in Figure 5.2b, where gray bars indicate positive gradients (pointing to the positive x axis) and black bars indicate negative gradients (pointing to the negative x axis). Once the index value at one edge of the sample along the x direction is given, the distribution of refractive index along this direction at 9.3 GHz can be calculated from the index gradients shown in Figure 5.2b. During the calculation, the refractive index is assumed to be continuous at the interfaces between every two neighboring subregions. After the index distribution is discretized along the x axis with the lattice size a_x, the dimension d of each I-shaped particle can be determined from the design curve shown in Figure 5.2c. This design curve is obtained using data fitting method after the effective index values for a set of I-shaped particles with different dimension d are extracted by standard retrieval process [2]. Note that the index value for the I-shaped particle at the left edge of the sample is 1.76.

5.1.2 Metasurface with Random Distribution of Reflection Phase

Such metasurface is an artificial composite constituted by reflecting elements each of which reflects local incident ray back into space with certain phase shift. One or several geometrical parameters of these elements vary along the metasurface such that the reflection phase along the surface is randomly distributed with respect to incident plane waves. Strictly speaking, such metasurface, which is referred to as random surface below, does not belong to the category of metamaterial since it cannot be regarded as effective medium. However, it is similar to metamaterial in that it is also a combination of artificial units.

Various radiators can be used as the reflecting elements of random surface. In particular, microstrip radiators such as metallic patches, dipoles, and loops are favorite reflecting elements when conformality and compactness are required [3–5]. Figure 5.3 illustrates a planar random surface consisting of three layers of square patches with variable side length above a conducting ground. The patches in each layer are arranged in square lattice and are supported by a piece of thin dielectric

Random Metamaterials ■ 151

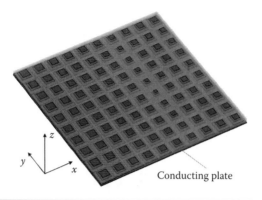

Figure 5.3 Visualization of random surface based on three-layer stacked square patches of variable size. The second and third substrate layers are set semitransparent so that the first and second patch layers could be partly shown.

substrate. This random surface can be viewed as a 2D array of reflecting elements each of which characterizes three-layer stacked square patches of variable size. As shown in Figure 5.4, each element occupies an $a \times a$ square area in the x–y plane. The side lengths for patches on the bottom, middle, and top layers are denoted by d, $r_1 d$, and $r_2 d$, respectively, where r_1 and r_2 are proportional factors smaller than 1. The patch thickness and substrate thickness for each layer are represented by t_m and t_s, respectively. It is well known that the microstrip patch over a ground plane responds resonantly to incident plane waves, leading to sensitive reflection

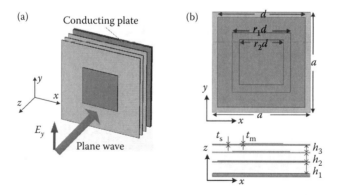

Figure 5.4 Reflecting element characterizing three-layer stacked square patches of variable size. (a) Oblique view. (b) Elevation view and side view. In the elevation view, the second and third substrate layers are omitted to show all the three patch layers clearly.

phase with respect to frequency variations near resonance. The resonant frequency mainly depends on the side length of patch and the reflection phase within certain frequency band can be effectively tuned by varying the side length. By stacking multiple patches, a multiresonant behavior is obtained and the tuning range of reflection phase can be expanded greatly [6,7]. For the case of three-layer stacked patches here, the phase range is greater than 2 times 360°. Figure 5.4b also indicates that every two neighboring layers (including the ground layer and the three patch layers) are separated by air, with the separation distances denoted by h_i ($i = 1, 2, 3$). Such configuration further broadens the frequency band of large effective phase range and results in a smoother and more linear phase variation versus side length.

As an example, a parametric study has been made on the stacked-patch element shown in Figure 5.4. Each patch layer of the element under study is fabricated from F4B copper-clad laminate with copper thickness of $t_m = 0.035$ mm, substrate thickness of $t_s = 0.43$ mm, and dielectric constant of $2.65 + i0.0053$. The lattice size, relative sizes of the stacked patches and the separation distances are considered fixed in the study ($a = 45$ mm, $r_1 = 0.68$, $r_2 = 0.5$, $h_1 = h_2 = h_3 = 4$ mm) and only the side length of bottom patch d is the independently tunable variable. Figure 5.5c demonstrates the relation curve of reflection phase versus the dimension d with respect to normal incidence at 4 GHz.[*] Note that the incident wave is x- or y-polarized and the reference plane just overlaps with the plane of top patch. Using the relation curve above, a random surface based on stacked-patch elements is designed and the corresponding sample is shown in Figure 5.5a. This sample is constituted by 10×10 elements and the dimension d of these elements are specified in such a way that the metasurface characterizes the reflection phase distribution shown in Figure 5.5b at 4 GHz.

5.2 Diffuse Reflections by Metamaterial Coating with Randomly Distributed Gradients of Refractive Index

It is known that when a plane wave impinges on a planar slab whose index varies linearly in a direction parallel to the plane of incidence, the transmitted wave will be uniformly deflected away from its original propagation direction [8]. If such a slab is further backed with a conducting plate, the beam reflected by the conducting plate will be deflected away from the specular direction. Hence, it is intuitive that if the simple linearly gradient index slab is replaced by a randomly gradient index slab with numerous subregions, the local beam of each subregion will be randomly deflected

---

[*] The reflection phase can be obtained via numerical simulations using commercial EM solvers such as Ansoft HFSS and CST MWS.

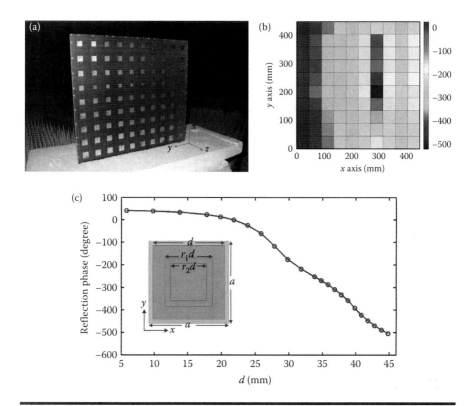

Figure 5.5 (a) Visualization of a random surface sample composed of reflecting elements based on three-layer stacked patches. Note that the bottom row of elements is buried in foams. (b) Two-dimensional distribution of reflection phase at 4 GHz for the sample. (c) A design curve relating reflection phase at 4 GHz with the dimension *d* for element of three-layer stacked patches with certain specifications. The reflection phase is associated with normally incident plane waves polarized along the *x* or *y* axis.

to different directions, leading to diffuse reflections instead of mirror reflection in front of the conducting plate [1]. In the rest of this section, we refer to such a substitute slab as randomly gradient index coating or random coating for brevity.

For simplicity, consider a 2D geometry shown in Figure 5.6, where a 1D randomly gradient index coating of thickness t is attached to a PEC plate. Both the coating and the PEC plate are infinite along the y axis. The subregion division of the coating is carried out along the x direction and the index distribution is continuous across the interface between every two adjacent subregions. The coating possesses diagonal permittivity and permeability tensors and is homogeneous along the y and z axis but inhomogeneous along the x axis. An infinite s-polarized plane wave with its electric field pointing to the y direction is incident on the coated PEC

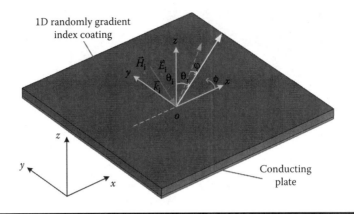

Figure 5.6 An infinite s-polarized plane wave incident on a 1D randomly gradient index coating backed with a PEC plate. The solid dark arrow represents the incident direction and the dashed dark arrow represents the specular direction.

plate in the *xoz* plane. Under such excitation, only ϵ_y, μ_x, and μ_z are relevant to the problem. We further assume that the coating is isotropic with respect to the incident wave, that is to say, $\sqrt{\epsilon_y \mu_x} = \sqrt{\epsilon_y \mu_z} = n_c$. Such hypothesis is reasonable to realistic random coatings made up of artificial "atoms" such as I-shaped and crossed-I particles. It is the refractive index denoted by n_c that is randomly gradient over the subregions. It is known from gradient index optics that each subregion could lead the local incident beam away from the specular direction by a deflection angle denoted as φ. Such a deflection angle mainly depends on the index gradient of each subregion and is restricted by

$$\sin(\theta_i + \varphi) - \sin(\theta_i) = 2 \cdot t \cdot g, \tag{5.1}$$

where θ_i is the incident angle and g is the local index gradient of each subregion. The factor "2" in Equation 5.1 implies that the beam undergoes twice the distance of the coating thickness. Since the index gradient for each subregion is randomly selected, the local beam related to each subregion will be randomly deflected so that diffuse reflections may be achieved in the *xoz* plane. The problem geometry shown in Figure 5.6 is used for subsequent research on diffuse reflections by 1D random coatings in the remaining part of this section.

As an example, a set of random index gradients, which corresponds to 15 subregions, has been assigned to the above 1D random coating with thickness of $t = 12$ mm and a 2D simulation has been made to validate the diffuse reflection performance of the resulting random coating. These index gradients, which are shown in Figure 5.7a, are calculated by Equation 5.1 on the assumption of normal incidence from a set of local-beam deflection angles, which is generated using MATLAB® code from a uniform random distribution in the range of $(-90°, 90°)$.

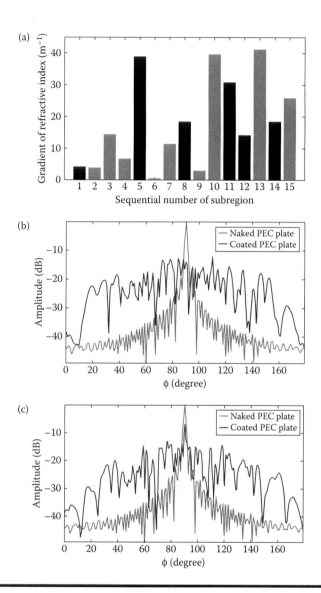

Figure 5.7 Comparison of diffuse reflection performances between 1D random coatings matched and unmatched to free space. (a) The spatial distribution of index gradients for both coatings. (b) Normalized backscattering patterns at 10 GHz for the random coating matched to free space. (c) Normalized backscattering patterns at 10 GHz for the random coating unmatched to free space. In both (b) and (c), the light-color line indicates the backscattering pattern of naked PEC plate with the same length as the coating. (X. M. Yang, Q. Cheng, and T. J. Cui, Research on the one-dimensional randomly gradient index coating. In *Proceedings of the 2010 International Conference on Microwave and Millimeter Wave Technology*, Chengdu, China, 1208–1211 © 2010 IEEE.)

We choose such a random fashion because a uniform distribution of deflection angles may intuitively lead to an equal allocation of scattered energy to various angular directions in the *xoz* plane. However, since Equation 5.1 is no longer correct when deflection angle is very large, the actual distribution of deflection angles is only approximately uniform. The simulation is performed using a commercial finite element solver, Comsol Multiphysics. In the simulation, each subregion of the random coating is 72 mm long and hence the length of the coating and the PEC plate is 1080 mm along the *x* axis. The start index value at the left edge of the coating is set as 3, with which the index distribution along the *x* axis is computed from the index gradients. Note that the refractive index is limited in the range of [1, 5] here.[*] Considering the 2π phase periodicity of wave transmission through the coating, the following algorithm is applied to update the index value in case that it exceeds the above range:

$$\begin{cases} n_c \leftarrow n_c - \dfrac{\pi}{k_0 \cdot t}, & \text{if } n_c > 5, \\ n_c \leftarrow n_c + \dfrac{\pi}{k_0 \cdot t}, & \text{if } n_c < 1, \end{cases} \quad (5.2)$$

where $\leftarrow$ is the assignment operator and k_0 is the wave number in free space at incident wave frequency (10 GHz here). The index distribution is also discretized by the spacing of $a_x = 3$ mm along the *x* axis. Since the discretization spacing is only the tenth of wavelength at 10 GHz, the influence of discretization on the scattering property of random coating can be neglected. Finally, the EM parameters of the coating are assumed lossless and matched to free space (i.e., $\epsilon_y = \mu_x = \mu_z$) in the simulation.

Figure 5.7b demonstrates the simulated 2D backscattering patterns of the random coating backed with PEC plate, with respect to the normally incident *s*-polarized plane wave at 10 GHz. For comparison, the 2D backscattering pattern of naked PEC plate with the same length is also shown. It can be seen that when the random coating is present, the power reflected by PEC plate in the specular direction is suppressed by more than 10 dB and there is no remarkable scattering lobe in the backscattering pattern, indicating the achievement of diffuse reflection.

To quantitatively measure the diffuse reflection performance of random coating, a new parameter "diffusion degree" is defined as the absolute difference between the peak backscattering power density values (in dB) for PEC plates with and without coating. For example, the backscattering pattern shown in Figure 5.7b reads a diffusion degree of 12.34 dB. The larger the diffusion degree, the more apparent the diffuse reflection phenomenon.

---

[*] The available index from both natural and artificial material technique is limited to a certain range in practice.

5.2.1 Role of Amount of Subregions or Length of Coating

In this section, we demonstrate the influence of coating length or amount of subregions on the diffuse reflection performance of 1D randomly gradient index coating via Comsol simulations. Similar to the example given above, a normally incident plane wave at 10 GHz is applied in these simulations and the coatings are assumed matched to free space. Besides, the coating thickness and the subregion length are fixed as 12 and 72 mm, respectively. The coating index is limited in the range of [1, 10] and the start index value at the left edge of the coating is set as 10. The discretization spacing is still 3 mm. Four cases of coating length are involved: 360, 720, 1080, and 1440 mm, which correspond to 5, 10, 15, and 20 subregions, respectively.

For each case of coating length, four procedures are conducted. First, 200 sets of index gradients are generated from the uniform distribution of local-beam deflection angles in the range of (−90°, 90°). Second, the backscattering patterns of both coated and naked PEC plates for each of the 200 sets of index gradients are simulated. Third, the diffusion degree corresponding to each set of index gradient is extracted from the backscattering patterns obtained in the last procedure. Finally, the proportions of index gradients sets with which the diffusion degree is located in different intervals are counted. The final statistical results of diffusion degree for all four cases of coating length are listed in Table 5.1.

Table 5.1 reveals that the proportion of index gradients sets with large diffusion degree grows as the amount of subregions grows. The possible reason for this fact is that the growth of subregion amount increases the randomicity of index gradients. Hence, it can be concluded in statistical sense that the larger the amount of subregions, the easier it is to get favorable diffuse reflection performance with the approach of random distribution of index gradients.

5.2.2 Influence of Impedance Mismatch

The random coatings involved in the above simulations are matched to free space and are hard to realize, as it is difficult to construct constituent particles whose

Table 5.1 Statistical Results of Diffusion Degree for 1D Random Coatings with Different Coating Lengths or Amount of Subregions

Interval	\multicolumn{4}{c}{Subregion Amount (%)}
	5	10	15	20
≥15 dB	0	0	0	1.67
≥10 dB and <15 dB	5.33	60.67	90	91.67
≥5 dB and <10 dB	86	36	8	5.83
<5 dB	8.67	3.33	2	0.83

relative permittivity and permeability responses are identical. As discussed previously, electric particles are preferred for implementing bulk randomly gradient index metamaterials with broadband behavior. With the absence of magnetic response, the engineering of effective refractive index for electric particles is achieved completely through the variation of effective permittivity, which makes the corresponding realistic randomly gradient index coating unmatched to free space. Hence, it is necessary to further investigate the influence of such impedance mismatch on the performance of random coating.

Figure 5.7c presents the simulated backscattering pattern of PEC plate covered with a 1D randomly gradient index coating unmatched to free space, with respect to the normally incident plane wave at 10 GHz. The EM parameters of this unmatched coating are set as $\epsilon_y = n_c^2$, $\mu_x = \mu_z = 1$, which is consistent with the property of electric particles. The other configurations of the coating are the same as those of the matched coating related to Figure 5.7b. Especially these two coatings share the same set of index gradients, which come from uniform random distribution of deflection angles. It can be seen from Figure 5.7 that the diffuse reflection performance of the unmatched coating is much worse than that of the matched coating. Though the mirror reflection of PEC plate is suppressed to some extent by the unmatched coating, it still appears large. Such large mirror reflection originates from primary reflection due to the impedance mismatch between the coating and free space. The energy related to primary reflection is reflected to the specular direction (i.e., the z direction) without the influence of gradient index. To allocate the total reflected energy to various angular directions as diffusely as possible, the index gradients of subregions should be designed to allow large deflection angles in high probability. Hence nonuniform distributions of local-beam deflection angles in the range of (−90°, 90°) with most samples far away from 0° may be a better choice for unmatched random coatings under normal incidence.

5.2.3 Influence of Random Distribution Mode

To verify the idea of nonuniform distribution of local-beam deflection angles, this section statistically compares the diffuse reflection performances of unmatched random coatings with uniformly and nonuniformly distributed local-beam deflection angles. Three kinds of nonuniform distributions of deflection angles are involved in the comparison. Figure 5.8 illustrates the probability density functions for these three nonuniform distributions together with the function for uniform distribution. In view of the normal incidence assumed here, the random variable (i.e., deflection angle) ranges from −90° to 90°. However, since the probability density function for each distribution is symmetric with respect to zero deflection angle, Figure 5.8 actually depicts the probability density curves against the absolute value of deflection angles for brevity. As is shown, all the three curves for nonuniform distributions rise as the deflection angle grows and such nonuniformity is strengthened as the marked number of distribution modes increases.

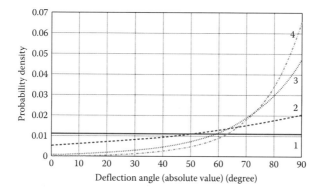

Figure 5.8 Probability density functions of local-beam deflection angles for different random distribution modes. The solid curve marked 1 represents the uniform distribution and the rest three curves correspond to three nonuniform distributions, respectively. (X. M. Yang, Q. Cheng, and T. J. Cui, Research on the one-dimensional randomly gradient index coating. In *Proceedings of the 2010 International Conference on Microwave and Millimeter Wave Technology*, Chengdu, China, 1208–1211 © 2010 IEEE.)

Table 5.2 presents the statistical results of diffusion degrees for four unmatched random coatings corresponding to the four random distribution modes shown in Figure 5.8. These results are obtained using analysis procedures analogous to those applied for Section 5.2.1. As before, a normally incident plane wave at 10 GHz is assumed in the statistical analysis and the coating thickness and length, the amount of subregions, the variation range of refractive index and the discretization spacing for coatings remain the same as those related to the last section. From Table 5.2, the proportion of index gradients for fairly large diffusion degrees ($\geqslant 10$ dB) apparently increases as the random distribution mode changes from mode 1 (uniform distribution) to mode 2, indicating that a degree of nonuniformity for random distribution of deflection angles does improve the diffuse reflection performance of unmatched random coatings. However, when the nonuniformity is strengthened further, the proportions of index gradients sets with fairly large diffusion degrees ($\geqslant 10$ dB) turns to decrease (see columns for modes 3 and 4 in Table 5.2). The reason for such a phenomenon is that as the nonuniformity is strengthened, the impedance mismatch between random coatings and free space becomes severe and the weight of primary reflections increases, both in the statistical sense. Because large nonuniformity would lead to high probabilities of large index gradients and hence large refractive indices are more likely to occur over random coatings.

5.2.4 Experimental Verification of Diffuse Reflections

To experimentally confirm the diffuse reflection feature of 1D randomly gradient index coatings, a metamaterial sample of 1D random coating is fabricated and

Table 5.2 Statistical Results of Diffusion Degrees for Unmatched 1D Random Coatings with Different Distribution Modes of Index Gradients

	Distribution Mode (%)			
Interval	1	2	3	4
≥10 dB and <15 dB	2	6	3	1
≥5 dB and <10 dB	73	77	84	84
<5 dB	25	17	13	15

2D near-field scanning measurements with 45° incident waves are conducted with regard to the sample.

Figure 5.9a shows the sample of 1D random coating used for experiment. It is implemented by a 3D crossed-I array with certain particle dimension varying along the x axis. The array is in cubic lattice and the lattice constants are $a_x = a_y = a_z = 3$ mm. The inset of Figure 5.9c shows the cubic unit cell for crossed-I particle printed on thin dielectric substrate. The crossed-I particle is capable of yielding electric response along the y axis and has almost no magnetic response along both the x and z axes with respect to the excitation manner indicated by the problem geometry shown in Figure 5.6. The electric response or effective permittivity ϵ_y can be tuned by adjusting the dimension s. Hence, the spatial distribution of effective refractive index $n_{yx} = \sqrt{\epsilon_y \mu_x}$ or $n_{yz} = \sqrt{\epsilon_y \mu_z}$ along the x axis could be achieved by variation of s along this axis. Note that in the presented design, the copper line widths w_x and w_y are fixed as 0.12 mm and the dimension d is not independent and is equal to $s + 4w_y$. All crossed-I particles are fabricated from FR4 copper-clad laminates with copper thickness of $t_m = 0.035$ mm, substrate thickness of $t_s = 0.19$ mm, and dielectric constant of $3.9 + i0.039$. Figure 5.9c shows the design curve revealing the relationship between the dimension s and effective refractive index at 10 GHz. This curve could be used to determine the dimension s along the x axis as long as the index distribution is known. The whole sample is 180 mm long along the x axis, 12 mm thick along the z axis and 12 mm high along the y axis. Hence the metamaterial scale is 60 × 4 × 4 in view of the lattice size of 3 mm along each axis. Air-like foams are stuffed between the 4 rows of planar crossed-I array along the z axis to support the sample. The short length of the sample along the x axis is due to the limited spatial measuring range. The sample has 10 subregions and the index gradients randomly selected for these subregions are shown in Figure 5.9b. The choice of random distribution mode for these index gradients is associated with the 45° incident wave. As the mirror reflection is along the direction of $\phi = 45°$ (refer to Figure 5.6 for the definition of ϕ), more energy should be deflected from this direction in an anticlockwise manner to ensure that the total reflected energy is

Random Metamaterials ■ 161

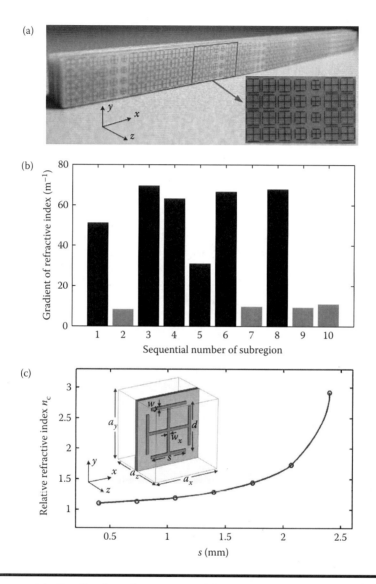

Figure 5.9 (a) Visualization of a metamaterial sample of 1D random coating composed of crossed-I particles. (b) Distribution of index gradients at 10 GHz for the coating sample. The sequential number of subregion increases toward the positive x direction. (c) A design curve relating refractive index n_c at 10 GHz with the dimension s for crossed-I particles with certain specifications, where n_c refers to $(n_{yx} + n_{yz})/2$. (X. M. Yang et al. Diffuse reflections by randomly gradient index metamaterials. *Opt. Lett.*, 35(6): 808, 2010. With permission of Optical Society of America.)

equally distributed in the whole angular range from $\phi = 0°$ to $\phi = 180°$. Hence most of the index gradients selected here are negative with large values.

The measurement is carried out in the 2D near-field mapping apparatus described in Section 4.2.2. The sample is placed in the planar waveguide of the apparatus as shown in Figure 5.10. A metal sheet that has the same length and height as the sample is tightly attached to the back side of the sample. As the field inside the planar waveguide is constrained to two dimensions, both the sample and the metal sheet can be regarded to be infinitely long along the y axis, which imitates the problem geometry shown in Figure 5.6. However, in order to detect the field right above the sample by the coaxial probe mounted on the upper metal plate, a tiny gap is kept between the upper metal plate and the sample. Hence the distance between the upper and lower metal plates is controlled to be 13 mm during the experiment. An approximate plane wave whose electric field is restricted to the y axis by the planar waveguide is produced to obliquely illuminate the sample and the metal sheet with a 45° incidence angle. Note that the incident beam originally emits from a waveguide adapter and has a narrow beam width. To ensure that the whole sample region is in a plane wave environment, a long and wide guiding channel made of absorbers is used to widen the wave front of incident beam before it encounters the sample. In order to acquire the scattered field of the coated metal sheet, two stages of measurement are performed. In these two stages, the incident field with the absence of scatterers and the total field surrounding the coated metal sheet are mapped in the X-band, respectively. The scattered field can then be calculated by simply subtracting the incident field data from the total field data. For the sake of comparison, the near-field scattered by the metal sheet alone is also measured, in a way similar to what is employed for coated metal sheet.

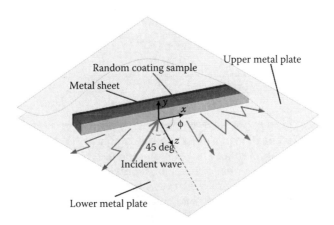

Figure 5.10 A sketch of experimental setup for measuring the near-field surrounding the coating sample and the metal sheet with a 45° incident wave.

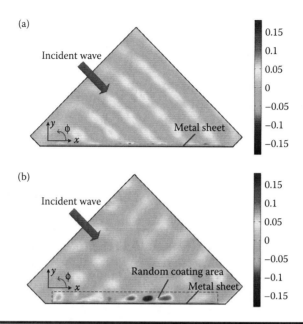

Figure 5.11 Measured scattered near-field distributions at 10.5 GHz for (a) naked metal sheet and (b) coated metal sheet. (X. M. Yang et al. Diffuse reflections by randomly gradient index metamaterials. *Opt. Lett.*, 35(6): 808, 2010. With permission of Optical Society of America.)

Figure 5.11 portrays the measured scattered near-field distributions at 10.5 GHz for coated and naked metal sheet, respectively. The diffuse reflection phenomenon is clearly observed in the presence of random coating (see Figure 5.11b). Using the near-to-far field transformation technique proposed in Reference 10, the normalized 2D backscattering patterns corresponding to the two scattering cases are further obtained in the X-band. It is found from these patterns that the peak backscattering intensity of the metal sheet is significantly reduced by the random coating from 10 to 11.2 GHz. As a proof, Figure 5.12 demonstrates the 2D backscattering patterns for both cases at 10.5 GHz.

In conclusion, the randomly gradient index metamaterial proves to be capable of creating diffuse reflections in front of conducting surfaces, which is beneficial to suppressing the remarkable scattering lobes of conducting objects. It is possible to become a new approach to radar stealth for conducting objects.

5.3 RCS Reduction by Metasurface with Random Distribution of Reflection Phase

Generally speaking, directional radiation is closely related to equiphase surface perpendicular to the radiation direction. For planar conducting plate that has uniform

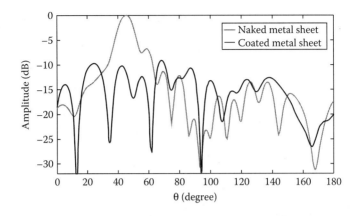

Figure 5.12 Comparison of measured backscattering patterns between naked and coated metal sheets at 10.5 GHz. (X. M. Yang et al. Diffuse reflections by randomly gradient index metamaterials. *Opt. Lett.*, 35(6): 808, 2010. With permission of Optical Society of America.)

reflection phase, the reflected field possesses a planar equiphase surface with respect to incident plane wave and hence the reflected energy is focused in specific direction (i.e., specular direction). If the conducting plate is covered with certain composite structures to form a metasurface that has randomly distributed reflection phase (abbr. random surface), the equiphase surface will be no doubt disturbed and the directive reradiation can be thereby suppressed or even eliminated, which is attractive for the application of radar cross section (RCS) reduction.

To verify the nondirectional reflection property of random surface, backscattering measurements have been conducted concerning the metasurface sample shown in Figure 5.5a. The measurements consists of two stages, which employ almost the same experimental setup. In the experimental setup for the first stage, a linearly polarized high-gain horn antenna is placed right in front of the metasurface sample with a distance of 3 m and the aperture of the horn antenna is in parallel with the sample surface, as is shown in Figure 5.13. The horn antenna is also connected to a single port of a VNA through a flexible coaxial cable. The whole setup is situated in an anechoic chamber environment to simulate a free space condition. During measurement, the VNA provides source microwave signals for the antenna and the main beam emitted from the horn antenna impinges on the sample surface. The back-scattered or reflected power from the sample is received by the same horn and then detected by the VNA. Note that the power reflected by the horn itself also enters the VNA in the meantime. By applying the gated-reflect-line (GRL) calibration technique provided by the time-domain analysis kit installed in the VNA, the influence of the power reflected by the horn is removed and the resultant reflection coefficient data read from the VNA is only related to the reflection from the

Random Metamaterials ▪ 165

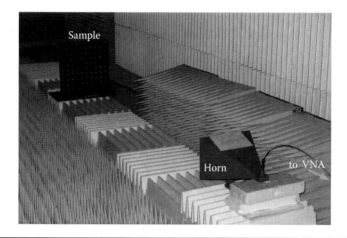

Figure 5.13 Experimental setup for measuring the reflectance of random surface under normal incidence.

sample. Two cases of incident beams are considered in the first stage. For the first case, the horn is arranged to allow a y-polarized incident beam with respect to the sample shown in Figure 5.5a. By simply rotating the horn antenna 90° around its axis of symmetry, the other case of incident beam with x polarization can be achieved afterward. The reflection coefficient data in decibels read from the VNA are recorded for both cases. As to the second stage of measurement, the metasurface sample is replaced by an ordinary square metallic plate, which owns the same size as the sample no matter whether the incident beam is y- or x-polarized. The reflection coefficient data in decibels given by the VNA at this stage are also recorded. The reflectance of the metasurface sample normalized to that of naked metallic surface for both incident polarizations can then be obtained by subtracting the newly recorded data from the two sets of data measured at the first stage, respectively. The final reflectance data can be approximately viewed as being associated with normally incident plane waves, considering that the sample is in the radiating near-field (Fresnel) region of the horn antenna, which indicates that the local incident rays for most of the reflecting elements are just roughly but not strictly perpendicular to the sample surface.

Figure 5.14 shows the measured normalized reflectance of the metasurface sample for both incident polarizations. It is observed from this figure that in the presence of the sample, the directional reflection or backscattering of the metallic surface in the band from 3.6 to 5.5 GHz is suppressed by more than 10 and 7 dB for the y- and x-polarized incident beams, respectively. Hence, the potential of random surface to reduce RCS of conducting objects is well demonstrated. We remark that although the phase distribution on the sample surface changes with frequency due to the frequency-dependent phase response of stacked-patch element, the random feature

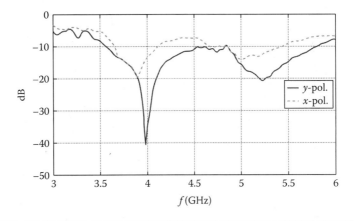

Figure 5.14 Experimental results of normalized reflectance of random surface with respect to both x- and y-polarized incident waves. The x and y directions are indicated by Figure 5.5a.

of phase distribution within a sufficiently large range exists in a fairly wide frequency band including the design frequency (4 GHz). This is the main reason why the frequency band with apparent backscattering suppression effect is also fairly wide from Figure 5.14.

In conclusion, we have introduced two categories of "super noncrystals" or random metamaterials in this chapter. One is the artificial medium characterizing randomly distributed gradients of refractive index, and the other is the metasurface characterizing random distribution of reflection phase along the surface. Both composite structures are capable of guiding or scattering EM waves irregularly due to their randomness. Diffuse reflection caused by randomly gradient index coating and RCS reduction achieved by random-phase surface, both in front of metallic plate, have been experimentally demonstrated.

References

1. X. M. Yang, X. Y. Zhou, Q. Cheng, H. F. Ma, and T. J. Cui. Diffuse reflections by randomly gradient index metamaterials. *Opt. Lett.*, 35(6): 808, 2010.
2. D. R. Smith, S. Schultz, P. Markos, and C. M. Soukoulis. Determination of effective permittivity and permeability of metamaterials from reflection and transmission coefficients. *Phys. Rev. B*, 65(19): 195104, 2002.
3. C. S. Malagisi. Microstrip disc element reflect array. *Electronics and Aerospace Systems Convention*, Arlington, VA, pp. 186–192, September 25–27, 1978.
4. J. P. Montgomery. A microstrip reflectarray antenna element. *Antenna Applications Symposium*, University of Illinois, Urbana, IL, p. 19, September 20–22, 1978.
5. D. M. Pozar and T. A. Metzler. Analysis of a reflectarray antenna using microstrip patches of variable size. *Electron. Lett.*, 29(8): 657, 1993.

6. J. A. Encinar. Design of two-layer printed reflectarray using patches of variable size. *IEEE Trans. Antenn. Propag.*, 49(10): 1403, 2001.
7. J. A. Encinar and J. A. Zornoza. Broadband design of three-layer printed reflectarrays. *IEEE Trans. Antenn. Propag.*, 51(7): 1662, 2003.
8. D. R. Smith, J. J. Mock, A. F. Starr, and D. Schurig. Gradient index metamaterials. *Phys. Rev. E*, 71(3): 036609, 2005.
9. X. M. Yang, Q. Cheng, and T. J. Cui. Research on the one-dimensional randomly gradient index coating. In *Proceedings of the 2010 International Conference on Microwave and Millimeter Wave Technology*, Chengdu, China, 1208–1211, 2010.
10. H. F. Ma, T. J. Cui, X. M. Yang, W. X. Jiang, and Q. Cheng. Far-field predictions of metamaterials from two-dimensional near-field measurement system. *Terahertz Sci. Technol.*, 3(2): 74, 2010.

Chapter 6

Inhomogeneous Metamaterials: Super Quasicrystals

Although homogeneous metamaterials have exhibited many unusual physical properties, such as negative refractions and zero-index refractions, inhomogeneous metamaterials provide a different way to control EM waves. In this chapter, we will mainly focus on the property and applications of inhomogeneous metamaterials. Inhomogeneity means that the EM parameters of the materials (e.g., the permittivity and permeability) vary with space position. The materials can be isotropic or anisotropic. Many novel devices have been proposed using inhomogeneous metamaterials, such as invisibility cloaks, concentrators, high-performance antennas, and illusion devices.

6.1 Inhomogeneous Metamaterials: Particularly Nonperiodic Arrays of Meta-Atoms

In our daily life, people usually use EM materials to control and direct the fields. For example, a glass lens in a camera is used to produce an image, a dielectric lens in a horn antenna is used to generate highly directional radiations, and so on. For homogeneous materials, optical design is mainly a matter of choosing the geometrical configuration of the device. However, inhomogeneous materials offer a different approach to control waves, and the introduction of specific gradients in material parameters can be used to form lenses and other EM devices. In this chapter, we will focus on the inhomogeneous metamaterials, which have been utilized to design many unusual artificial devices.

In the earlier work on negative refractions and zero-index refractions before 2005, metamaterials have been constructed from periodic unit cells, which means that the meta-atoms are identical in spatial distribution. Such periodic media are considered as homogeneous metamaterials because the averaged EM response is almost the same for each unit cell. Although many interesting EM properties have been revealed and presented by homogeneous metamaterials, such as negative-index refractions [1,2] and subwavelength imaging [3], inhomogeneous metamaterials have more possibilities and capabilities to control EM waves, due to the flexibility in designing the unit particles and their spatial arrangements, producing many exciting new physical phenomena and real applications. Inhomogeneous metamaterials imply that the unit cells are not necessarily periodically distributed, and may be particularly nonperiodic arrays. Such materials can be utilized in many applications, including artificial lenses [4–6], invisibility cloaks [7,8], and other interesting functional devices [9–11].

As we have known, Fermat's principle is a very important rule in optics, which was formulated in 1662 by Pierre de Fermat, and was destined to shape geometrical optics [12]. Fermat's principle describes the shortest optical path: light rays passing between two spatial points chose the optically shortest path. The optical path length s is related to the refractive index n and is defined as

$$s = \int n dl. \tag{6.1}$$

Fermat's principle has profoundly influenced modern optics. The principle governs the path between two spatial points if it is known that light travels from one point to another. If the refractive index varies in space, for example, in inhomogeneous materials, the shortest optical path is no longer a straight line but a curved one. This bending of waves may generate many optical illusions.

Here, we take a planar focusing lens as an example to illustrate inhomogeneous metamaterials [4]. Gradient-index metamaterials, as a kind of inhomogeneous metamaterials, have been experimentally confirmed in References 5 and 6. To describe the principle, we consider two normally incident rays entering a planar gradient-index slab of thickness t, as shown in Figure 6.1. The rays will acquire different optical path lengths as they propagate through the slab. However, in order to focus at the same point, their optical path lengths must be the same. Assuming that two rays enter at locations x and $x + \Delta x$ along the slab surface, the acquired optical path lengths of the two beams traversing the slab will be

$$n(x)t + l = n(x + \Delta x)t + \sqrt{l^2 + (\Delta x)^2}, \tag{6.2}$$

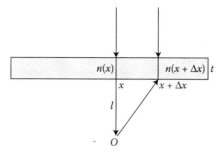

Figure 6.1 Design of the gradient-index focusing lens which consists of inhomogeneous metamaterials.

where l is the distance from the focusing point to the planar lens. From the above equation, we have

$$n'(x) = \lim_{\Delta x \to 0} \frac{n(x + \Delta x) - n(x)}{\Delta x} = -\lim_{\Delta x \to 0} \frac{\sqrt{l^2 + (\Delta x)^2} - l}{t \Delta x}, \quad (6.3)$$

which shows that the material parameters vary only in one direction for a planar focusing lens. It should be noted that such a simplified analysis applies strictly to thin samples, as the phase fronts will otherwise not be uniform within the material. With such a planar lens, a plane wave will be focused on one point after propagating through the lens. According to the reciprocity principle, cylindrical waves radiated by a line source at the focal point, in 2D case, will be transformed into a plane wave after penetrating the lens. Such a lens has been applied in high-gain lens antenna design [5].

6.2 Geometric Optics Method: Design of Isotropic Metamaterials

In this section, we will first show how to calculate the material parameters for isotropic but inhomogeneous metamaterial devices designed by geometrical optics. According to geometrical optics [13], when the light propagates in an isotropic material, the light rays are always normal to the Eikonal curves. Hence, if the light rays in one medium is given, it is possible to get the Eikonal curves. Mathematically, Eikonal function S can be obtained through analytic or numerical approaches. Since $|\nabla S|^2 = n^2$, the refractive index of the material can be obtained.

In the first example, a simple device is designed [14], which is a 2D metallic waveguide bend with an angle of θ. The light rays propagate in the circumferential directions. Hence, $\theta = C$ represents the curves of the isophase, and the Eikonal function should be

$$S = f(\theta), \tag{6.4}$$

where f is an arbitrary function of θ. For simplicity, a linear function is chosen,

$$S = A\theta + C, \tag{6.5}$$

in which A and C are two constants. We remark that other functions are also possible to choose. For this linear function, the refractive index for the bend structure is

$$n = \sqrt{\epsilon_r \mu_r} = A/r. \tag{6.6}$$

Clearly, the refractive index is a function of radius, but independent of bending angle. The material inside the metallic waveguide bend is isotropic, and the range of refractive index is determined by the parameter A. Obviously, it can be realized using inhomogeneous metamaterials.

To design an arbitrary waveguide bend, the property of an analytic function may be useful. For an arbitrary analytic function $w(z) = u(x, y) + iv(x, y)$, the families of curves corresponding to $u(x, y) = C$ and $v(x, y) = C$ are orthogonal. If one family of curves, for example, $v(x, y) = C$, is considered as the light rays, the other group, that is, $u(x, y) = C$, is naturally the Eikonal curves. Following the similar process, the refractive index can be obtained for the arbitrary bends

$$n = |\nabla u(x, y)| = |\overline{W'(z)}|. \tag{6.7}$$

As an example, we choose an elliptical bend structure, in which,

$$w(z) = A \sin^{-1}(z/p), \tag{6.8}$$

where A and p are two positive constants to be determined. Then it is clear that $v(x, y) = C$ corresponds to elliptic curves in the Z plane and $u(x, y) = C$ the hyperbolic ones. Using Equation 6.7, the refractive index is $n = A/((p^2 - x^2 + y^2)^2 + 4x^2 y^2)^{1/4}$.

For isotropic material parameters designed by geometrical optics, the refractive index can also be calculated based on the analogy between light rays and electric force lines, as well as the equivalence between Eikonal curves and isopotential lines [15]. Supposing that the ray trajectories in the device are given, and they are considered as electric force lines in electrostatics, then the original problem is converted into a boundary value problem in electrostatics. The refractive index distribution in the device is obtained through the evaluation of the electric field by $n = |\nabla S| = |\nabla \Phi| = |\mathbf{E}|$, where Φ is the electric potential and $\mathbf{E}$ is the electric field intensity in the static problem. Such a method can be used to design an isotropic directional invisibility cloak [15].

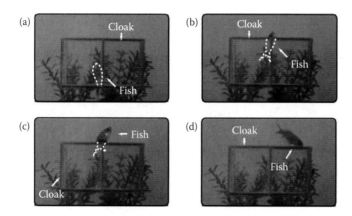

Figure 6.2 Experimental observation of fish with a geometrical-optics-based cloak. (a) The main fish body inside the cloak is invisible; only the tail outside of the cloak is visible. (b) Only the fish head outside of the cloak is visible. (c) The main body of the fish comes out of the cloak and thus becomes visible. (d) The whole fish has come out of the cloak. (Reprinted by permission from Macmillan Publishers Ltd. *Nat. Comm.* H. Chen, B. Zheng, L. Shen, H. Wang, X. Zhang, N. I. Zheludev, and B. Zhang. Ray-optics cloaking devices for large objects in incoherent natural light. 4: 2652. Copyright 2013.)

Recently, an invisibility cloak based on geometrical optics was demonstrated for natural light at multiple observation angles [16]. Such a cloaking device operating within the ray-optics approximation offers good performance for hiding macroscopic objects much larger than the wavelength of light. Their experimental demonstration with only isotropic materials was the first simplified test of a non-phase-preservation cloak. Such a method can be extended in multiple directions for arbitrary polarization over a broad range of optical frequencies, and therefore may simplify the construction of a cloak in many real applications, in which only a certain number of detectors or observers are involved. Using the optical glass available in nature, polarization-insensitive cloaks can be constructed and used to hide a fish in a fish tank. The experimental observation is shown in Figure 6.2, which provides the dynamic process when a fish wanders through the cloak.

6.3 Quasi-Conformal Mapping: Design of Nearly Isotropic Metamaterials

In recent years, quasi-conformal mapping has been widely used to design nearly isotropic metamaterial devices, such as ground-plane cloak (or carpet cloak) [17,18], flattened Luneburg lens [19,20], and so on. Compared to spatial invisibility cloak, which we will discuss in the next section, the ground-plane cloak does not require singular values for the material parameters. Furthermore, the range of permittivity and permeability is much smaller. By choosing a suitable spatial transformation, the anisotropy of ground-plane cloak can be minimized and the magnetic response is not necessary. As a result, to construct a ground-plane cloak, only isotropic dielectrics are needed. Moreover, the ground-plane cloak can be low-loss and broadband.

For simplicity, a 2D problem with **E**-field polarized in the z direction is considered. Supposing an object be placed on a ground plane, which is regarded as a perfect conductor here, a cloak is covered on the object so that the system is perceived as a flat ground plane again. The object is concealed between the cloak and the original ground plane as shown in Figure 6.3a. In the physical space with coordinate (x, y), we assume that the cloak is a rectangle of size s, w, and h except the cloaked region for the object. The virtual space is what the configuration observer perceives with coordinate (ξ, η), shown in Figure 6.3b. To design a ground-plane cloak, a coordinate transformation will be constructed, which maps the rectangle $(0 \leq x \leq w, 0 \leq y \leq h)$ in the virtual space to the irregular region in the physical space.

If the virtual space is filled with isotropic homogeneous medium of permittivity ε_{ref} and unit permeability, then the corresponding physical medium induced by the

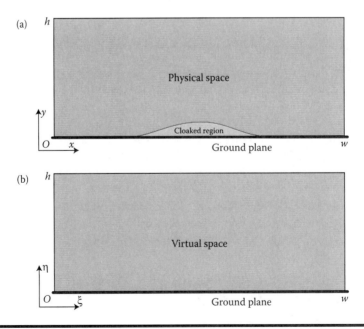

Figure 6.3 Illustration of ground-plane cloak. (a) Physical space and (b) virtual space.

coordinate transformation is given by

$$\varepsilon = \varepsilon_{ref}/|\Lambda|, \quad \mu = \Lambda \cdot \Lambda^T/|\Lambda|, \qquad (6.9)$$

where Λ is the Jacobian matrix of the transformation. To discuss the anisotropy of the transformed medium, μ_T and μ_L are defined as the principal values of the permeability tensor in the physical medium, and hence the corresponding refractive indices are $n_T = \sqrt{\mu_L \epsilon}$ and $n_L = \sqrt{\mu_T \epsilon}$. An anisotropy factor α is defined to indicate the extent of anisotropy

$$\alpha = \max(n_T/n_L, n_L/n_T). \qquad (6.10)$$

If there is a very fine rectangular grid in the virtual domain with tiny cells of size $\delta \times \delta$, every such tiny square is transformed to a parallelogram in the physical domain with two sides $\xi_1 \delta$ and $\xi_2 \delta$. Li and Pendry's approach is to minimize the induced anisotropy by choosing a suitable coordinate transform. If the anisotropy is small enough, we can simply drop it (by assigning 1) and only keep the refractive index n. The minimal of the anisotropy occurs at the quasi-conformal map [21]. Every small cell in the transformed grid in the physical domain is now a rectangle of constant aspect ratio $M : m$ where M is called the conformal module of the physical domain and $m = w/h$ is the conformal module of the virtual domain, that is,

$$|\xi_1|/|\xi_2| = M/m, \quad \sqrt{\det(g)} = |\xi_1||\xi_2|. \qquad (6.11)$$

By substituting Equation 6.11 into 6.10, we have

$$\alpha = \max\left(\frac{M}{m}, \frac{m}{M}\right),$$

independent of position. We note that, the quasi-conformal map approaches the conformal map [8] in the limit M/m. In general, a quasi-conformal map is required for an unchanged topology of space without creating additional singular points in the coordinate transformation.

For example, to design a ground-plane cloak in the microwave band [18], the area of a rectangle bounded by $0 \leq \xi \leq 250\,\text{mm}$, and $0 \leq \eta \leq 96\,\text{mm}$ in the virtual space is mapped to the same rectangle in the physical space but with the bottom boundary specified by

$$y(x) = \begin{cases} 12 \cos^2((x-125)\pi/125), & 62.5 \leq x \leq 187.5, \\ 0, & \text{otherwise}. \end{cases} \qquad (6.12)$$

To test the effectiveness of the designed ground-plane cloak, a few experiments have been completed at different frequencies. The first experimental realization of a ground-plane cloak was reported in Reference 18, in which a perturbation on a flat conducting plane was concealed. To match the complex spatial distribution of the required constitutive parameters, a metamaterial sample consisting of thousands of elements was constructed. The ground-plane cloak was realized with the use of nonresonant metamaterial elements, and hence had a broad operational bandwidth and exhibited extremely low loss. This ground-plane cloak can work from 13 to 16 GHz. Inspired by the work reported in the microwave band, the experimental demonstration of optical cloaking was soon delivered. The optical ground-plane cloak was designed using quasi-conformal mapping so as to conceal an object that was placed under a curved reflecting surface by imitating the reflection of a flat surface [22]. The cloak consists only of isotropic dielectric materials, which enables broadband and low-loss invisibility at a wavelength range of 1400–1800 nm. Soon after, a cloak operating in the near infrared at a wavelength of 1550 nm was demonstrated [23]. The cloak concealed a deformation on a flat reflecting surface, under which an object can be hidden. The device has an area of 225 mm^2 and hides a region of 1.6 mm^2. It was composed of nanometer-sized silicon structures with spatially varying densities across the cloak. The density variation was defined using effective medium theory so as to get the effective index distribution of the cloak.

6.4 Optical Transformation: Design of Anisotropic Metamaterials

Using the invariance of Maxwell's equations, Pendry et al. proposed a design strategy to redirect EM fields at will, the so-called "transformation optics" [7]. One exciting functional device, the invisibility cloak, was taken as an example to describe the transformation optics theory. To date, such a theory has been developed to design many kinds of invisibility cloaks and a lot of transformation devices. With this theory, the flexibility of metamaterials can be engineered to achieve new EM devices.

Before 2006, research on metamaterials had been mainly focused on negative-index refraction, which is an impressive material property that does not exist in nature but has been empowered by metamaterials. Now, metamaterials are employed to provide much more freedom about material parameters. We are able to construct a material whose permittivity and permeability values may vary independently and arbitrarily, taking positive or negative values as desired.

Based on the idea of transformation optics, if the appropriate metamaterials are given, the conserved field quantities—the electric displacement field **D**, the magnetic field intensity **B**, and the Poynting vector **S**—can all be directed at will. In particular, these fields can be focused as required or made to avoid objects and propagate around them smoothly, and finally return to their original path. These conclusions are not confined to a ray approximation. They encompass all forms of field phenomena in principle.

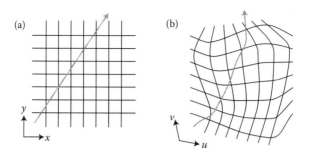

Figure 6.4 The principle of transformation optics. (a) The original virtual space and (b) the transformed real space.

The design process of transformation optics is as follows. We first choose an arbitrary configuration, as the virtual space, regardless of the sources. The virtual space is filled with an arbitrary dielectric and magnetic medium. The original virtual space, would be chosen to have the same topology as the designed device, the physical space. For example, to design an invisibility cloak, we might start with a uniform electric field and require that the field lines be moved to avoid a given cloaked region. We can record the initial virtual space on a Cartesian mesh, while the physical space is distorted by some pulling and stretching process, shown in Figure 6.4. The distortions can now be recorded as a coordinate transformation between the original Cartesian mesh and the distorted mesh

$$u(x, y, z), \quad v(x, y, z), \quad w(x, y, z),$$

where (u, v, w) is the location of the new point with respect to the x, y, and z axes. Maxwell's equations have exactly the same form in any coordinate system, but the permittivity ε and permeability μ are scaled by a common factor. In the new coordinate system, the values of the permittivity and permeability are renormalized as

$$\varepsilon' = \frac{\Lambda \varepsilon \Lambda^T}{|\Lambda|}, \quad \mu' = \frac{\Lambda \mu \Lambda^T}{|\Lambda|}. \tag{6.13}$$

As usual, $E' = \Lambda \cdot E$, $H' = \Lambda \cdot H$, $\mathbf{B} = \mu_0 \mu' \mathbf{H}'$, $\mathbf{D} = \varepsilon_0 \varepsilon' \mathbf{E}'$.

It should be noted that for orthogonal coordinate systems, the formulae are particularly simple. The general case is given in Reference 24 and the equivalence of coordinate transformations and changes to ε and μ has also been referred to in Reference 25.

6.5 Examples

In this section, we will discuss some unusual EM devices based on inhomogeneous metamaterials.

6.5.1 Invisibility Cloaks

As mentioned in Section 6.4, invisible cloaks show the unique feature of inhomogeneous metamaterials to realize new phenomena. The idea of perfect invisibility cloak in free space was proposed by Sir J. B. Pendry, together with transformation optics [7]. We would like to conceal an object contained in a given volume of space. The idea is to achieve concealment by cloaking the object with a metamaterial layer whose function is to guide the waves around the object, and return them to their original direction. Hence, the waves attempting to penetrate the concealed volume are smoothly guided around it by the cloak and finally propagate in the original direction as if it had passed through the empty volume of space. As a result, an object may be hidden because it remains untouched by external detecting waves. A perfect EM cloaking is generated and the propagating waves are excluded from the concealed region.

Without the loss of generality, a sphere of radius a was chosen as the hidden object and the cloaking region is contained within the annulus $a \leq r \leq b$ [7]. A simple coordinate transformation for achieving the desired cloak can be found by compressing the fields from the region $r \leq b$ into the region $a \leq r \leq b$ with

$$r' = a + (b-a)r/b, \quad \varphi' = \varphi, \quad \theta' = \theta. \tag{6.14}$$

Applying the transformation optics theory, the material parameters for cloaking layer $a \leq r \leq b$ are

$$\epsilon_r = \mu_r = \frac{b}{b-a}\frac{(r-a)^2}{r^2}, \tag{6.15}$$

$$\epsilon_\varphi = \mu_\varphi = \frac{b}{b-a}, \tag{6.16}$$

$$\epsilon_\theta = \mu_\theta = \frac{b}{b-a}. \tag{6.17}$$

In the design process of invisibility cloak, the distortion of the fields is represented by a spatial transformation. Then, the electrical permittivity and magnetic permeability of the cloak are generated by the spatial transformation based on form invariability of Maxwell's equations. Anisotropic inhomogeneous metamaterials make realization of invisibility cloaks a practical possibility. As Pendry et al. pointed out, the principle of transformation optics can be used to static fields, as shown in Figure 6.5b. Recently, the invisibility cloak in static magnetic fields and

Inhomogeneous Metamaterials ■ 179

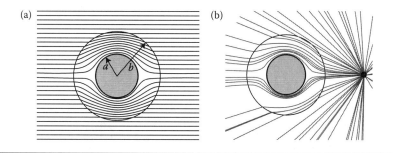

Figure 6.5 The principle of the invisibility cloak. (a) Time-harmonic field and (b) static field.

static electric fields were also proposed and investigated, which will be discussed in Chapter 9.

After the proposal of invisibility cloaks, many theoretical works are devoted to their exciting properties. Chen et al. analytically investigated the interactions of EM wave with a general class of spherical cloaks [26]. The total scattering cross section of an ideal cloak is absolutely zero, but for a cloak with loss, only the backscattering is exactly zero, which indicates the cloak can still be rendered invisible with a monostatic detection. Furthermore, for a cloak with imperfect parameters the bistatic scattering performance is more sensitive to the material parameters of the cloak. Ruan et al. studied the scattering field for an ideal 2D cylindrical invisibility cloak using a cylindrical wave expansion method [27]. Their results showed that a cloak with the ideal material parameters is a perfect invisibility cloak by systematically studying the change of the scattering coefficients from the near-ideal case to the ideal one. However, because of the slow convergence of the zeroth-order scattering coefficients, a tiny perturbation on the cloak would induce noticeable field scattering and penetration.

In principle, the circularly cylindrical or spherical cloaks can wrap up any object to be invisible. But they are very wasteful and inconvenient for long and thin objects. Hence, some work has been devoted to the shape variety of the invisibility cloaks. An arbitrarily elliptical–cylindrical cloak was proposed in 2008 [28]. The material parameters in elliptical cloaking are singular at only two points, instead of on the whole inner circle for circular cloaking, and hence are much easier to be realized in actual applications. The near-field distribution of a perfect elliptical cloak is shown in Figure 6.6a. After that, to design an arbitrarily shaped cloak, the nonuniform rational B-spline (NURBS) was used to represent the geometrical modeling of the arbitrary object [29]. The near-field distribution of a heart-shaped cloak is shown in Figure 6.6b. The powerflow lines of incoming EM waves are bent smoothly in the cloak and returned to their original propagation directions after propagating around the object. The numerical results show that the scattered field from a metallic object coated with the invisible cloak is much smaller than that from the metallic core.

180 ■ *Metamaterials*

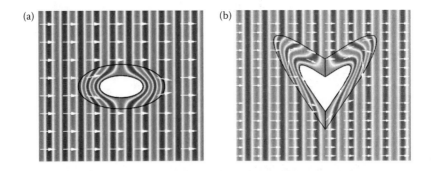

Figure 6.6 Near-field distributions of (a) elliptical–cylindrical cloak and (b) hear-shaped invisibility cloak.

6.5.2 Concentrators

Aside from the invisibility cloaks, the EM concentrators are another kind of interesting transformation devices [10]. The first-proposed concentrator is a circular one. Due to its cylindrical symmetry, it is convenient to describe the transformation equations in a cylindrical coordinate system. The transformation equations for the optical design of the cylindrical concentrator are denoted as

$$r' = \begin{cases} \dfrac{a}{b} r & 0 \leq r \leq b; \\ \dfrac{c-a}{c-b} r - \dfrac{b-a}{c-b} c & b \leq r \leq c; \end{cases} \quad (6.18)$$

$$\phi' = \phi \quad (6.19)$$
$$z' = z. \quad (6.20)$$

From the transformation, it is obvious that the space is compressed into a cylindrical region with radius a at the expense of an expansion of the space between a and c, as shown in Figure 6.7a. The transformation is continuous to free space at c. The relative permittivity and permeability tensors are expressed in the cylindrical coordinates as

$$\overline{\varepsilon} = \overline{\mu} = \begin{cases} \mathrm{diag}\left(1,\ 1,\ \left(\dfrac{b}{a}\right)^2\right), & 0 \leq r \leq a, \\ \mathrm{diag}\left(t,\ t^{-1},\ \left(\dfrac{c-b}{c-a}\right)^2 t\right), & a \leq r \leq c, \end{cases} \quad (6.21)$$

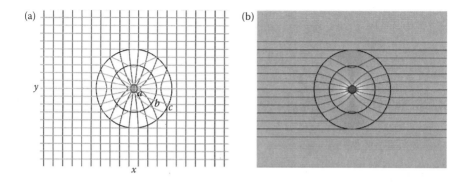

Figure 6.7 (a) Visualization of the space transformation for concentrator and (b) normalized power flow distribution of a circular concentrator.

where $t = (b-a)/(c-b)\, c/r + 1$. Equation 6.21 represents the material properties of the cylindric concentrator in the original space (r, ϕ, z). If the incident EM wave is polarized with the electric field in z direction, three components are of interest: the radial and azimuthal magnetic components μ_r and μ_ϕ and the electric z component ε_z, all of which are gradient functions of the radial position r. Hence, the experimental implementation of the concentrator requires independent control of the local values of all three parameters.

Figure 6.7a displays the equi-phase surfaces and the power flow lines of the electric field for a z-polarized TE plane wave. The horizontal gray lines represent the direction of the power flow. As can be seen, the fraction of the plane wave is completely focused by the concentrator into the center region. Figure 6.7b illustrates the normalized intensity distribution of the TE wave. It is obvious that the field intensities are strongly enhanced in the inner region of radius a within the concentrator material. The intensity enhancement factor for the chosen structure can be computed as the ratio between the maximal value of the field intensities inside the concentrator region with radius a and that outside the circular region with radius c. Significantly stronger enhancements can be achieved by increasing the ratio b/a. As can be seen, the enhancement theoretically diverges to infinity as a goes to zero. Due to the rotational symmetry around the axis perpendicular to the $x - y$ plane, the concentrator focuses waves impinging from arbitrary directions.

The concentrator is reflectionless due to inherent impedance matching in the transformation-optical design method. Arbitrarily shaped concentrator has been designed in Reference 30. The real part of the electric field distribution of a heart-shaped concentrator is shown in Figure 6.8a. The fraction of the plane wave is focused by the concentrator into the center region regardless of the shape. Figure 6.8b illustrates the normalized intensity distribution of the TE wave. Again, the field intensities are strongly enhanced in the center region within the concentrator material.

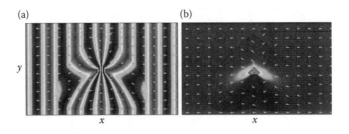

Figure 6.8 (a) Real part of E-field distribution and (b) normalized power flow distribution of a heart-shaped concentrator.

6.5.3 High-Performance Antennas

The principle of many high-performance antennas is based on the conversion from cylindrical waves to plane waves. An alternative method of the conversion from cylindrical waves to plane waves will be discussed here, in a short range through a metamaterial layer based on an embedded optical transformation. The first work on conversion from cylindrical waves to plane waves in a short range was proposed by using the finite embedded optical transformation [11,31]. The principle of the conversion is shown in Figure 6.9a. The radii of inner and outer circles are a and b, respectively, and the sidelength of the outer square is $2c$. The square domain is divided into four triangles and in the triangle OAB, the following transformation is constructed,

$$r' = \begin{cases} br/a, & 0 \leq r \leq a, \\ \dfrac{cr-bx}{(b-a)x}(r-a)+b, & a \leq r \leq b, \end{cases} \quad (6.22)$$

where (x, y) is the coordinate variables in the original space and (x', y') is the corresponding variables in the transformed space, and $r = \sqrt{x^2+y^2}$, $r' = \sqrt{x'^2+y'^2}$. Obviously, $0 \leq r' \leq b$ if $0 \leq r \leq a$; and $b \leq r'$, $x' \leq c$ if $a \leq r \leq b$.

There are two transformations in Equation 6.22. The first extends the space in sector $\widehat{OCD}$ to a bigger space in sector $\widehat{OA_1B_1}$, and the second transforms the arc CA_1B_1D to the trapezia-like region A_1ABB_1. Hence, if cylindrical waves are excited at the origin and propagate through the domain A_1ABB_1, the phase front will change from circular to plane.

The relative permittivity and permeability tensors of the materials for the single triangular domain OAB in the transformed space would be calculated by transformation optics. By using rotation operators with angles of $\pi/2$, π and $3\pi/2$ around the z axis, the corresponding material parameter tensors will be obtained.

Figure 6.9c illustrates the distributions of electric fields inside and outside the conversion materials when the cylindrical waves are excited by a line current source. In this example, we set $a = 0.05$ m, $b = 0.1$ m, and $c = 0.15$ m. When the

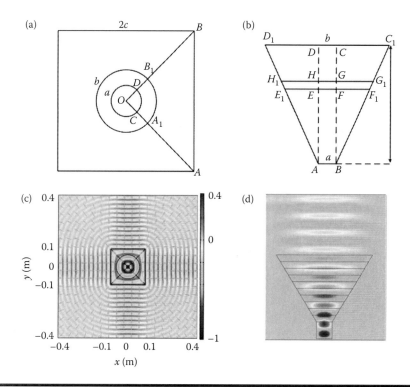

Figure 6.9 (a) Configuration of cylindrical-to-plane-wave conversion. (b) The 2D layered EM lens structure. (c) The distributions of electric fields and power flow lines for the conversion from cylindrical waves to plane waves. (d) The distribution of electric fields inside and outside the lens structures, which are placed in a metallic horn antenna, under the excitation of a line-current source.

cylindrical waves propagate through the metamaterial conversion layer, four beams of plane waves emerge in the surrounding space. Hence, in a very short range ($c = 0.15$ m), cylindrical waves emitted from the line source are converted to plane waves in four directions. The converter provides the high intensity of the waves, and hence, it is able to modify the directivity of the infinite line source in comparison with the unmodified source.

Next, we review a highly directive lens antenna based on layered embedded optical transformation [32], which makes all layers of the lens antenna homogeneous and uniaxially anisotropic. Such an optical-transformation device can amplify the local properties of fields in a small region to a large region. Consider a 2D structure in the Cartesian coordinate system, as shown in Figure 6.9b, in which the virtual space is a rectangle $ABCD$ and the physical space is a trapezia ABC_1D_1. We divide the physical and virtual spaces into n layers in the same way. For the kth ($1 \leq k \leq n$)

layer, we define an elongating mapping using the following equations:

$$x' = x + \frac{(k-0.5)(b-a)x}{na}, \quad y' = y, \quad z' = z \quad (1 \leq k \leq n), \quad (6.23)$$

where (x, y) is an arbitrary point in the virtual space, and (x', y') is the transformed point in the physical space. We have assumed the lengths of AB and $C_1 D_1$ to be a and b, respectively, and the height of the trapezia $ABC_1 D_1$ to be L. We remark that the discrete transformation implied in Equation 6.23 maps rectangles to wider rectangles. Therefore, the physical space is actually the staircase approximation of the trapezium obtained by stacking rectangles on top of each other. From this point of view, the larger the layer number n, the more accurate the transformation.

Hence the relative permittivity and permeability tensors of the transformation medium in the kth layer ($1 \leq k \leq n$) are expressed as

$$\varepsilon_{xx}^k = \mu_{xx}^k = \alpha_k, \quad \varepsilon_{yy}^k = \mu_{yy}^k = 1/\alpha_k, \quad \varepsilon_{zz}^k = \mu_{zz}^k = 1/\alpha_k, \quad (6.24)$$

in which

$$\alpha_k = 1 + (k - 0.5)(b-a)/(na). \quad (6.25)$$

For the TE wave incidence, the electric fields are polarized along the z axis, and only ε_{zz}^k, μ_{xx}^k, and μ_{yy}^k are required in Equation 6.24. The dispersion relations of the transformed media remain unchanged as long as the products of $\mu_{xx}^k \varepsilon_{zz}^k$ and $\mu_{yy}^k \varepsilon_{zz}^k$ are kept the same in the above equations. Hence one advantageous choice is to select

$$\mu_{xx}^k = 1, \quad \varepsilon_{zz}^k = 1, \quad \mu_{yy}^k = 1/\alpha_k^2, \quad 1 \leq k \leq n \quad (6.26)$$

due to the practical reason. In this set of parameters, only μ_{yy}^k is different in each layer of the lens structure, which makes the whole structure very easy to realize. We remark that the reduced media parameters provide the same wave trajectory inside each layer of the lens antenna, and there will be no reflection for the normal incidence because of the wave-impedance matching at the upper and lower boundaries. However, for other angles of incidence, there will be tiny reflections.

Figure 6.9d illustrates the distribution of electric fields inside and outside the lens antenna, in which we choose $a = 0.05\,\text{m}$, $b = 0.3\,\text{m}$, and $L = 0.2\,\text{m}$. We divide the whole structure into 10 layers. Although there are some reflections inside the lens antenna, almost all EM powers are concentrated as a beam of plane waves in the front of lens antenna. The transformation-optics lens antennas have much better performance than the conventional lens.

A broadband transformation optics lens has been designed and fabricated [33]. Such a lens can convert the radiation from an embedded isotropic source into any desired number of highly directive beams pointing at arbitrary directions. The sample is shown in Figure 6.10a. The impedance bandwidth of the embedded antenna

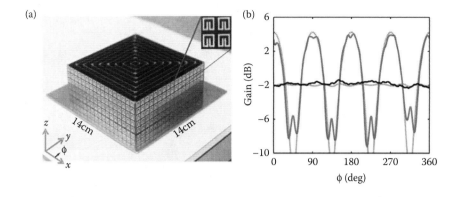

Figure 6.10 (a) Photograph of the fabricated broadband transformation optics lens. (b) Simulated and measured gain patterns of the monopole with and without the transformation optics lens. (Reprinted with permission from Z. H. Jiang, M. D. Gregory, and D. H. Werner, *Phys. Rev. B*, 84: 165111. Copyright 2009 by the American Physical Society.)

was greatly enhanced by exploiting the dispersive properties of the metamaterial unit cells. Simulated and measured gain patterns of a monopole with and without the transformation optics lens have been shown in Figure 6.10b.

6.5.4 Illusion-Optics Devices

The last example of inhomogeneous metamaterials is illusion-optics device. Inhomogeneous and anisotropic metamaterials can not only make an object invisible, but also make an object look exactly like another with different shape and material makeup as one desires. Illusion optics was first proposed by Lai et al. [34]. The principle behind the illusion device is not to make light bend, but rather to cancel and restore the EM signature within the virtual boundary. It should be noted that the illusion device can work at a distance from the object.

An interesting schematic illustration of illusion device is shown in Figure 6.11. In the real space, there is a man with an illusion device, as shown in Figure 6.11a. Such a device makes any observer outside the virtual boundary see the image of the virtual object a woman instead. A blueprint for the device is shown in Figure 6.11c. The device can be divided into two parts. Part 2 includes the complementary medium, which is used to cancel the EM signature of the real object, a man, while Part 1 includes the restoring medium, which is used to generate the signature of the virtual object, a woman. The spatial transformation of Lai's illusion is explained as follows: the complementary medium is formed by a coordinate transformation of folding Region 3 (Figure 6.11c), which contains the actual object into Region 2. The restoring medium is formed by a coordinate transformation of compressing

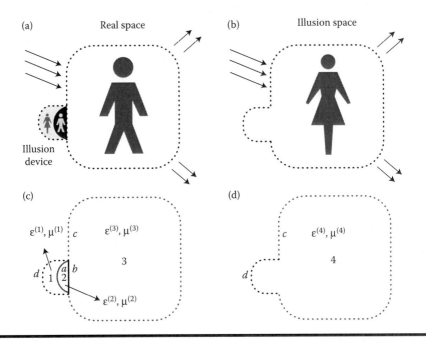

Figure 6.11 The schematic illustration of illusion device. (a) The real space, (b) the virtual space, (c, d) the blueprint of illusion device. (Reprinted with permission from Y. Lai, J. Ng, H. Chen, D. Han, J. Xiao, Z.-Q. Zhang, and C. T. Chan, *Phys. Rev. Lett.*, 102: 253902. Copyright 2009 by the American Physical Society.)

Region 4 (Figure 6.11d), which contains the virtual object into Region 1. The material parameter tensors of both media in the illusion device are obtained by transformation optics.

The above idea of illusion device is very fascinating, but the realization of complementary media with both negative permittivity and negative permeability in free space is difficult [35], and hence, the free-space illusion device based on such an idea has not been reported until now. Fortunately, an alternative method of illusion optics device was presented soon. Unlike the above illusion devices which are composed of left-handed materials [34], all permittivity and permeability components of the new-proposed illusion media are finite and positive [36]. Hence, such illusion media are fairly realizable.

An intuitive example to illustrate the method is that, in the real space, the actual object, a metallic sphere, is enclosed with an illusion medium layer which makes any detector outside the virtual boundary perceive the scattering fields of the virtual objects, two dielectric spheres, instead of one metallic sphere. The illusion medium layer has two functions: concealing the optical signature of the metallic sphere and generating the image of two dielectric spheres.

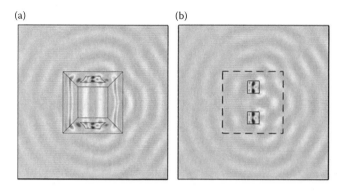

Figure 6.12 The scattering field of (a) a metallic square object with an illusion device in real space and (b) two dielectric square objects in virtual space. (Reprinted with permission from W. X. Jiang, H. F. Ma, Q. Cheng, and T. J. Cui. Illusion media: Generating virtual objects using realizable metamaterials. *Appl. Phys. Lett.*, 96: 121910. Copyright 2010, American Institute of Physics.)

For example, an illusion medium layer, which transforms a metallic square cylinder to two dielectric square cylinders, is considered in 2D case. Figure 6.12 illustrates the scattering electric fields of such an illusion device, which is better to illustrate the scattering character of objects than the total fields. The plane waves are incident horizontally from left to right. Figure 6.12a and b shows the scattered patterns of the metallic square object enclosed by the illusion device and two dielectric square objects with $\varepsilon_r = 3$ and $\mu_r = 2$, respectively. By comparing the scattering-field pattern of the illusion-device-coated metallic object in Figure 6.12a with that of two dielectric objects in Figure 6.12b, the field patterns have been proved to be exactly the same outside the virtual boundary.

The illusion device in Figure 6.12a is composed of one layer of inhomogeneous and anisotropic medium. The spatial transformation for such an illusion device in the Cartesian coordinates is expressed as

$$x' = \left(\frac{R_2 - R_1}{R_2} + \frac{R_1}{r}\right)x, \quad y' = \left(\frac{R_2 - R_1}{R_2} + \frac{R_1}{r}\right)y, \quad z' = z, \quad (6.27)$$

where $R_1 = ar/(k_1 x + k_2 y)$, $R_2 = br/(k_1 x + k_2 y)$, and $r = \sqrt{x^2 + y^2}$, a is the distance between the origin and the inner boundary, b is the distance between the origin and the outer boundary if the boundary is vertical or parallel to the x axis. For the vertical case, $k_1 = 1$ and $k_2 = 0$; for the parallel case, $k_1 = 0$ and $k_2 = 1$. For other cases, $k_1 = 1$ and $k_2 = -1/k$, in which k is the slope of the boundary, and a and b are abscissa of the intersection between the x axis and the inner and outer boundaries or their prolongations, respectively. In such an example, $a = 0.025$ m,

$b = 0.05$ m, and the side length of each virtual object is 0.02 m. Based on the above transformation, the electric permittivity and magnetic permeability tensors of the illusion medium layer can be calculated.

The illusion objects are assumed to be isotropic with $\bar{\varepsilon} = \varepsilon_r \bar{I}$ and $\bar{\mu} = \mu_r \bar{I}$ for each illusion object. For two trapezia regions in Figure 6.12a, $\varepsilon_r = 3$, $\mu_r = 2$, which creates the illusion of two dielectric square cylinders in Figure 6.12b; for other region in the illusion device, $\varepsilon_r = 1$, $\mu_r = 1$, which is the same as the free-space cloak. It is noted that the transformation in this illusion device is a compressing mapping, instead of the folding of geometry. Hence any parameters of the illusion device is positive. We will discuss the realization of such illusion device in Chapter 9.

In this chapter, we have discussed some methods of designing inhomogeneous devices, such as geometrical optics, quasi-conformal mapping, and transformation optics. Metamaterials provide many possibilities to realize these devices. In the following chapters, we will discuss some more specific inhomogeneous devices, isotropic and anisotropic.

References

1. R. A. Shelby, D. R. Smith, and S. Schultz. Experimental verification of a negative index of refraction. *Science*, 292: 77, 2001.
2. D. R. Smith, J. B. Pendry, and M. C. K. Wiltshire. Metamaterials and negative refractive index. *Science*, 305: 788, 2004.
3. N. Fang, H. Lee, C. Sun, and X. Zhang. Sub-diffraction-limited optical imaging with a silver superlens. *Science*, 308: 534, 2005.
4. D. R. Smith, J. J. Mock, A. F. Starr, and D. Schurig. Gradient index metamaterials. *Phys. Rev. E*, 71: 036609, 2005.
5. X. Chen, H. F. Ma, X. Y. Zou, W. X. Jiang, and T. J. Cui. Three-dimensional broadband and high-directivity lens antenna made of metamaterials. *J. Appl. Phys.*, 110: 044904, 2011.
6. H. F. Ma, X. Chen, H. S. Xu, X. M. Yang, W. X. Jiang, and T. J. Cui. Experiments on high-performance beam-scanning antennas made of gradient-index metamaterials. *Appl. Phys. Lett.*, 95: 094107, 2009.
7. J. B. Pendry, D. Schurig, and D. R. Smith. Controlling electromagnetic fields. *Science*, 312: 1780, 2006.
8. U. Leonhardt. Optical conformal mapping. *Science*, 312: 1777, 2006.
9. M. Rahm, D. A. Roberts, J. B. Pendry, and D. R. Smith. Transformation-optical design of adaptive beam bends and beam expanders. *Opt. Express*, 16: 11555, 2008.
10. M. Rahm, D. Schurig, D. A. Roberts, S. A. Cummer, D. R. Smith, and J. B. Pendry. Design of electromagnetic cloaks and concentrators using form-invariant coordinate transformations of Maxwell's equations. *Phot. Nano. Fund. Appl.*, 6: 87, 2008.
11. M. Rahm, S. A. Cummer, D. Schurig, J. B. Pendry, and D. R. Smith. Optical design of reflectionless complex media by finite embedded coordinate transformations. *Phys. Rev. Lett.*, 100: 063903, 2008.

12. U. Leonhardt and T. G. Philbin. Chapter 2: Transformation optics and the geometry of light. *Progr. Optics*, 53: 69–152, 2009.
13. M. Born and E. Wolf. *Principles of Optics*. Cambridge University, Cambridge, 1999.
14. Z. L. Mei and T. J. Cui. Arbitrary bending of electromagnetic waves using isotropic materials. *J. Appl. Phys.*, 105: 104913, 2009.
15. Z. L. Mei, J. Bai, T. M. Niu, and T. J. Cui. Design of arbitrarily directional cloaks by solving the Laplace's equation. *J. Appl. Phys.*, 107: 124502, 2010.
16. H. Chen, B. Zheng, L. Shen, H. Wang, X. Zhang, N. I. Zheludev, and B. Zhang. Ray-optics cloaking devices for large objects in incoherent natural light. *Nat. Comm.*, 4: 2652, 2013.
17. J. Li and J. B. Pendry. Hiding under the carpet: A new strategy for cloaking. *Phys. Rev. Lett.*, 101: 203901, 2008.
18. R. Liu, C. Ji, J. J. Mock, J. Y. Chin, T. J. Cui, and D. R. Smith. Broadband ground-plane cloak. *Science*, 323: 366, 2009.
19. N. Kundtz and D. R. Smith. Extreme-angle broadband metamaterial lens. *Nat. Mater*, 9: 129, 2010.
20. H. F. Ma and T. J. Cui. Three-dimensional broadband and broad-angle transformation-optics lens. *Nat. Commun.*, 1: 124, 2010.
21. J. F. Thompson, B. K. Soni, and N. P. Weatherill. *Handbook of Grid Generation*. CRC Press, Boca Raton, 1999.
22. J. Valentine, J. Li, T. Zentgraf, G. Bartal, and X. Zhang. An optical cloak made of dielectrics. *Nat. Mater*, 8: 568, 2009.
23. L. H. Gabrielli, J. Cardenas, C. B. Poitras, and M. Lipson. Silicon nanostructure cloak operating at optical frequencies. *Nat. Photon.*, 3: 461, 2009.
24. A. J. Ward and J. B. Pendry. Refraction and geometry in Maxwell's equations. *J. Mod. Opt.*, 43: 773, 1996.
25. U. Leonhardt. Notes on waves with negative phase velocity. *IEEE J. Sel. Topics Quantum Electron.*, 9: 102, 2003.
26. H. Chen, B.-I. Wu, B. Zhang, and J. A. Kong. Electromagnetic wave interactions with a metamaterial cloak. *Phys. Rev. Lett.*, 99: 063903, 2007.
27. Z. Ruan, M. Yan, C. W. Neff, and M. Qiu. Ideal cylindrical cloak: Perfect but sensitive to tiny perturbations. *Phys. Rev. Lett.*, 99: 113903, 2007.
28. W. X. Jiang, T. J. Cui, G. X. Yu, X. Q. Lin, Q. Cheng, and J. Y. Chin. Arbitrarily elliptical cylindrical invisible cloaking. *J. Phys. D: Appl. Phys.*, 41: 085504, 2008.
29. W. X. Jiang, J. Y. Chin, Z. Li, Q. Cheng, R. Liu, and T. J. Cui. Analytical design of conformally invisible cloaks for arbitrarily shaped objects. *Phys. Rev. E*, 77: 066607, 2008.
30. W. X. Jiang, T. J. Cui, Q. Cheng, J. Y. Chin, X. M. Yang, R. Liu, and D. R. Smith. Design of arbitrarily shaped concentrators based on conformally optical transformation of nonuniform rational B-spline surfaces. *Appl. Phys. Lett.*, 92: 264101, 2008.
31. W. X. Jiang, T. J. Cui, H. F. Ma, X. Y. Zhou, and Q. Cheng. Cylindrical-to-plane-wave conversion via embedded optical transformation. *Appl. Phys. Lett.*, 92: 261903, 2008.
32. W. X. Jiang, T. J. Cui, H. F. Ma, X. M. Yang, and Q Cheng. Layered high-gain lens antennas via discrete optical transformation. *Appl. Phys. Lett.*, 93: 221906, 2008.
33. Z. H. Jiang, M. D. Gregory, and D. H. Werner. Experimental demonstration of a broadband transformation optics lens for highly directive multibeam emission. *Phys. Rev. B*, 84: 165111, 2009.

34. Y. Lai, J. Ng, H. Chen, D. Han, J. Xiao, Z.-Q. Zhang, and C. T. Chan. Illusion optics: The optical transformation of an object into another object. *Phys. Rev. Lett.*, 102: 253902, 2009.
35. C. Li, X. Meng, X. Liu, F. Li, G. Fang, H. Chen, and C. T. Chan. Experimental realization of a circuit-based broadband illusion-optics analogue. *Phys. Rev. Lett.*, 105: 233906, 2010.
36. W. X. Jiang, H. F. Ma, Q. Cheng, and T. J. Cui. Illusion media: Generating virtual objects using realizable metamaterials. *Appl. Phys. Lett.*, 96: 121910, 2010.

Chapter 7
Gradient-Index Inhomogeneous Metamaterials

Gradient-refractive-index (GRIN) materials play an important role in inhomogeneous metamaterials. As we all know, the refractive index $n = \sqrt{\epsilon\mu}$. This means that the materials can be realized by tailoring their permittivity and/or permeability. In most cases, these materials are isotropic and nonmagnetic, that is, their electromagnetic (EM) parameters can be characterized by a scalar dielectric constant. However, as the technology evolves and more magnetic materials are involved in this area, GRIN materials employing inhomogeneous and magnetic materials can also be envisioned. In this chapter, however, we only focus on nonmagnetic materials.

In homogeneous materials, light rays always travel in straight lines, in line with Fermat's principle. However, in inhomogeneous materials, light rays may travel in very strange ways. This fact explains the formation of mirage effect in deserts, where airs at different levels are heated differently by the heated earth, leading to the gradient of refractive indexes for the atmosphere. Based on the same idea, various lenses can be designed, which have different light manipulation capabilities. In this chapter, we will introduce some representative GRIN devices in this area, including half Maxwell fisheye lens, full and half Luneburg lens, the artificial "black hole," etc. Both 2D and 3D cases will be explained in detail.

These GRIN devices can be designed with the method of geometrical optics (GO) [1], conformal mapping [2–5], quasi-conformal mapping [6–10], and other methods. They can be analyzed through ray tracing or full-wave method. The former is an approximate method and the latter is an accurate one. The flowchart for the theoretical analysis and experimental verification of GRIN device is given in Figure 7.1. Below are some details:

- Index profile: For known devices, the index profile is given; while for new GRIN devices, it is obtained using various methods. Ray tracing can be used in this step to roughly analyze the performance of the device.
- Discretization: The GRIN device with continuous medium is discretized into subwavelength blocks, which can then be mapped into metamaterial unit

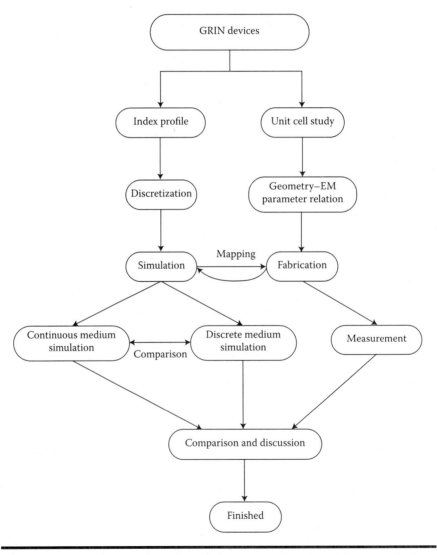

Figure 7.1 Flowchart for the design, analysis, and realization of the GRIN devices.

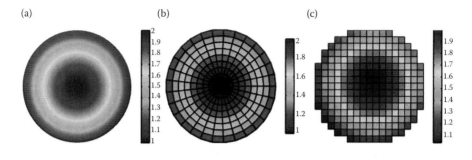

Figure 7.2 Refractive index discretization. **(a) Continuous refractive index distribution, (b) discretization in the polar grids, (c) discretization in the rectangular grids.**

cells. Discretization can be realized with various methods. For example, in the 2D case, both polar grid and rectangular grid can be adopted to discretize the continuous medium. In Figure 7.2, we show the refractive index profile for a 2D spherical lens in (a), its polar gird discretization in (b), and the rectangular discretization in (c).

- Simulation: Rigorous full-wave simulations, which can be divided into two types, that is, continuous-medium simulation based on the original refractive index and the discrete simulations based on metamaterial unit cells (sometimes also called full structure simulations). The comparison between these two simulations will decide whether the device is realizable or not; hence it is a very important step in the flowchart. However, the discrete simulation is a time-consuming task, sometimes even impossible to run.
- Unit cell study: Choose the appropriate unit cell for the GRIN device realization. Generally speaking, they should be broadband, less lossy, and have good impedance matching with the background.
- Geometry–EM parameter relation: Through unit cell simulation or rigorous analysis, one should obtain the relationship between the effective EM parameters, that is, the effective refractive index and the geometrical size of the unit cell structure. Usually, this is done by the well-known retrieval method, as has been shown in the previous chapters [11]. The results can then be utilized to realize the mapping between the unit cell and the previously discretized, subwavelength blocks. Due to the nature of GRIN devices, each unit cell in the device usually has different geometrical sizes.
- Fabrication: The device is fabricated using the mapped unit cell structures. In the microwave frequency, it can usually be implemented using the PCB technology.
- Measurement: The GRIN device is measured.
- Comparison: The measured data are compared with the full-wave simulation results.

In the following sections, we will strictly follow this chart and analyze several GRIN lenses. However, before we start, we will introduce several representative GRIN metamaterials.

7.1 Several Representative GRIN Metamaterials

GRIN metamaterials can be realized using various methods. In this section, we mainly focus on three representative metamaterials, that is, the hole-array material [12,13], "I-shaped" material [7,9,14], and the "waveguide" material [15].

7.1.1 Hole-Array Metamaterial

This kind of material may be one of the simplest metamaterials studied. The working principle can be easily understood using the parallel capacitor model, which is shown in Figure 7.3.

In the effective medium theory, since the working wavelength is larger compared with the hole dimensions, the problem can be approximately treated as a quasistatic problem. Then, the total capacitor between the two conducting plates can be expressed as

$$C = \frac{\epsilon S}{d}, \tag{7.1}$$

where S is the area for one plate, ϵ is the effective material permittivity between the two conducting plates, and d is their distance. However, if we treat the capacitor as the parallel connection of many small capacitors, it is quite clear that the following equation holds

$$C = \sum C_i, \tag{7.2}$$

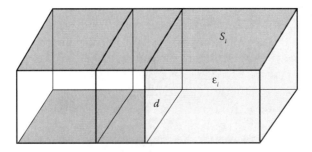

Figure 7.3 Parallel capacitor model.

which can be further expressed as

$$\frac{\epsilon S}{d} = \sum \frac{\epsilon_i S_i}{d}. \quad (7.3)$$

Then, it is easily proved that the effective material parameter can be represented as

$$\epsilon = \sum \epsilon_i f_i, \quad (7.4)$$

where $f_i = S_i/S$ is the filling ratio for each constitutive material. By adjusting the filling ratio, which is easily realizable in practice, the effective material parameters can be flexibly adjusted. In most cases, two materials are used to obtain the required effective parameters, and for most of the time, one of the two materials is air. Then, it is easy to obtain the following equation: $\epsilon = f + \epsilon_d(1-f)$, where ϵ_d is the dielectric constant of the material (substrate). Using the current PCB technology, it is very simple to adjust the filling ratio through different via-holes, and hence, various effective materials. Because of its simplicity and controllability, hole-array materials have been widely used in various devices [12,13].

7.1.2 I-Shaped Metamaterial

I-shaped metamaterial was first proposed for the experimental realization of ground-plane cloak (or carpet cloak) [7], and it represents a typical broadband metamaterial. As we all know, for metamaterials with negative permittivity or permeability, the constitute unit cells always work at the resonant region, which means the working band is very narrow. However, if one does not focus on the negative parameters, the unit cell can be detuned from the resonance frequency, leading to large working bandwidth and low loss. In this regard, the I-shaped metamaterial is a very good example.

Figure 7.4 shows the equivalent material parameters for an I-shaped unit cell, which is obtained using the retrieval method [11]. As can be clearly seen in the figure, the effective refractive index varies with the arm length of the cell. Hence, we can flexibly realize the required refractive index profile using different I-shaped unit cells, as will be shown in this chapter.

7.1.3 Waveguide Metamaterial

When EM waves are propagating in certain wave guiding structures, their phase propagation constants (β) will change, leading to fast or slow waves compared with those in an unbounded medium (k_0). Equivalently, this can also be interpreted by introducing an effective refractive index, the ratio of β/k_0. In 2011, Falco and coworkers demonstrated a fully functional Luneburg lens in silicon photonics that can be used in integrated imaging devices [15]. In the design, they exploit the 3D

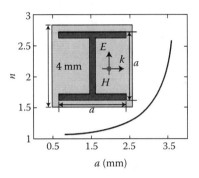

Figure 7.4 Effective refractive index of an I-shaped unit cell structure. The inset shows the geometry of the structure.

nature of the waveguide to create an effective index profile for 2D wave propagation on the chip. Hence we can call it "waveguide material."

Figure 7.5a shows a typical planar dielectric waveguide. For the TE mode illustrated in the figure, the dispersion equation can be easily obtained, which is

$$\tan(\tau_1 d) = \frac{\tau_1(\tau_2 + \tau_3)}{\tau_1^2 - \tau_2 \tau_3}, \tag{7.5}$$

where $\tau_1^2 = k_0^2 n_1^2 - \beta^2$, $\tau_2^2 = \beta^2 - k_0^2 n_2^2$, and $\tau_3^2 = \beta^2 - k_0^2 n_3^2$. Based on this equation, we find that the propagation constant β has an implicit relation

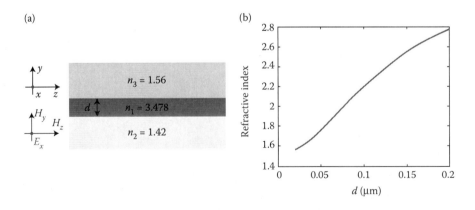

Figure 7.5 A planar dielectric waveguide (a) and the effective refractive index variation with the thickness d (b). Note that the working wavelength is about 1.55 μm.

with the thickness d. If we take $n_{eff} = \beta/k_0$, then it is apparent that the effective refractive index in the planar dielectric waveguide depends on its geometric dimensions. As a result, the structure actually provides a very flexible method for the creation of required index profiles by changing the thickness d. Figure 7.5b shows the corresponding refractive index for the planar waveguide in (a).

Using the above-mentioned GRIN metamaterials, various EM devices with certain functionalities can be realized, that is, the so-called GRIN devices. In the following sections, we will follow the flowchart in the previous section and analyze several GRIN devices.

7.2 2D Planar Gradient-Index Lenses

In this section, we will show some 2D planar lenses with GRIN profiles. Based on the quasi-static theory, such materials at the microwave band can be realized by drilling hole arrays on ordinary dielectric materials. It can also be applied in the optical band as long as quasi-static conditions are satisfied. These lenses can realize beam steering or focusing functions and have wide applications in various sectors. The method and the devices may find applications in integrated circuit systems.

7.2.1 Derivation of the Refractive Index Profile

Figure 7.6 shows the working principle of the lenses discussed in this section. For the first two lenses, the incident beams are plane waves, while for the third one, it is a cylindrical wave. We will use the method of geometrical optics to derive the index profile for each lens. Here, the essential idea is to concentrate on the shape of the equi-phase curves (surfaces), whose normal directions represent the wave propagation direction.

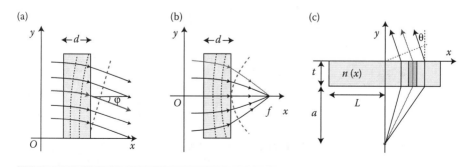

Figure 7.6 Schematic diagram of the lenses. (a) The deflection lens, (b) the focusing lens, (c) the beam scanning lens.

For the first deflection lens, let us pay attention to a typical horizontal ray at position *y*, which is denoted as the center gray line in the figure. Then, the phase delay when it emanates from the lens is $k_0 dn(y)$. Since the ray is deflected with an angle φ from the horizontal direction, the outward rays should share the same equiphase curve, which is denoted by the dashed line outside the lens. This means the phase delay between the incident surface and the dashed line will be the same for all rays, which leads to

$$k_0 dn(y) + k_0 y \sin \varphi = k_0 dn(0) = \text{Constant.} \tag{7.6}$$

Hence,

$$n(y) = n(0) - (\sin \varphi / d) y. \tag{7.7}$$

Using the similar process, the refractive index for the second focusing lens is

$$n(y) = n(0) - \left(\sqrt{f^2 + y^2} - f\right) / d. \tag{7.8}$$

Here, we make an approximation that the wave vector $\vec{k}$ is in the horizontal direction in the lens structure. Theoretically, the ray travels along a curved path in inhomogeneous media. But our approximation is good enough to estimate the optical length, as long as the thickness of the lens is small.

The same method applies to the third lens, which is used to transform a cylindrical wave into plane waves propagating in a predesigned direction (in the figure, with an angle θ from the vertical direction). However, for this case, the phase delay between the point source and the equi-phase curve (the dashed line outside the lens) of the outward rays should be a constant, which means

$$k_0 \sqrt{a^2 + x^2} + k_0 n(x) t + k_0 x \sin \theta = k_0 a + k_0 n(0) t. \tag{7.9}$$

So, we have

$$n(x) = n_0 - \left[x \sin \theta + \left(\sqrt{a^2 + x^2} - a\right)\right] / t. \tag{7.10}$$

This is in line with the result given by Ma and coworkers [16]. The readers are strongly encouraged to repeat the derivation process so as to grasp the core of the problem. In the following part, however, we only study the performance of the first two lenses due to the similarity of the devices.

7.2.2 Full-Wave Simulations (Continuous Medium)

To test the performance of the two lenses, we first make full-wave simulations with the finite element method, and the simulation is made using Comsol Multiphysics. Figure 7.7 shows the corresponding results. In the simulation, the working

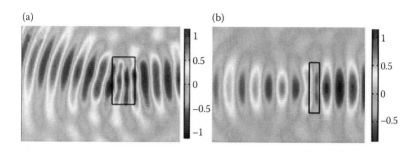

Figure 7.7 Simulated electric field distributions for the deflection lens (a) and focusing lens (b).

frequency is set to 10 GHz, and the refractive index profile is set according to the above two equations. Two Gaussian beams propagate from the right side to the left side and they experience focusing and deflection effect due to the presence of the two lenses. According to the figure, it is very clear that the two lenses work very well. Since the refractive profile of the two lenses are different from the ambient air, they are not impedance matched with surrounding materials and this will give rise to unwanted reflections. The reflection can be clearly observed in Figure 7.7a, where the incident beams are obviously distorted due to its interference with the reflected waves.

7.2.3 Hole-Array Metamaterials

As mentioned before, when two nonmagnetic materials are mixed together, whose permittivities are ϵ_1 and ϵ_2, respectively, then in the quasi-static regime, we have $\epsilon = \epsilon_1 f_1 + \epsilon_2 f_2$, where f_1 and f_2 are the filling ratio of the two materials, ϵ is the resulting material parameter, and $f_1 + f_2 = 1$. In a very special case, subwavelength hole arrays can be drilled on ordinary dielectric materials, which give the following parameters: $\epsilon = f + \epsilon_d(1-f)$, where ϵ_d is the dielectric constant of the substrate, and $n = \sqrt{f + \epsilon_d(1-f)}$. This scheme is used in our design.

Figure 7.8 shows the geometry of the unit cell and the variation of the effective refractive index with r, the radius of the hole. According to this figure, it is clear that the effective parameter can be easily and flexibly controlled by adjusting the radius of the hole. The parameters are obtained using the standard retrieval process (solid curve with dots) with simulated S parameters from the Ansoft HFSS environment [11]. Parameters calculated using the effective medium theory are also given at 10 GHz (solid curve). The retrieved parameters agree well with the calculated ones, a further proof of the validity of the quasi-static applicability. If the working wavelength is large compared with the hole size, the resulting effective material can be broad band and have very low losses, this is demonstrated by the dispersion curve shown in Figure 7.9.

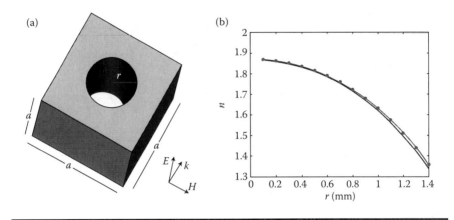

Figure 7.8 Geometry of the unit cell (a) and the effective refractive index variation along with the radius of the hole (b).

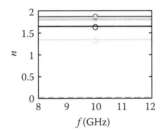

Figure 7.9 Dispersion of the hole-array material. Curves from top to bottom represent the real part of n with hole radii of 0, 0.7, 1.3, and 1.9 mm, respectively, while the corresponding dashed lines mean the imaginary part and circles indicate parameters calculated with effective medium theory. (The source of the material Z. L. Mei, J. Bai, and T. J. Cui, Gradient index metamaterials realized by drilling hole arrays. *J. Phys. D: Appl. Phys.*, 43: 055404, 21 January 2010, IOP Publishing is acknowledged.)

7.2.4 Full-Wave Simulations (Discrete Medium)

By discretizing the continuous medium of the two lenses, we can obtain two discretized lenses, which can be easily realized with the hole-array materials. We then make full-wave simulations for these two resulting lenses, so as to test the feasibility of the scheme. The deflection lens has a deflection angle of 16°. It consists of 20 by 10 holes, and each hole occupies a 3 by 3 mm^2 cell. The radii of the array are obtained using the effective medium theory. For the focusing lens, a 19 by 4 array is used, with a 3 by 3 mm^2 unit cell. Again, the simulations are made in Comsol Multiphysics. Figure 7.10 shows the simulated electric field distributions.

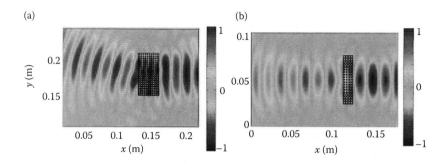

Figure 7.10 Simulated electric field distributions for the discrete lens structure. (a) The deflection lens. (b) The focusing lens. (The source of the material Z. L. Mei, J. Bai, and T. J. Cui, Gradient index metamaterials realized by drilling hole arrays. *J. Phys. D: Appl. Phys.*, 43: 055404, 21 January 2010, IOP Publishing is acknowledged.)

Careful observations of the continuous medium simulation and the discrete medium simulation show that our scheme works perfectly well, meaning the lenses are practically realizable with hole-array materials.

7.2.5 Experimental Realization

We finally fabricate the beam-steering device and focusing lens by drilling hole arrays on ordinary dielectrics. In both lenses, F4BK is used for realization, which has a dielectric constant of 3.5 (10 GHz), a loss tangent of less than 0.001 (10 GHz), and a thickness of 4 mm. The fabricated lenses are tested using the 2D field-mapping apparatus and Figure 7.11 shows the measured electric field distribution. Comparisons between the simulated and measured data show good agreements, which firmly confirm the correctness of the lenses.

7.3 2D Luneburg Lens

The Luneburg lens antennas have been widely used in many commercial applications [14,15,17–19]. The lens has a spherical geometry and can transform the spherical waves from a point source on its surface into plane waves on the diametrically opposite side of the lens. In the 2D case, the lens is infinitely long in the z direction, however, the index profile in the transverse direction is the same as that in the cross section of a 3D Luneburg lens. In 2009, our group presented a 2D Luneburg design with nonresonant metamaterials [14]. The near-field distributions of the metamaterial lens antenna are measured using the 2D field-mapping apparatus, and agree well with the simulation results. The far-field radiation patterns

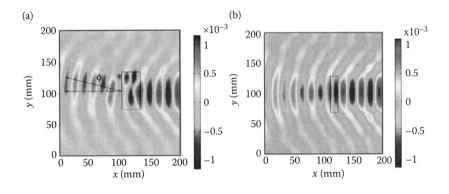

Figure 7.11 Measured electric field distributions for the fabricated lens structure at 10 GHz. (a) The deflection lens. (b) The focusing lens. (The source of the material Z. L. Mei, J. Bai, and T. J. Cui, Gradient index metamaterials realized by drilling hole arrays. *J. Phys. D: Appl. Phys.*, 43: 055404, 21 January 2010, IOP Publishing is acknowledged.)

of the antenna are also presented to show the high-gain performance in the broad bandwidth.

7.3.1 Refractive Index Profile

For Lunegurg lens, the refractive index profile is given by the following equation:

$$n = \sqrt{2 - r^2/R^2}, \tag{7.11}$$

where R is the radius of the lens. The distribution is also given in Figure 7.12 for illustration.

7.3.2 Ray Tracing Performance

In the Hamiltonian formulation of geometrical optics, rays are the solutions to a Hamiltonian system of equations [20,21].

$$\frac{dx_i}{dt} = -\frac{\partial H}{\partial p_i}, \quad \frac{dp_i}{dt} = \frac{\partial H}{\partial x_i}, \quad i = 1, 2, 3, \tag{7.12}$$

where x_1, x_2, x_3 are Cartesian coordinates, p_1, p_2, p_3 are the optical direction cosines, and H is the Hamiltonian function. Note that different forms of the Hamiltonian function are related with different parameterizations along the ray trajectories.

Gradient-Index Inhomogeneous Metamaterials ■ 203

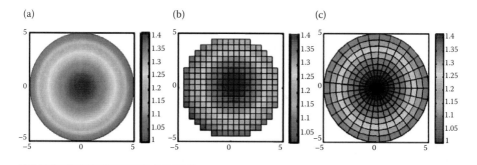

Figure 7.12 Refractive index profile for 2D Luneburg lens. (a) Continuous distribution, (b) discrete distribution in the rectangular grid, (c) discrete distribution in the polar grid.

The Hamiltonian function for an isotropic medium can be written as [21]

$$H(x_1, x_2, x_3, p_1, p_2, p_3) = \left(n^2(x_1, x_2, x_3) - p_1^2 - p_2^2 - p_3^2\right)/2, \quad (7.13)$$

where $n(x_1, x_2, x_3)$ is the refractive index distribution.

Based on the above theory, when the refractive index profile is given, it is not difficult to trace the light using Equation 7.12. Corresponding results are demonstrated in Figure 7.13. Then it is clear that a point source at the surface of lens can produce highly directive radiations after passing through the spherical lens.

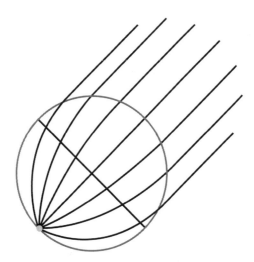

Figure 7.13 Ray trajectories of the 2D Luneburg lens.

7.3.3 Full-Wave Simulations (Continuous Medium)

Using Comsol Multiphysics, we make full-wave simulations with the finite element method. Note that in our simulation, a perfectly electric conducting (PEC) reflector is put behind the point source to greatly reduce the backward radiation, which is clearly not wanted in practical situations. Figure 7.14 shows the simulated electric field distribution, which firmly validates our argument.

7.3.4 Metamaterials Utilized

In order to realize the refractive index of the Luneburg-like lens, the nonresonant I-shaped structures with broadband features are chosen [7]. The response of the I-shaped metamaterial structure is shown in Figure 7.15 in terms of the vertical arm length g. In this design, the PCB is the copper-clad F4B with a thickness of 0.25 mm, the thickness of the copper layer is 0.018 mm, and the line width is 0.2 mm. As shown in the figure, when g is varied from 0.2 to 2.4 mm, the corresponding relative permittivity ϵ_r is gradually increased from 1.31 to 2.38, and the relative permeability μ_r is decreased gradually from 1 to 0.98. It is obvious that the index profile can be realized with the chosen cells.

7.3.5 Experiments

We fabricate the metamaterial cylindrical lens at the center frequency of 8 GHz, the radius is chosen as $R = 49.4$ mm and the height is set to 12 mm. In order to realize the refractive index of the Luneburg-like lens, the rectangular grid is used for discretization. The lens is first divided horizontally into many rectangle slabs with the same thickness, that is, 3.8 mm. The length of each slab is the chord length of the circle and its height equals the height of the lens, which is 12 mm. Then, a series of rectangular slabs with different lengths can be assembled to approximate the cylindrical lens antenna, and all can be fabricated using the PCB technology.

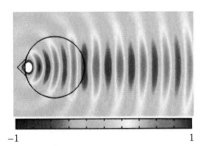

Figure 7.14 Simulated electric field distribution of the 2D Luneburg lens.

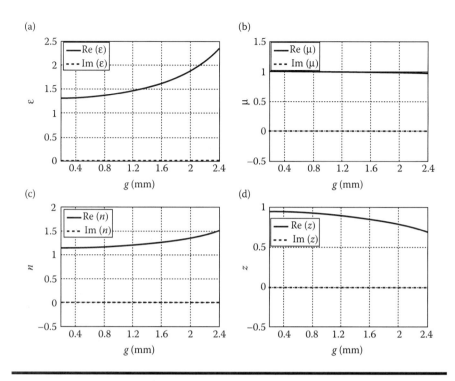

Figure 7.15 Effective EM parameters of the I-shaped unit cell. (a) Permittivity, (b) permeability, (c) refractive index, (d) wave impedance. (With kind permission from Springer Science + Business Media: *Chinese Sci. Bull.*, A broadband metamaterial cylindrical lens antenna, 55, 2010, 2066, H. F. Ma et al., Figure 2.)

The cylindrical lens is measured using the 2D field mapping apparatus, which is actually a parallel plate waveguide system.

The comparison between the simulation and measurement results for the near-field electric distributions of the metamaterial lens antenna is illustrated in Figure 7.16. Figure 7.16a and b shows the simulated and measured near fields of the lens at 8 GHz, and they agree very well. Clearly, the designed Luneburg-like lens antenna transforms the cylindrical waves from the line source into plane waves on the opposite part of the lens, and radiates to the far-field region. Figure 7.16c and d are electric field distributions of the lens antenna at 7 and 8.5 GHz, respectively, and they also have good performances. Hence the proposed metamaterial Luneburg-like lens antenna can work in broadband frequencies.

Figure 7.17a illustrates the comparison between the simulated and measured far-field radiation patterns of the lens antenna at the frequency of 8 GHz. For both curves, the sidelobes are below −15 dB, and the beams are both directed exactly to 90°. Obviously, the measured main lobe is wider than its simulation counterpart.

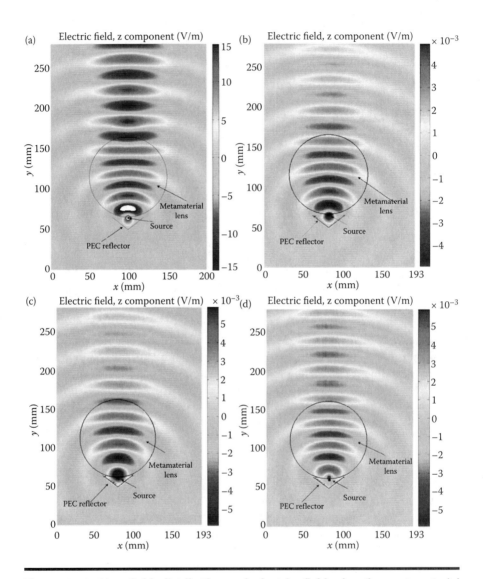

Figure 7.16 Near-field distributions of electric fields for the metamaterial Luneburg-like lens antenna. (a) The simulated result at 8 GHz. (b) The measured result at 8 GHz. (c) The measured result at 7 GHz. (d) The measured result at 8.5 GHz. (With kind permission from Springer Science + Business Media: *Chinese Sci. Bull.*, A broadband metamaterial cylindrical lens antenna, 55, 2010, 2066, H. F. Ma et al., Figure 5.)

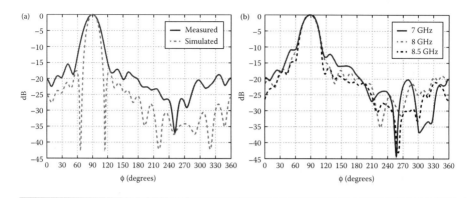

Figure 7.17 Far-field pattern for the Luneburg lens at different frequencies. (a) Simulated and measured data at 8 GHz. (b) Measured data at 7, 8, and 8.5 GHz. (With kind permission from Springer Science + Business Media: *Chinese Sci. Bull.*, A broadband metamaterial cylindrical lens antenna, 55, 2010, 2066, H. F. Ma et al., Figure 6.)

This is understandable since the antenna is completely assembled by hand, which will definitely lead to performance degradations. Considering the relatively small electric size of the aperture ($2R = 2.63\,\lambda$), the metamaterial lens antenna has a good performance with relatively high gain. Figure 7.17b demonstrates the measurement results of the radiation pattern for the metamaterial lens antenna working at 7, 8, and 8.5 GHz, respectively. The results show that the proposed lens antenna can work in broadband frequencies with high gain and good directive behavior.

7.4 2D Half Maxwell Fisheye Lens

Like the Luneburg lens, the Maxwell fisheye (MFE) lens is a typical GRIN device proposed in 1854 [22]. This lens is usually given as an example of a GO perfect imaging instrument, which can transform a point source at the lens surface into a focus at the diametrically opposite side of the lens. In recent years, Fuchs et al. suggested to use half of the MFE lens as a highly directive antenna, which is termed as half Maxwell fisheye lens (HMFE) [23–25]. Using the same idea, we proposed a 2D HMFE design using the I-shaped metamaterials [26].

7.4.1 Refractive Index Profile

The refractive index for the MFE lens is well known, which is

$$n(r) = \frac{n_0}{1 + (r/R)^2}, \qquad (7.14)$$

where n_0 is the refractive index at the lens center and R is the radius of the lens. Obviously, the refractive index gradually decreases to $n_0/2$ from the lens center to the spherical surface. Then HMFE is obtained by cutting the sphere through its center, and the refractive index on that flat surface does not match to the ambient air. Then it is expected that the lens will cause unwanted reflections at its surface.

7.4.2 Ray Tracing Performance

To roughly predict the performance of the 2D HMFE lens, we apply the Hamilton–Jacobi equation to perform the geometric ray tracing in the lens region. As shown in Figure 7.18, the lens can transform rays from a point source into parallel rays which are perpendicular to its flat surface. In contrast, a beam of parallel light rays incident upon the flat side of the 2D HMFE lens will be focused on a point at the spherical lens surface. The ray tracing performance suggests that the lens can be used in antenna systems for collimating EM waves.

7.4.3 Full-Wave Simulations (Continuous Medium)

In this section, we present the simulation results for the HMFE lens antenna. The simulation is made using finite-element-based commercial software, Comsol Multiphysics. Figure 7.19a shows the simulation result for a 10-GHz cylindrical wave passing through the designed HMFE lens. Undoubtedly, the cylindrical wave is transformed to the plane wave according to our prediction. In the simulation, the lens is fed by a current line source at the center part of the curved surface, and the whole structure is surrounded by perfect matching layers (PMLs) in three directions except in the main radiation direction. The configuration is similar to our experimental setup. The observation suggests that HFME can produce a highly directive radiation. In real applications, we can place a PEC corner reflector behind the line source to restrain the back radiation and suppress the side lobes, so that the lens

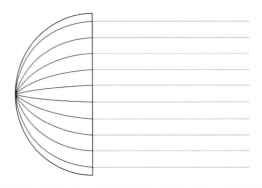

Figure 7.18 Ray tracing performance of the HMFE lens.

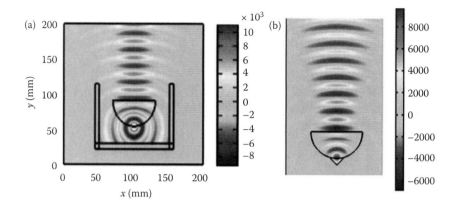

Figure 7.19 Full-wave simulation results for HMFE lens antenna. (a) The antenna is fed by a point source and surrounded by PMLs. (b) The antenna is fed by a point source, which is further covered by a corner reflector.

antenna will achieve both high directivity and radiation efficiency. The simulated near-field distribution is given in Figure 7.19b, which confirms our arguments.

7.4.4 Metamaterials Utilized

The HMFE lens antenna can be realized using I-shaped unit cells [7]. Figure 7.20a shows the retrieved EM parameters for different arm lengths at 10 GHz [11]. It can be seen that the refractive index varies from 1.06 to 2 when a varies from 0.8 to 3.44 mm. Therefore, the I-shaped unit cells are sufficient to satisfy the requirement of the HMFE lens ($n_0 = 2$). Moreover, the imaginary part of material parameters are nearly 0; this means the material loss is very small. Note that in the simulation, we fix $p = 4$ mm, $w = 0.3$ mm. The geometry of the I-shaped cell is shown in Figure 7.20c.

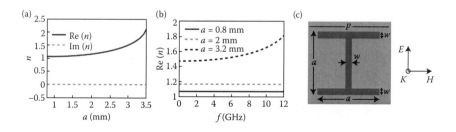

Figure 7.20 Variation of retrieved EM parameters with the geometrical size (a), dispersion of the I-shaped unit cell (b), and the cell structure (c).

Figure 7.20b shows the dispersion curves for unit cells with different geometrical sizes. We can see clearly that the effective constitutive parameters vary slowly with the frequency. That is to say that the constitutive parameters are nearly constants at the lower frequencies. These features will be helpful for designing the broadband EM devices using metamaterials.

7.4.5 Experiments

The HMFE is then fabricated with I-shaped unit cells. The fabricated HMFE lens antenna has a radius about 48 mm with a thickness of 12 mm (3 unit cells in the z direction). As we have described for the 2D Luneburg lens, the lens is (logically) divided into many $4 \times 4 \times 4$ mm^3 small blocks, each of which has a specific refractive index and can be realized using I-shaped cell. Physically speaking, in order to simplify the implementation process, the lens is divided into many parallel slabs with a 4 mm thickness, and then each is mapped to a PCB strip with I-shaped arrays on it. The PCB strips are fabricated using the lithography technology, and the supporting frame is a handmade foam structure with permittivity close to one. Figure 7.21 shows the picture of the fabricated lens antenna.

The HMFE lens antenna is measured in our 2D field mapping equipment, which is actually a parallel plate waveguide. The bottom plate can slowly move in two directions using two-stage motors, and a small dipole is placed near the plate center for the excitation of cylindrical waves. Another probe is implemented on the top plate for the field measurement. Microwave-absorbing materials are inserted between two plates around the measuring area to emulate an infinite environment. The measured electric field distributions at 8, 9, 10, 11, and 12 GHz are shown in Figure 7.22. Similar transformation effects from cylindrical waves to plane waves

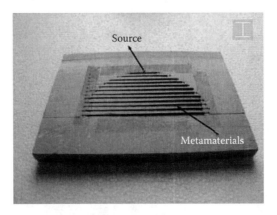

Figure 7.21 Picture of the fabricated HMFE lens antenna. A unit cell is also shown in the picture.

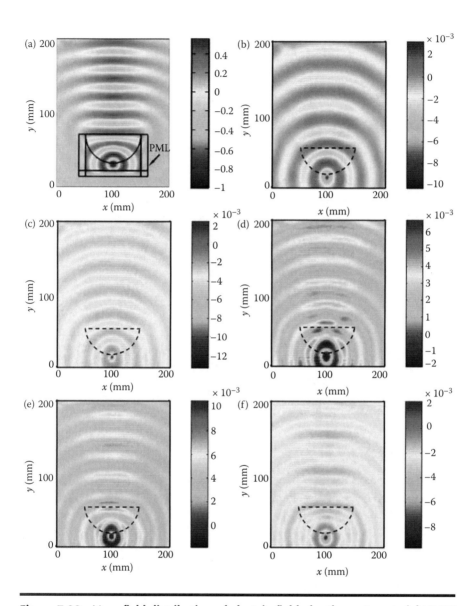

Figure 7.22 Near-field distribution of electric fields for the metamaterial HMFE lens antenna. (a) The simulation result at 10 GHz and the HMFE lens is surrounded with PML. (b)–(f) The measured results using the 2D mapper at 8, 9, 10, 11, and 12 GHz.

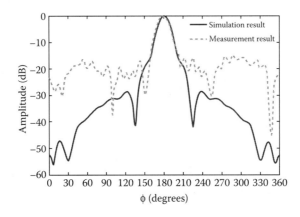

Figure 7.23 Normalized far-field radiation patterns for HMFE lens antenna at 12 GHz. Solid line represents the ideal parameter distribution of HMFE lens antenna and dashed line means the measured result.

are clearly observed at different frequencies. However, the "flattening" effect is more obvious at higher frequency, which implies higher directivity and smaller main-lobe width.

The far-field radiation patterns at 12 GHz are also calculated using the measured data, shown in Figure 7.23. In the peak radiation direction, the measured and simulated results agree well with each other. However, large disparities are observed in other directions. This can be attributed to energy leakage in the experimental setup. In our simulation, PMLs are used to surround the point source except for the peak direction. While in the measurement, we use microwave-absorbing materials. Since the height of the material is slightly smaller than the distance between top and bottom plates (so that the two can move smoothly for the scanning process), energy will surely leak into the ambient environment. That explains why the side lobes in the measured data are higher than the simulation results. Careful examination for the measured result shows that the side lobes are below -15 dB and the half-power beam width is about $20°$.

7.5 3D Planar Gradient-Index Lens

We have shown some 2D devices with GRIN metamaterials. In the following sections, we will present additional 3D devices. In 2011, we proposed a high-directivity lens antenna made of metamaterials [27]. The lens is composed of multilayer microstrip square-ring arrays. These elements are arranged on the planar substrate and can transform the spherical wavefront into a planar one. Moreover, two impedance matching layers (IML) are designed to minimize the reflection loss.

In fact, we have used the lens antennas in the measurements of a 3D invisibility cloak due to the high directivity of the proposed antenna [8].

7.5.1 Refractive Index Profile

Figure 7.24a shows the cross section of the designed lens. As the first step, no matching layers are adopted in the figure. A point source is put at the axis of the lens, which will radiate in all directions. According to geometrical optics, if rays emanating from the lens are parallel with the lens axis, they should have equal phases after passing through the lens structure, that is, every optical path from the source to the upper surface of the GRIN lens should have the same phase delay; this will lead to the following refractive index profile [27]:

$$n(r) = n_0 - \frac{\sqrt{r^2 + S^2} - S}{T}, \qquad (7.15)$$

in which $n(r)$ is the radial function of the index of refraction, S is the distance between the source and the lens, and T is the thickness of the lens. Note that the expression given in the original paper is wrong due to printing errors.

Though the profile of refractive index obtained can give rise to highly directive beams, the reflection is also very large. This is due to the fact that the refractive index is different from 1, the value for the ambient air. To reduce the reflections, two additional impedance layers (coat) can be adopted at the two surfaces of the lens. In this section, the lens without the coats is named as Lens A, while that with the matching layers is named as Lens B. Both lenses are illustrated in Figure 7.24. Then,

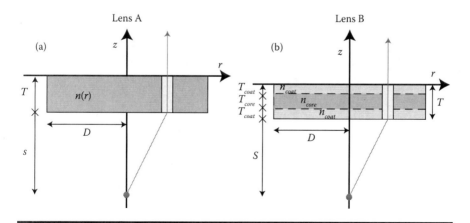

Figure 7.24 Schematic illustration of the proposed lens, where the cross section along its rotational axis is given. (a) Lens A, without the matching layers. (b) Lens B, with the matching layers.

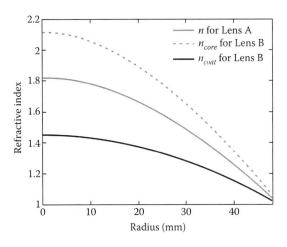

Figure 7.25 Refractive index needed for the two lenses.

for Lens B, the requirement of equal phase will lead to the following equation [27]:

$$2T_{coat}n_{coat} + T_{core}n_{core} = Tn, \qquad (7.16)$$

where T's and n's represent the thickness and refractive index for the core and coating layers, and $T = 2T_{coat} + T_{core}$. This requirement will ensure that the two lenses will lead to same phase accumulations seen from outside the lenses, and hence Equation 7.15 holds for both A and B. Using the geometrical sizes given in the paper, that is, $T_{coat} = 6$ mm, $T_{core} = 15$, it is easy to obtain the following equation, which shows the relationship between the three refractive indexes [27].

$$4\sqrt{n_{core}} + 5n_{core} = 9n. \qquad (7.17)$$

We remark that in the original paper, the right side of the equation is wrong. And the following expressions can be obtained [27]:

$$n_{core} = \frac{8}{25} + \frac{9}{5}n - \frac{4}{25}\sqrt{4 + 45n}. \qquad (7.18)$$

Here, the impedance matching is realized using the quarter-wavelength transformer, which suggests that $n_{coat} = \sqrt{n_{core}}$. The refractive index for the designed lens is given in Figure 7.25.

7.5.2 Full-Wave Simulations (Continuous Medium)

In order to demonstrate the performance of the lens, we first make full-wave simulations in the computer simulation technology (CST) Microwave Studio. Both

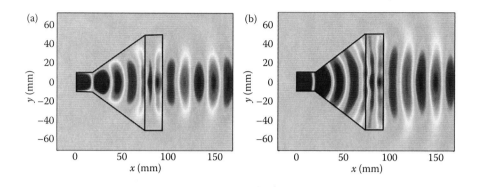

Figure 7.26 Near-field distributions inside and outside the GRIN lens B and horn. (a) E plane. (b) H plane. (Reprinted with permission from X. Chen et al. Three-dimensional broadband and high-directivity lens antenna made of metamaterials. *J. Appl. Phys.*, 110: 044904. Copyright 2011, American Institute of Physics.)

impedance matched lens (B) and not matched lens (A) are installed in a designed metallic conical horn, which is fed by a 25-mm-diameter cylindrical waveguide. The cylindrical waveguide is excited to propagate TE11 mode only. The simulation results of the near-field distribution and far-field directivity patterns for Lens B are demonstrated in Figures 7.26 and 7.27, respectively. Both near-field distribution and far-field pattern show that the lens can transform spherical-like waves into quasi-plane waves, and hence have high directivity at the axial direction. Simulations data for Lens A show similar performance; however, they are not given in the book. Note that in our simulation model, both isotropic and anisotropic factors are considered. At first, the GRIN lens is modeled as an isotropic index medium with varying refractive index along the radius, shown in Figure 7.25. Later, the anisotropic effect of the used metamaterial is considered in the simulation, in which ϵ_{xx} and ϵ_{yy} follow the curve of Figure 7.25, while ϵ_{zz} is fixed at value 1.1. The two cases give similar far-field performances, as is clearly shown in Figure 7.27.

7.5.3 Metamaterials Utilized

For the lens antenna design, we choose electrically small closed square ring (CSR) structures. In this cell structure, a metallic square ring with side length $dx(dy)$ and line width w is coated on an ordinary dielectric substrate, and each cell has an area of $ax * ay$. When in the working state, the metallic CSR pattern on the dielectric substrate is perpendicular to the wave vector, as given in Figure 7.28. According to the figure, we see that the proposed CSR supports two electric field polarizations, one for the x direction and the other for the y direction, because of its 90° rotational symmetry around the z axis.

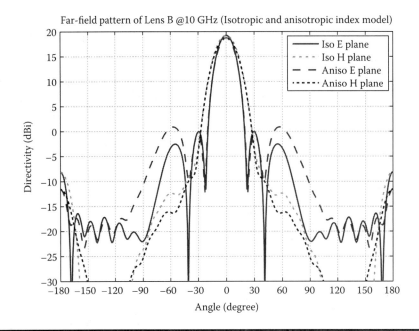

Figure 7.27 Far-field pattern of lens antenna B at 10 GHz. (Reprinted with permission from X. Chen et al. Three-dimensional broadband and high-directivity lens antenna made of metamaterials. *J. Appl. Phys.*, 110: 044904. Copyright 2011, American Institute of Physics.)

The effective parameters of the unit cell can be obtained using the retrieval method [11] and the results are demonstrated in Figure 7.29. From this figure, we clearly see that the effective permittivity of the CSR metamaterial can be controlled by tuning the geometry of the CSR element. At the frequency of 10 GHz, we can tune the refractive index from 1.05 to 2.0, with the CSR size (dx) being varied

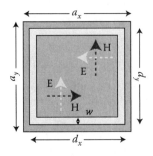

Figure 7.28 Structure of the closed square ring.

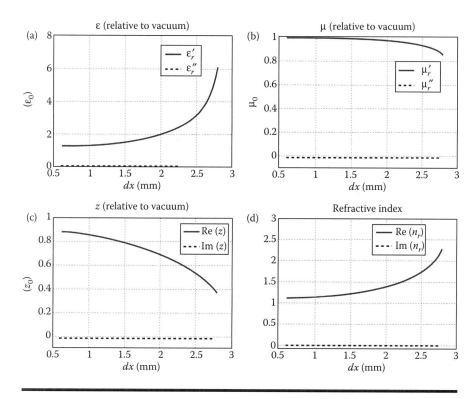

Figure 7.29 Effective parameters of the CSR structure in terms of the geometrical size. (a) Permittivity, (b) permeability, (c) wave impedance, (d) refractive index. (Reprinted with permission from X. Chen et al. Three-dimensional broadband and high-directivity lens antenna made of metamaterials. *J. Appl. Phys.*, 110: 044904. Copyright 2011, American Institute of Physics.)

from 0.6 to 2.7 mm. The ability to control index of refraction over a broadband and support dual polarizations lays the solid foundation of the metamaterial GRIN lens antenna design.

7.5.4 Experiments

To further verify the design, a GRIN lens with IMLs (Lens B) and a metallic horn are manufactured and tested. The conical horn is connected to a cylindrical–rectangular transition waveguide and a waveguide coaxial adaptor. The lens (B) is put at the open end of the horn surface to collimate the outgoing waves. And the measured far field is given in Figure 7.30.

The E plane far-field patterns of the lens antenna over the whole X-band maintains a very highly directive beam with the HPBW around 18° and the side lobe level (SLL) below −17 dB, as demonstrated in Figure 7.30. The measured far-field

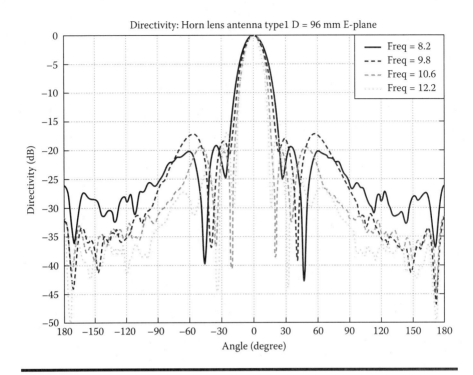

Figure 7.30 Measured E plane far-field directivity of the antenna with lens B at 8.2, 9.8, 10.6, and 12.2 GHz. (Reprinted with permission from X. Chen et al. Three-dimensional broadband and high-directivity lens antenna made of metamaterials. *J. Appl. Phys.*, 110: 044904. Copyright 2011, American Institute of Physics.)

data agree well to the simulated results. The H plane far-field patterns are even better with the HPBW around 23° and the SLL below −24 dB (not given in this book). We also measured the empty horn antenna as a comparison, and significant improvement of the antenna performance is observed after using the GRIN lens [27].

7.6 3D Half Luneburg Lens

As has been demonstrated in the previous section, both 2D Luneburg and fisheye lenses have many applications in various sectors. However, for practical applications, 3D lenses are more desired than their 2D counterparts. In 2013, our group designed and realized two functional devices in the microwave frequencies, a half-spherical Luneburg lens and an HMFE lens based on 3D GRIN metamaterials [28]. The measurement results show that both devices have very good performance, and illustrate the great application potentials of the 3D GRIN metamaterials. In this and the next section, we will briefly show these two designs.

7.6.1 Refractive Index Profile

As has been given in the 2D case, the refractive index profile for the 3D Luneburg lens is

$$n = \sqrt{2 - r^2/R^2}, \qquad (7.19)$$

where r is the radial coordinate in the spherical coordinate system, and R is the radius of the lens. Note that in our realization, only half of the lens is considered. This will reduce the volume of the lens, and at the same time, change the performance of the original lens. To compensate for this factor, a perfectly conducting plate is coated on the flat surface of the half Luneburg lens. Then, using the knowledge of geometrical optics, it is obvious that the modified lens still has the property to change rays from a point source source into parallel beams.

7.6.2 Ray Tracing Performance

The ray tracing performance can also be obtained with the Hamiltonian formalism, as has been discussed in the previous section. Figure 7.31 shows the corresponding results. In the figure, both full and half Luneburg ray trajectories are demonstrated, which are represented by the dashed and solid lines, respectively. Obviously, they are mirror-symmetric with respect to the PEC coating. And it is apparent that the half lens structure can also transform rays from a point source into parallel ones.

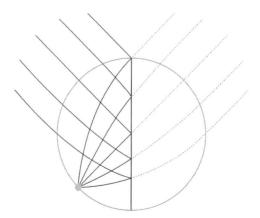

Figure 7.31 Ray tracing for the 3D half Luneburg lens.

7.6.3 Full-Wave Simulations

To further test the performance of the half Luneburg lens, we then carry out a full-wave simulation using the commercial software, CST Microwave Studio. In our simulation, the radius of Luneburg lens is set to 70 mm, whose index of refraction is isotropic and follows the profile in Equation 7.19. The lens is discretized by using 35 spherical shells with different material parameters (refractive indexes), and the thickness of each shell is chosen to be 2 mm. A Ku-band coax-to-waveguide transition with a cross section of $16 \times 8 \, mm^2$ is used to feed the antenna. In the simulation, the spherical coordinate system is utilized, and the flat lens surface is put on the *xoy* plane with the lens axis in the *z*-direction.

In Figure 7.32, the simulated far-field radiation patterns are given. Here, the lens is fed with the magnetic field parallel to the flat PEC plate (i.e., the broad side of the waveguide cross section parallel to the PEC plate, termed as H-field parallel polarization, or HPP for short). According to the figure, we see that the half Luneburg lens antenna has one main lobe for each fixed excitation, and this corresponds to the highly directive radiation direction. When the excitation is moved along the spherical surface, the main lobes will change accordingly. The main lobes of the half-spherical Luneburg lens antenna appear in the directions of $-15°$, $-45°$, and $-75°$ when the incident angle is $15°, 45°$, and $75°$, respectively. The observation shows that the proposed half Luneburg lens can transform a quasi-point source into a parallel beam. Moreover, the main radiation direction can be flexibly controlled by changing the position of the feed.

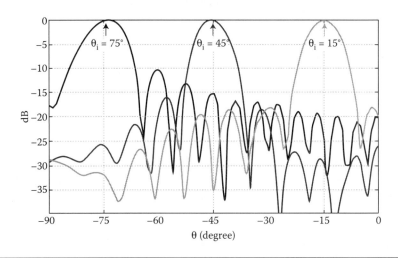

Figure 7.32 Simulated far-field radiation patterns for different incident angles.

7.6.4 Metamaterials Utilized

An easy way to realize nearly isotropic GRIN metamaterials is to use the drilled hole dielectric unit cells, as demonstrated in the previous sections. Such a unit cell can be treated as approximately isotropic and is easy to fabricate using the PCB technology. Figure 7.33 shows the geometries for these unit cells, and the corresponding refractive index variations with the PCB dielectrics and via-hole diameters. The geometry of the unit cell is shown in Figure 7.33a. In our design, a, b, and c are all set to 2 mm to obtain the required indices of refraction. The three kinds of unit cells are as follows. U1 is a $2 \times 2 \times 2$ mm^3 unit cell based on FR4 substrate; U2 has the same geometry but with a F4B substrate; and U3 is a $2 \times 2 \times 1$ mm^3 unit cell using F4B substrate. In the figure, the left panel represents U1 and U2, and the right panel is for U3. Figure 7.33b illustrates the indices of refraction for

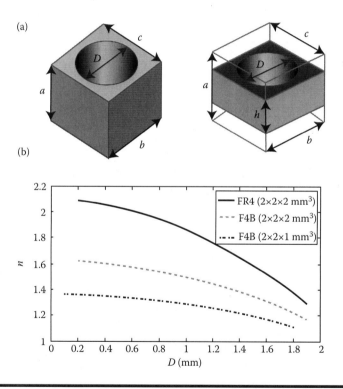

Figure 7.33 Unit cells and effective medium parameters for the 3D half Luneburg lens. (a) Unit cells with two different geometries. In the experiment, three unit cells, that is, U1, U2, and U3 are utilized. U1 is a 2 × 2 × 2 mm FR4 dielectric structure, U2 is a 2 × 2 × 2 F4B structure, and U3 is a 2 × 2 × 1 F4B structure. The left panel represents U1 and U2, and the right panel is for U3. (b) Refractive index variation with the diameter of the hole for three unit cells.

different unit cells with different via-hole diameters. Obviously, we find that the refractive index (n) decreases gradually as the diameter increases. Our studies show that though different polarizations will lead to effective refractive indexes with different values (anisotropic effect); however, the difference between each polarization is negligible, especially if one combines the three unit cells appropriately for the realization of different material parameters. That is to say, the unit cells can be roughly treated as isotropic structures. Using the above three kinds of unit cells, we can design arbitrarily 3D nearly isotropic and inhomogeneous metamaterials with the refractive index changing from 2.1 to 1.

7.6.5 Experiments

According to the distribution of refractive index given in Equation 7.19, the permittivity of Luneburg lens varies from 2 to 1 continuously from the spherical centre to surface. Hence, we only need the U2 and U3 unit cells to realize the corresponding index of refraction. In our experiments, the half Luneburg lens has a radius of

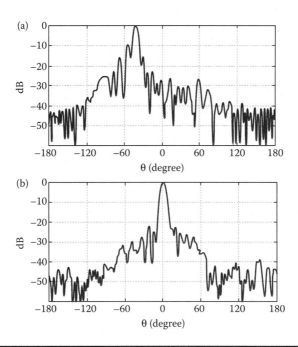

Figure 7.34 Measured far-field pattern for the 3D half Luneburg lens. (a) E plane with $\phi = 0$, HPP, and the wave is incident at $\theta = 45°$. (b) H plane with $\theta = -45°$, HPP polarization.

70 mm. A Ku-band (12.3–18 GHz) coax-to-waveguide transition with an aperture size of 16 × 8 mm is used as the feeding source, and a PEC plate is attached to the half Luneburg lens as the reflector. Two polarizations have been considered in our measurements by adjusting the polarizations of feeding device: HPP, in which the magnetic field is parallel to the PEC plate, and E-field parallel polarization (EPP), in which the electric field is parallel to the PEC plate. The measured far-field radiation patterns under HPP polarizations are illustrated in Figure 7.34. Here, Figure 7.34a shows the E plane patterns at 18 GHz. As expected, the waves from the feed ($\theta = 45°, \phi = 0$) is transformed into plane waves, and emit from the direction ($\theta = -45°, \phi = 0$) by propagating through the half Luneburg lens and be reflected by the PEC plate. The radiation patterns on the E plane demonstrate that the HPBWs are narrow, and the side lobes are lower than 10 dB within the whole Ku band. Figure 7.34b depicts the H plane far-field radiation patterns at 18 GHz, which also show the narrow HPBWs and low side lobes (less than 20 dB in the whole Ku band). Similar to HPP polarization, we also measure both E and H plane radiation patterns under the EPP polarization. Corresponding results are omitted in this part. We remark that the sidelobes in E and H planes are lower than 10 and 17 dB in the whole Ku band, which shows the good performance of the proposed lens antenna.

7.7 3D Maxwell Fisheye Lens

We have studied the 2D HMFE lens antenna in the previous section, which is realized using I-shaped metamaterials. In this section, a 3D HMFE lens antenna is briefly described. The lens antenna is realized using the same unit cells as the 3D Luneburg lens, that is, 3D hole-array materials.

7.7.1 Refractive Index Profile

Similar to the 2D case, the refractive index distribution for the 3D MFE lens is as follows:

$$n = n_0/(1 + r^2/R^2), \qquad (7.20)$$

where R is the radius of the lens and r is the radial variable in the spherical coordinate system. Note that refractive index profile given in Ma's work is wrong due to printing errors [28].

7.7.2 Ray Tracing Performance

Figure 7.35 shows the 3D ray tracing performance of the 3D HMFE lens with the finite element method. It is clearly shown that rays from a point source at the

224 ■ *Metamaterials*

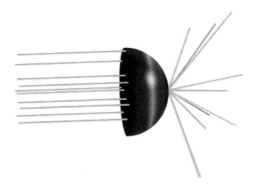

Figure 7.35 Ray tracing performance for the 3D HMFE lens.

spherical surface radiate isotropically in all directions. However, when they are captured by the lens, they will change directions accordingly. As a result, a bundle of highly directive rays can be found along the axis direction. This is the working principle for the HMFE lens.

7.7.3 Full-Wave Simulations and Experiments

We also make the corresponding full-wave simulations using CST Microwave Studio. In the simulation, the radius of the half-spherical fisheye lens is 60 mm, and the index of refraction follows Equation 7.20. Since one cannot use inhomogeneous materials directly in the CST environment, the lens is again first discretized into 30 spherical shells in the simulation model, and the thickness of each shell is about 2 mm.

The simulated far-field radiation pattern is presented in Figure 7.36. It is quite obvious that the 3D HMFE lens antenna have high directivities in both the E and H plane, and the side lobes are very low.

We then fabricate the half-spherical fisheye lens, whose radius is about 60 mm. The metamaterials used are the same as those used for the 3D half Luneburg lens. The feeding source used here is also the same as that in the Luneburg lens. The readers are kindly referred to the previous Section 7.6.4 for more information. However, for the fisheye lens, all three kinds of unit cells U1, U2, and U3 are required to realize the indices of refraction from 1 to 2.

The measured far-field radiation patterns of the fisheye lens on the E and H planes are also illustrated in Figure 7.36. It is obvious that the measured results also show high directivities. Moreover, compared with 3D half Luneburg lens in the previous section, the fisheye lens shows better agreements between simulations and measurements.

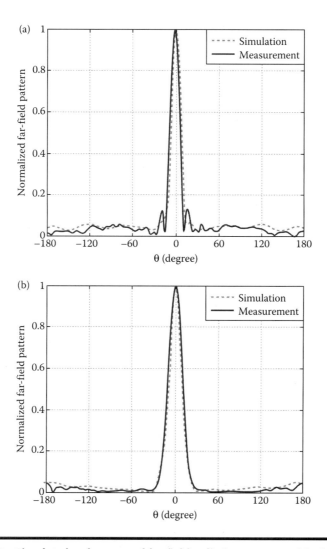

Figure 7.36 Simulated and measured far-field radiation patterns of the half fisheye lens antenna under HPP polarization at 15 GHz. (a) E plane. (b) H plane.

7.8 Electromagnetic Black Hole

In 2009, Narimanov and Kildishev proposed a theoretical design for an optical omnidirectional light absorber, which they termed as "optical black hole" [29]. They showed that all optical waves hitting the absorber were trapped and absorbed. Following this seminal work, Cheng and coworkers gave the first experimental demonstration of an omnidirectional EM absorber at the microwave band [30]. The fabricated device was composed of nonresonant and resonant metamaterial structures, which can trap and absorb EM waves coming from all directions without any reflections. Such a device could be used as a thermal emitting source and to harvest EM waves. In this section, we will show how their EM "black hole" works.

7.8.1 Refractive Index Profile

The optical "black hole" is composed of spherically or cylindrically symmetric inhomogeneous dielectric materials, where the incident light is guided and trapped in the device center. The core of the problem is to design the dielectric profile so as to guide and trap the incident light. Using the Hamiltonian description, the light propagation is governed by the following Hamiltonian function:

$$H = \frac{p_r^2}{2\epsilon(r)} + \frac{p_\theta^2}{2\epsilon(r)r^2}, \quad (7.21)$$

where p_r and p_θ are the light momentum in the radial and angular directions, respectively. If we treat a light ray as the trajectory of a point particle with unitary mass, then the above Hamiltonian can also be interpreted to describe the classical motion of such a particle in a central potential:

$$V_{\it eff}(r) = \frac{1}{2}(\omega/c)^2[\epsilon_0 - \epsilon(r)], \quad (7.22)$$

where c is the speed of light and ω is the angular frequency. Using the analogy between light propagation and classical mechanics, it is not difficult to obtain the expression for the potential, which is not unique for the current problem. As a result, the dielectric profile is not unique either. In terms of practical realization, however, the following dielectric profile is adopted [29]:

$$\epsilon(r) = \begin{cases} \epsilon_0, & r > R \\ \epsilon_0 (R/r)^2, & R_c < r < R \\ \epsilon_c + i\gamma, & r < R_c \end{cases} \quad (7.23)$$

where R and R_c are the exterior and interior radius of the shell, respectively, and the dielectric permittivity inside the shell center is deliberately designed to have losses, which are used to absorb the trapped light. Matching the parameters at R_c leads to the following expression:

$$R_c = R\sqrt{\frac{\epsilon_0}{\epsilon_c}}. \quad (7.24)$$

Figure 7.37 shows the dielectric profile for an EM black hole, where $\epsilon_0 = 1$, $R = 0.2$ m, $R_c = 0.05$ m, and $\epsilon_c = 16 + 20i$.

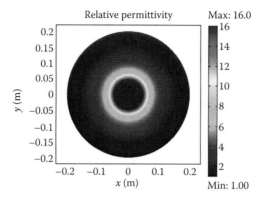

Figure 7.37 Dielectric permittivity distribution for the black hole.

7.8.2 Ray Tracing Performance

When the Hamiltonian is known, as is shown in the above equation, it is easy to trace the light rays inside the device. Figure 7.38 shows the ray tracing performance for different rays. It is very clear that when the rays hit the device, they are attracted into the device center and finally absorbed.

7.8.3 Full-Wave Simulations (Continuous Medium)

To demonstrate the performance of the "black hole" in the microwave frequency, we then make full-wave simulations of the device in the case of Gaussian-beam incidence. Figure 7.39a and b illustrates the magnitudes of simulated electric fields $|Ez|$ at 8 GHz when a Gaussian beam is incident on the device on-center and off-center,

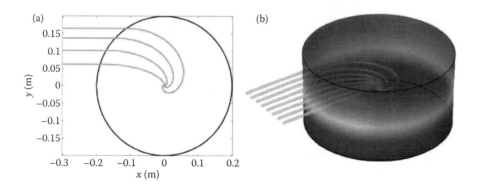

Figure 7.38 Ray tracing for the "black hole" with the dielectric profile given in Equation 7.23. (a) Home-made code calculation. (b) Comsol simulation.

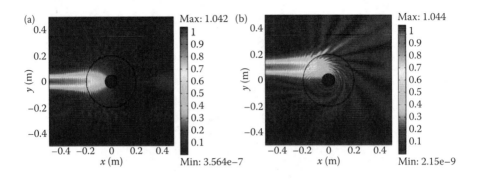

Figure 7.39 Simulated electric field distributions when a Gaussian beam is incident on the device on-center (a) and off-center (b). In the simulation, the working frequency is 8 GHz, $\epsilon_c = 16 + i20$, $R = 0.2$ m, $R_c = 0.05$ m, and $\epsilon_0 = 1$.

respectively. It is obvious that all on-center rays are directly attracted by the device with small reflections, and nearly all off-center rays bend in the shell region spirally and are trapped by the core. The ray tracing performance and full-wave simulations suggest that we can use the device as an excellent absorber at microwave frequencies.

7.8.4 Metamaterials Utilized

The EM black hole was first experimentally verified by Cheng and coworkers [30]. In their design, the continuous dielectric profile is first discretized into small blocks with the cylindrical grids, and each block is then realized using different metamaterial unit cells. In the design, the nonresonant I-shaped structure is chosen as the basic unit for the outer shell of the omnidirectional absorbing device [7], and the electric-field-coupled (ELC) resonator as the basic unit for the inner core, which has large permittivity and large loss tangent simultaneously near the resonant frequency [31]. Figure 7.40 shows the geometry of the two unit cells and the equivalent parameters. In Figure 7.40a, where the frequency is 18 GHz, it is shown that by changing the height of the I-shaped unit, the real part of permittivity $\text{Re}(\epsilon_z)$ ranges from 1.27 to 12.64 at 18 GHz, while the permeability components, $\text{Re}(\mu_r)$ and $\text{Re}(\mu_\phi)$, are always close to unity. From Figure 7.40b, we see that when the operating frequency is close to the resonant frequency of the ELC structure, that is, 18 GHz, a very lossy permittivity $\epsilon_z = 9.20 + i2.65$ can be obtained, and the permeability components are $\mu_r = 0.68 + i0.01$ and $\mu_\phi = 0.84 + i0.14$.

7.8.5 Experiments

Using the above unit cells, a laboratory EM black hole is fabricated by Cheng and coworkers using the PCB technology. In the realization, the radii of the absorber and

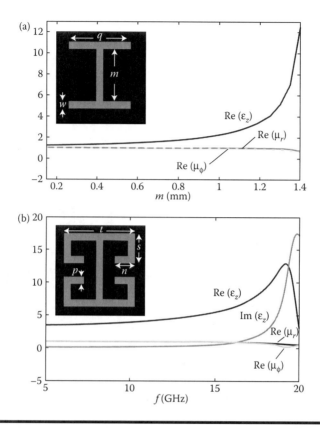

Figure 7.40 Effective material parameters for the I-shaped unit cell in the shell (a) and the ELC unit cell in the center (b). Note that $w = 0.15$, $q = 1.1$, $t = 1.6$, $g = 0.3$, $p = 0.15$, and $s = 0.65$ mm. (The source of the material Q. Cheng and T. J. Cui, An omnidirectional EM absorber made of metamaterials. *New J. Phys.*, 12: 063006, 3 June 2010, IOP Publishing & Deutsche Physikalische Gesellschaft is acknowledged.)

the lossy core are $R = 108$ mm and $Rc = 36$ mm, respectively. And the experimental prototype is measured in the 2D field mapping apparatus. The measurement results are presented in Figure 7.41. Note that since it is difficult to generate the Gaussian beam, the authors use a monopole probe with a corner reflector to produce the narrow beam. The simulation results are also demonstrated in Figure 7.41a and c. And this is in line with the experimental setup. As expected, the simulation and experimental results agree well. From Figure 7.41, we can draw the conclusion that the device is a good attractor and absorber of microwaves. Since the lossy core can transfer EM energies into heat energies, it is expected that the proposed device could find important applications in thermal emitting and EM wave harvesting.

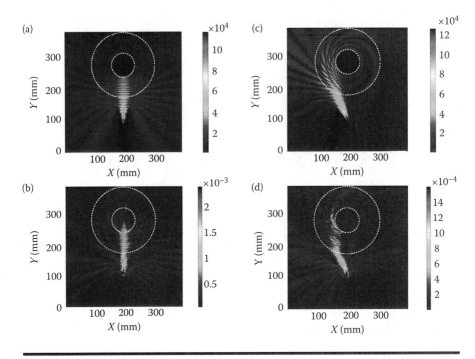

Figure 7.41 Distributions of electric fields $|E_z|$ for the designed omnidirectional absorbing device at the frequency of 18 GHz. An electric monopole is placed inside a corner reflector to produce the desired incident beam with finite width. The two circles stand for boundaries of the outer shell and the inner core. (a) The full-wave simulation result under the vertical incidence of the produced narrow beam. (b) The experimental result under the vertical incidence of the produced narrow beam. (c) The full-wave simulation result under the oblique incidence of the produced narrow beam. (d) The experimental result under the oblique incidence of the produced narrow beam. (The source of the material Q. Cheng and T. J. Cui, An omnidirectional EM absorber made of metamaterials. *New J. Phys.*, 12: 063006, 3 June 2010, IOP Publishing & Deutsche Physikalische Gesellschaft is acknowledged.)

References

1. M. Born and E. Wolf. *Principles of Optics*. Cambridge University Press, Cambridge, 1999.
2. U. Leonhardt. Optical conformal mapping. *Science*, 312: 1777, 2006.
3. P. A. Huidobro, M. L. Nesterov, L. Martin-Moreno, and F. J. Garcia-Vidal. Moulding the flow of surface plasmons using conformal and quasiconformal mapping. *New J. Phys.*, 13: 033011, 2011.
4. Y. G. Ma, N. Wang, and C. K. Ong. Application of inverse, strict conformal transformation to design waveguide devices. *J. Opt. Soc. Am. A*, 27: 968, 2010.

5. K. Yao and X. Jiang. Designing feasible optical devices via conformal mapping. *J. Opt. Soc. Am. B*, 28: 1037, 2011.
6. J. Li and J. B. Pendry. Hiding under the carpet: A new strategy for cloaking. *Phys. Rev. Lett.*, 101: 203901, 2008.
7. R. Liu, C. Ji, J. J. Mock, J. Y. Chin, T. J. Cui, and D. R. Smith. Broadband ground-plane cloak. *Science*, 32: 323, 2009.
8. H. F. Ma and T. J. Cui. Three-dimensional broadband ground-plane cloak made of metamaterials. *Nat. Commun.*, 1: 21, 2010.
9. Z. L. Mei, J. Bai, and T. J. Cui. Experimental verification of a broadband planar focusing antenna based on transformation optics. *New J. Phys.*, 13: 063028, 2011.
10. Z. L. Mei, J. Bai, and T. J. Cui. Illusion devices with quasi-conformal mapping. *J. Electromagn. Waves Appl.*, 24: 2561, 2010.
11. D. R. Smith, S. Schultz, P. Markos, and C. M. Soukoulis. Determination of effective permittivity and permeability of metamaterials from reflection and transmission coefficients. *Phys. Rev. B*, 65: 195104, 2002.
12. Z. L. Mei, J. Bai, and T. J. Cui. Gradient index metamaterials realized by drilling hole arrays, *J. Phys. D: Appl. Phys.*, 43: 055404, 2010.
13. J. Valentine, J. Li, T. Zentgraf, G. Bartal, and X. Zhang. An optical cloak made of dielectrics. *Nat. Mater.*, 8: 568, 2009.
14. H. F. Ma, X. Chen, X. M. Yang, H. S. Xu, Q. Cheng, and T. J. Cui. A broadband metamaterial cylindrical lens antenna. *Chinese Sci. Bull.*, 55: 2066, 2010.
15. A. D. Falco, S. C. Kehr, and U. Leonhardt. Luneburg lens in silicon photonics. *Opt. Express*, 19: 5156, 2011.
16. H. F. Ma, X. Chen, H. S. Xu, X. M. Yang, W. X. Jiang, and T. J. Cui. Experiments on high-performance beam-scanning antennas made of gradient-index metamaterials. *Appl. Phys. Lett.*, 95: 094107, 2009.
17. R. K. Luneburg. *Mathematical Theory of Optics*. Brown University Press, Providence, Rhode Island, pp. 189–213, 1944.
18. Q. Cheng, H. F. Ma, and T. J. Cui. Broadband planar Luneburg lens based on complementary metamaterials. *Appl. Phys. Lett.*, 95: 181901, 2009.
19. B. Zhou, Y. Yang, H. Li, and T. J. Cui. Beam-steering Vivaldi antenna based on partial Luneburg lens constructed with composite materials. *J. Appl. Phys.*, 110: 084908, 2011.
20. D. Schurig, J. B. Pendry, and D. R. Smith. Calculation of material properties and ray tracing in transformation media. *Opt. Express*, 14: 9794, 2006.
21. J. C. Miñano, P. Benítez, and A. Santamaría. Hamilton-Jacobi equation in momentum space. *Opt. Express*, 20: 9083, 2006.
22. J. C. Maxwell. Solutions of problems. *Camb. Dublin Math J.*, 8: 188, 1854.
23. B. Fuchs, Q. Lafond, S. Rondineau, M. Himdi, and L. L. Coq. Design and characterisation of half-Maxwell fish-eye lens antenna in 76–81 GHz band. *Electron. Lett.*, 42: 261, 2006.
24. B. Fuchs, O. Lafond, S. Rondineau, and M. Himdi. Design and characterization of half Maxwell fish-eye lens antennas in millimeter waves. *IEEE Trans. Microw. Theory Tech.*, 54: 2292, 2006.
25. B. Fuchs, O. Lafond, S. Palud, L. L. Coq, M. Himdi, M. C. Buck, and S. Rondineau. Comparative design and analysis of luneburg and half Maxwell fish-eye lens antenna. *IEEE Trans. Antennas. Propag.*, 56: 3058, 2008.
26. Z. L. Mei, J. Bai, T. M. Niu, and T. J. Cui. A half Maxwell fish-eye lens antenna based on gradient-index metamaterials. *IEEE Trans. Antenn. Propag.*, 60: 398, 2012.

27. X. Chen, H. F. Ma, X. Y. Zou, W. X. Jiang, and T. J. Cui. Three-dimensional broadband and high-directivity lens antenna made of metamaterials. *J. Appl. Phys.*, 110: 044904, 2011.
28. H. F. Ma, B. G. Cai, T. X. Zhang, Y. Yang, W. X. Jiang, and T. J. Cui. Three-dimensional inhomogeneous microwave metamaterials and their applications in lens antennas. *IEEE Trans. Antenn. Propag.*, 61: 2561, 2013.
29. E. E. Narimanov and A. V. Kildishev. Optical black hole: Broadband omnidirectional light absorber. *Appl. Phys. Lett.*, 95: 041106, 2009.
30. Q. Cheng and T. J. Cui. An omnidirectional electromagnetic absorber made of metamaterials. *New J. Phys.*, 12: 063006, 2010.
31. D. Schurig, J. J. Mock, and D. R. Smith. Electric-field-coupled resonators for negative permittivity metamaterials. *Appl. Phys. Lett.*, 88: 041109, 2006.

Chapter 8
Nearly Isotropic Inhomogeneous Metamaterials

Optical transformation devices realized with anisotropic inhomogeneous metamaterials are challenging in real design and fabrication. Alternatively, nearly isotropic inhomogeneous metamaterials can be adopted for a series of transformation devices. In these designs, quasi-conformal mapping is applied between the virtual space and the physical space. In this chapter, we will discuss some interesting broadband devices and antennas using nearly isotropic inhomogeneous metamaterials.

8.1 2D Ground-Plane Invisibility Cloak

In Chapter 6, we have talked about how to design invisibility cloaks using optical transformation. A cloak of invisibility is essentially a physical space that is filled with inhomogeneous media. It wraps an object to be concealed, and mimics a virtual space—the empty free space. Inside the cloak, an incident wave is controlled to travel around the object without any disturbance, and then recompose into its original manner at the other side of the cloak. Therefore, the incident wave will see the whole cloaked area as the free space. A 2D cloak was experimentally verified at microwave frequencies by Smith et al. in 2006 [1], followed by more explorations at different frequencies [2–6].

In practice, however, the theoretically fantastic cloak can never be perfect. One big challenge is that usually there are some singular points in the physical space. For example, in the spherical cloak shown in Figure 6.5, when the virtual space $r < b$ is transformed into the physical space $a < r' < b$, the center point is mapped to a sphere on the inner boundary. This means the parameters of transformation media can achieve infinite values (0 in this case) on the sphere of $r' = a$. This property is extremely difficult to be realized by existing materials or even artificial materials. Although carefully designed metamaterials are able to approach infinite values, in practice they exhibit extremely narrow band and potentially large reflections. In addition, the transformation media usually have anisotropic and inhomogeneous permittivity and permeability.

A new strategy for a special kind of invisibility cloak, the ground-plane cloak (or called as carpet cloak), is therefore proposed to circumvent these problems. In Reference 7, Li and Pendry reported the method of designing nearly isotropic and inhomogeneous metamaterials for a ground-plane cloak. Essentially, a ground-plane cloak is an EM device that conceals objects on the ground plane, as illustrated in Figure 8.1. In the physical space, the bulk covering the car represents the ground-plane cloak. If there exist conducting objects on the ground (e.g., a car here), the ground-plane cloak can cover the objects and decrease scatterings from them. As a result, to an outside observer, the reflection from the ground-plane cloak would appear to be the same as that from the virtual space, an empty ground. In this way, objects on the ground are undetectable.

In Section 6.5, we have studied the fundamentals of ground-plane cloak. Here, we will demonstrate in detail that generating a nearly isotropic ground-plane cloak is possible whenever the physical space and the virtual space have no strongly curved boundaries. As an example, a ground-plane cloak is designed at microwave frequencies to conceal a conducting bump on the ground plane. 2D section views of the physical space and the virtual space are shown in Figure 8.2. In the physical space in Figure 8.2a, a perfect electric conducting bump, represented with the curved

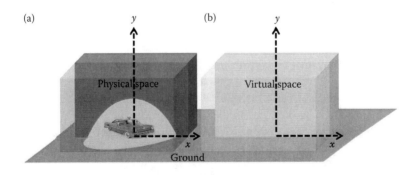

Figure 8.1 Design of a ground-plane cloak. Optical transformation is operated between the physical space (a) and the virtual space (b).

bold line, sits on the ground plane. In the virtual space in Figure 8.2b, however, the ground plane is empty and the space is filled with the air.

Optical transformation is carried out to map the physical space to the virtual space. Due to the presence of ground plane, a ground-plane cloak brings in significantly less bending and distortion of the incident waves, and therefore, a proper grid with near-orthogonal small cells can be generated in both the physical space and the virtual space. In each cell, the local coordinate system ((x, y) or (x', y')) is defined, as shown in Figure 8.2. Thereby, the two spaces are discretized and described with near-orthogonal local coordinates instead of global coordinates, and the optical transformation between them are based on a quasi-conformal coordinate mapping. In Reference 7, Li et al. used the modified Liao functional upon slipping boundary condition to generate the quasi-conformal mapping grids [8,9]. In

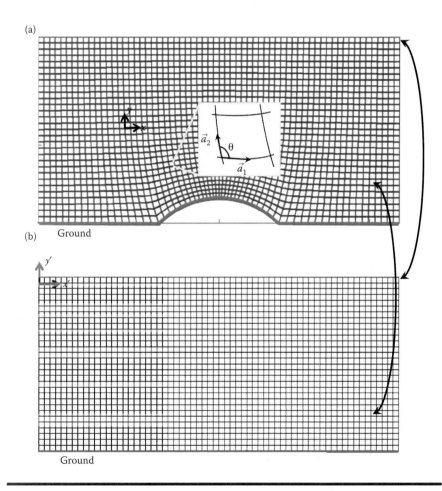

Figure 8.2 2D section view of (a) the physical space and (b) the virtual space.

practice, quasi-conformal mapping grids can be generated using Matlab, or grid-generating softwares such as "GenGrid" and "PointWise." Note that according to the Nyquist–Shannon sampling theorem, when converting a signal from a continuous space into a discrete space, to fully reconstruct the signal, the sampling rate should be no less than twice the maximum operating frequency. In other words, in the discretized space, the dimension of each cell should be less than half the wavelength inside the media.

Once the quasi-conformal mapping grids, as well as the local coordinates, are decided, the discrete optical transformation, also termed as the discrete coordinate transformation (DCT), is applied. Note that the DCT is operated between each pair of the local coordinate system in the physical space and the virtual space. Two pairs of the local coordinate systems are marked with arrow in Figure 8.2 as examples. Now, let us suppose that the coordinate transformation between the virtual space and the physical space is $x = x(x', y', z')$, $y = y(x', y', z')$, $z = z(x', y', z')$, where (x, y, z) are local coordinates in the physical space and (x', y', z') are local coordinates in the virtual space, as described in the grids in Figure 8.2. Based on the form invariant nature of Maxwell equations, there exists a relation that

$$\bar{\bar{\varepsilon}} = \frac{J\bar{\bar{\varepsilon}}'J^T}{det(J)}, \quad \bar{\bar{\mu}} = \frac{J\bar{\bar{\mu}}'J^T}{det(J)}, \quad (8.1)$$

where

$$J = \begin{pmatrix} \frac{\partial x}{\partial x'} & \frac{\partial x}{\partial y'} & \frac{\partial x}{\partial z'} \\ \frac{\partial y}{\partial x'} & \frac{\partial y}{\partial y'} & \frac{\partial y}{\partial z'} \\ \frac{\partial z}{\partial x'} & \frac{\partial z}{\partial y'} & \frac{\partial z}{\partial z'} \end{pmatrix}. \quad (8.2)$$

Since the virtual space is filled with the air, we have

$$\bar{\bar{\varepsilon}}' = \varepsilon_0 \mathbf{I},$$
$$\bar{\bar{\mu}}' = \mu_0 \mathbf{I}, \quad (8.3)$$

where $\mathbf{I}$ is the unitary matrix.

To simplify the problem, we now assume that the transformation is 2D, and thus the device is infinite in the z direction, normal to the $x - y$ plane defined in Figure 8.1. The bold reduction from 3D to 2D makes the design process much simpler. In practice, because many devices are symmetric, or work at a specified polarization, 3D transformation devices can be obtained by rotating or extending the corresponding 2D models to an axis. Examples will be presented in Sections 8.5 and 8.6.

In the 2D case, the Jacobian matrix has a simpler form:

$$J = \begin{pmatrix} \frac{\partial x}{\partial x'} & \frac{\partial x}{\partial y'} & 0 \\ \frac{\partial y}{\partial x'} & \frac{\partial y}{\partial y'} & 0 \\ 0 & 0 & 1 \end{pmatrix}, \qquad (8.4)$$

and for an E-polarized incident wave (electric field along the z direction), only six components (μ'_{xx}, μ'_{xy}, μ'_{yy}, μ'_{yx}, and ε'_{zz}) contribute, and the permittivity and the permeability become

$$\varepsilon_z \equiv \varepsilon_{zz} = \varepsilon_0 det(J)^{-1}, \qquad (8.5)$$

$$\bar{\bar{\mu}} = \frac{\mu_0}{det(J)} \begin{pmatrix} \left(\frac{\partial x}{\partial x'}\right)^2 + \left(\frac{\partial x}{\partial y'}\right)^2 & \frac{\partial x}{\partial x'}\frac{\partial y}{\partial x'} + \frac{\partial x}{\partial y'}\frac{\partial y}{\partial y'} \\ \frac{\partial y}{\partial x'}\frac{\partial x}{\partial x'} + \frac{\partial x}{\partial y'}\frac{\partial y}{\partial y'} & \left(\frac{\partial y}{\partial x'}\right)^2 + \left(\frac{\partial y}{\partial y'}\right)^2 \end{pmatrix}, \qquad (8.6)$$

respectively.

Instead of using the permittivity ε and the permeability μ to describe the transformation media in the physical space, we now use the anisotropy factor α and the averaged refractive index n, which have geometrical meanings in terms of the metric. On the one hand, the anisotropy factor is represented by the cell aspect ratio in the grids. Since the grids in both the virtual space and the physical space are near-orthogonal ones consisting of quadrilateral cells, the anisotropy factor α can be approximately considered as 1. On the other hand, the averaged refractive index n is defined by [7]

$$n = \sqrt{n_L n_T}, \qquad (8.7)$$

where n_L and n_T are the principal values of the refractive index tensor of the transformation media, which can be defined from Equations 8.5 and 8.6 as

$$n_L = n_{xx} = \sqrt{\mu_{yy} \varepsilon_{zz}} / \sqrt{\varepsilon_0 \mu_0} \qquad (8.8)$$

and

$$n_T = n_{yy} = \sqrt{\mu_{xx} \varepsilon_{zz}} / \sqrt{\varepsilon_0 \mu_0}. \qquad (8.9)$$

Consequently, the averaged refractive index becomes

$$n^2 = \sqrt{\mu_{yy} \mu_{xx}} \varepsilon_z / (\varepsilon_0 \mu_0). \qquad (8.10)$$

Equation 8.10 indicates that if $\mu_{xx}\mu_{yy} = \mu_0^2$, that is, if there is no magnetic dependence, then the refractive index of the transformation media, which determines the trace of the wave, can be realized by the permittivity alone, leading to an all-dielectric device. Next, we shall show that this condition is approximately satisfied if a certain grid is properly chosen in the distorted space, the physical space in Figure 8.2.

The explicit value of $\mu_{xx}\mu_{yy}$ from Equation 8.6 is

$$\mu_{xx}\mu_{yy} = \mu_0^2 \frac{[(\partial x/\partial x')^2(\partial y/\partial x')^2 + (\partial x/\partial x')^2(\partial y/\partial y')^2 + (\partial x/\partial y')^2(\partial y/\partial x')^2 + (\partial x/\partial y')^2(\partial y/\partial y')^2]}{[(\partial x/\partial x')^2(\partial y/\partial y')^2 - 2(\partial x/\partial x')(\partial y/\partial y')(\partial x/\partial y')(\partial y/\partial x') + (\partial x/\partial y')^2(\partial y/\partial x')^2]}.$$

(8.11)

According to Equation 8.11, the approximate condition $\mu_{xx}\mu_{yy} \simeq \mu_0^2$ is satisfied when at the same time

$$\frac{\partial x}{\partial y'} \simeq 0, \qquad (8.12)$$

$$\frac{\partial y}{\partial x'} \simeq 0. \qquad (8.13)$$

Since x and y are functions of both x' and y', Equations 8.12 and 8.13 can be also written using the chain rule as

$$\frac{\partial x}{\partial y'} = \frac{\partial x}{\partial y}\frac{\partial y}{\partial y'} \simeq 0, \qquad (8.14)$$

$$\frac{\partial y}{\partial x'} = \frac{\partial y}{\partial x}\frac{\partial x}{\partial x'} \simeq 0. \qquad (8.15)$$

It is easy to understand that the above condition can indeed be satisfied because we can generate a grid in the physical space with near-orthogonal cells such that

$$\frac{\partial x}{\partial y} \simeq 0, \qquad (8.16)$$

$$\frac{\partial y}{\partial x} \simeq 0. \qquad (8.17)$$

To illustrate how this orthogonality restriction can be approximately satisfied, let us look at a sample distorted cell in the inset of Figure 8.2. The 2 × 2 covariant metric g is defined to characterize the distortion as [8,10,11]

$$g = \begin{pmatrix} g_{11} & g_{12} \\ g_{21} & g_{22} \end{pmatrix}, \qquad (8.18)$$

$$g_{i,j} = \vec{a}_i \cdot \vec{a}_j \quad (i,j = 1, 2). \qquad (8.19)$$

$\vec{a}_1$ and $\vec{a}_2$ are the covariant base vectors defined in Figure 8.2, and θ is the angle between them. We quantify the orthogonality of the grid using the parameter θ for each cell, defined through

$$\cos\theta = \sqrt{\frac{g_{12}g_{21}}{g_{11}g_{22}}}. \qquad (8.20)$$

From a straightforward point of view, for a perfectly orthogonal grid, g is a unit matrix, $\cos\theta = 0$, $\theta = 90°$, and ultimately $\mu_{xx}\mu_{yy} = \mu_0^2$. In practice, as long as the angle parameter θ is distributed around $90°$, most of the local coordinates are considered as near-orthogonal. For example, the grid in Figure 8.2a is a near-orthogonal one. The condition of $\theta \simeq 90°$, $\mu_{xx}\mu_{yy} \simeq \mu_0^2$ is approximately satisfied, yielding an all-dielectric device with very minor sacrifice in performance.

Furthermore, the near-orthogonal property ensures an approximation of Equation 8.6 that

$$\bar{\bar{\mu}} = \frac{\mu_0}{\det(J)} \begin{pmatrix} \left(\frac{\partial x}{\partial x'}\right)^2 & 0 \\ 0 & \left(\frac{\partial y}{\partial y'}\right)^2 \end{pmatrix}. \qquad (8.21)$$

Because all cells are generated to be approximately square-shaped for conformal mapping, μ_{xx} and μ_{yy} have very similar values that are close to the unity. Accordingly, the relative permeability of the ground-plane cloak can be assumed to be isotropic and unity, and the effective relative refractive index in Equation 8.10 is only dependent on ε_z as

$$n^2 \simeq \varepsilon_z/\varepsilon_0 = \frac{1}{\det(J)}. \qquad (8.22)$$

Note that under the orthogonal condition of Equations 8.12 and 8.13, the refractive index profile of the ground-plane cloak can be directly retrieved from each cell within the grid, using Equation 8.4:

$$n^2 \simeq \frac{1}{(\partial x/\partial x')(\partial y/\partial y')} \simeq \frac{\Delta x' \Delta y'}{\Delta x \Delta y}, \qquad (8.23)$$

where Δx, Δy, $\Delta x'$, $\Delta y'$ are the dimensions of cells in the physical space and the virtual space, respectively. In this way, the refractive index map, or the permittivity map of a ground-plane cloak, is obtained.

It should be pointed out that a ground-plane cloak in free space includes less-than-unity permittivities. From a straightforward point of view, the squeezing of a space leads to higher permittivity while the expanding of a space results in lower permittivity when compared to that of the air. Usually, for the ground-plane cloak, materials with less-than-unity permittivities are required around the two ends of conducting objects. Since these areas are electrically small at operating frequencies, and those permittivity values are often very close to one, it is safe in practice to neglect these materials. We will discuss more on this issue in following sections.

The first ground-plane cloak was proposed and designed in 2008 [7]. In numerical demonstration, a wide-band Gaussian pulse was launched to stimulate a reflecting surface on the ground plane. The incident beam was deflected and split into two different angles due to the existence of the surface. In contrast, when the ground-plane cloak was added on the conducting surface, the field outside the cloak resembled the field as if there is only a flat ground plane.

One year later, the ground-plane cloak was experimentally verified at microwave frequencies [12]. In order to avoid less-than-unity permittivities, this ground-plane cloak is embedded in a higher dielectric region with refractive index $n = 1.331$. Figure 8.3a illustrates the whole refractive index distribution. The coordinate transformation region is shown within the box outline in black. Outside the box is the background material. We also observe an impedance matching layer (IML) surrounding the entire structure. Figure 8.3b shows the expanded view of the transformation region. This design leads to refractive index values for the ground-plane cloak ranging from 1.08 to 1.67. These values can be achieved with the use of nonresonant metamaterial elements, in particular, the I-shaped structures here. We have studied on nonresonant particles in Chapter 3. Here, Figure 8.4 includes photograph of the fabricated metamaterial sample, and the relation between the geometry of the I-shaped structure and the effective index. The ground-plane cloak is finally fabricated on copper-clad printed circuit board with FR4 substrate whose dielectric constant is $3.85 + i0.02$. The completed sample is 500 by 106 mm with a height of 10 mm at X band. This 2D ground-plane cloak is tested in a near-field microwave scanning system inside a planar waveguide, with the polarization of the waves being transverse electric.

A large-area electric field map is shown in Figure 8.5. A nearly collimated microwave beam is launched as the incidence. A flat ground plane produces a nearly perfect reflection of the incident beam, as shown in Figure 8.5a. Once a conducting perturbation is located on the ground plane, strong scatterings are observed, as shown in Figure 8.5b. However, when the ground-plane cloak is added to surround the perturbation, the reflected beam is restored, as if the ground plane were flat, as shown in Figure 8.5c. This impressive performance is observed at 13, 14, 15,

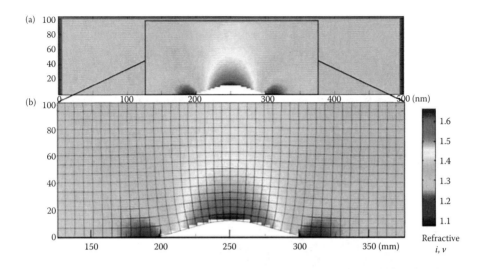

Figure 8.3 (a) Metamaterial refractive index distribution. The coordinate transformation region is shown within the box outlined in black. The surrounding material is the higher index embedding region and the IMLs. (b) Expanded view of the transformation optical region in which the mesh lines indicate the quasi-conformal mapping. (From R. Liu et al. Broadband ground-plane cloak. *Science*, 323: 366, 2009. Reprinted with permission of AAAS.)

and 16 GHz, respectively, proving that the ground-plane cloak can exhibit a broad frequency bandwidth owing to the use of nonresonant metamaterial elements.

Although the first prototype of ground-plane cloak has been proved successful in experiment, this design is rather complicated. Specifically, a large number of elements are required, and background material is involved. Therefore, compact-sized ground-plane cloaks in the free-space background have been explored. We will show some examples in the next section.

8.2 2D Compact Ground-Plane Invisibility Cloak

2D ground-plane cloak has the merits of broadband performance and low loss when compared to arbitrary free-space cloaks. It is highly expected to be involved in practical applications such as the stealth radomes. Although the experimental demonstration at microwave frequencies is successful [12], there exist two restrictions. First, the ground-plane cloak is embedded in a background material with a refractive index of 1.331. In other words, such a ground-plane cloak cannot work in the air. Second, the ground-plane cloak is electrically large. For example, the fabricated ground-plane cloak in Reference 12 is 500 by 106 mm with a height of

242 ■ *Metamaterials*

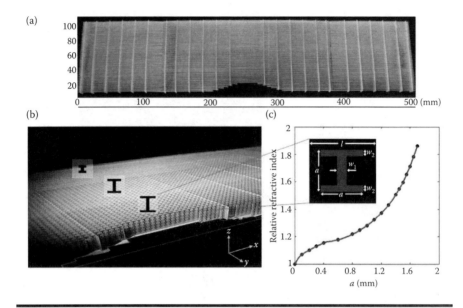

Figure 8.4 (a) Photograph of the fabricated metamaterial sample. (b) The model of the fabricated carpet cloak. (c) The design of the nonresonant elements and the relation between the unit cell geometry and the effective index. (From R. Liu et al. Broadband ground-plane cloak. *Science*, 323: 366, 2009. Reprinted with permission of AAAS.)

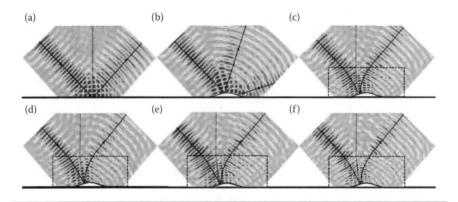

Figure 8.5 Measured field mapping (E field) of the ground, perturbation, and ground-plane cloaked perturbation. (a) Collimated beam incident on the ground plane at 14 GHz. (b) Collimated beam incident on the perturbation at 14 GHz. (c) Collimated beam incident on the ground-plane cloaked perturbation at 14 GHz. (d–f) Collimated beam incident on the ground-plane cloaked perturbation at 13, 15, and 16 GHz, respectively. (From R. Liu et al. Broadband ground-plane cloak. *Science*, 323: 366, 2009. Reprinted with permission of AAAS.)

10 mm at X band. From an engineering point of view, these restrictions make the ground-plane cloak null in real applications.

Several compact-sized ground-plane cloaks in the free-space background have therefore been proposed. The following simplifications are carefully applied to the full-parameter ground-plane cloak so as to obtain easier design and less expensive fabrication.

First of all, in order to reduce the size of a ground-plane cloak, regions with less-than-unity refractive index are treated as free space. Let us take a look at an example reported in Reference 13. A ground-plane cloak is designed in free space based on the technique presented in Reference 7, and refractive index distribution is plotted in Figure. 8.6a. The complete ground-plane cloaking region is 250 by 100 mm.

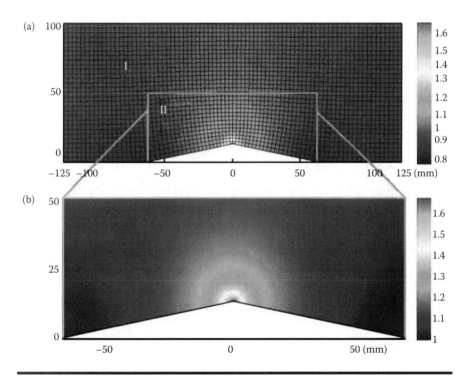

Figure 8.6 Design for a compact-sized ground-plane cloak in free-space background. (a) Metamaterial refractive index distribution of the complete carpet cloaking region in which the mesh lines indicate the quasi-conformal mapping. The compact-sized cloaking region is shown within the box. (b) Expanded view of the compact-sized cloaking region in which the refractive indices below one are all designed as 1. (H. F. Ma et al. Compact-sized and broadband carpet cloak and free-space cloak. *Opt. Express*, 17: 19947, 2009. With permission of Optical Society of America.)

Since the ground-plane cloak is embedded in the air, the optical transformation generates a region (region *I* here) where refractive indices are close to 1. Such a region can be simplified as free space, and in this way the cloak is reduced to region *II*, at the price of slightly increased reflection from the cloak surface. This simplification has been validated numerically in Reference 13 by comparing the near-electric-field distributions and far-field patterns of the full-parameter cloak and the simplified cloak at 10 GHz. Simulated results prove that the simplified ground-plane cloak has very similar performance as the full-parameter cloak to conceal the hidden object, and only slightly improved side lobes are observed in the far field. This fact has also been demonstrated in Reference 14. Note that since the ground-plane cloak is designed in free space, IMLs are no longer needed.

Second, the ground-plane cloak is simplified to be made of only a few blocks of isotropic all-dielectric materials. A ground-plane cloak created by quasi-conformal mapping can have a very detailed permittivity map if the transformation space is discretized with a fine grid. Although all required permittivity values can be realized using metamaterials, it is rather complicated to fabricate many elements and distribute them into the assigned profile. A postprocessing is to simplify the permittivity map, as is studied in Reference 15. The fundamental limitation of resolution for a transformation device is based on the Nyquist–Shannon sampling theorem [16]. Sampling is the process of converting a signal (e.g., a function of continuous time or space) into a numeric sequence (a function of discrete time or space) [17]. Shannon's version of the theorem states: "If a function $f(t)$ contains no frequencies higher than W cps, it is completely determined by giving its ordinates at a series of points spaced 1/2W seconds apart." Figure 8.7 interprets this theorem in the spatial domain. When propagating in the physical space, at a fixed time, the EM wave is a function of continuous space. Transformation media with a high resolution discretize the space using a high sampling rate while low-resolution ones discretize the space using a low sampling rate. If the electromagnetic wave operates in a frequency band from f_L to f_H, to accurately reconstruct information of the wave, the sampling

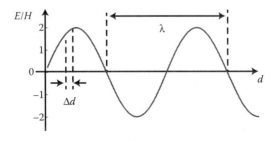

Figure 8.7 EM wave propagates in the spatial domain at a fixed time.

rate f_s should satisfy

$$f_s > 2f_H. \tag{8.24}$$

Because

$$f_s = \frac{1}{\Delta d}, \quad f = \frac{1}{\lambda}, \tag{8.25}$$

the resolution of the discrete space should satisfy

$$\Delta d < \frac{1}{2}\lambda_H. \tag{8.26}$$

Δd is the dimension of blocks in the permittivity map of a transformation device, and λ_H is the minimum wavelength within the frequency band. According to Equation 8.26, as long as the dielectric blocks are smaller than half a wavelength at the operating frequency, the simplified device can maintain the property of a high-resolution one. When the operating frequency goes higher, the resolution of dielectric map should increase accordingly.

For the first ground-plane cloak reported in Reference 12, the entire sample region is divided into 2-by-2-mm squares, requiring more than 10,000 elements, about 6000 of which are unique. For the compact ground-plane cloak reported in Reference 13, the sample is divided to 3-by-3-mm squares at the same operating frequency. Although the resolution of the compact ground-plane cloak is lowered, it has been proved that the sample with a low-resolution map retains similar performance as those with high-resolution ones within a limited bandwidth of operation, while accompanied by predictable degradations of the near-field distribution and far-field pattern above a "cut-off" frequency. A very simple 2D compact ground-plane cloak composed of 34.25-by-30-mm dielectric blocks has also been reported [14]. A down-sampled selection of six dielectric blocks, with refractive indexes of 1.08, 1.14, and 1.21, are involved in this design. Polyurethane foam mixed with different ratios of barium titanate is used to produce the required range of refractive indexes, as illustrated in Figure 8.8.

After these two steps of simplification, a 2D compact ground-plane cloak is achieved. In Reference 13, a prototype with the size of 125 by 50 mm is fabricated to conceal a triangular metallic bump with the height of 13 mm and the bottom of 125 mm. I-shaped resonators working off the resonance are chosen as composing elements. They are printed on F4B substrate whose thickness is 0.25 mm, dielectric constant is 2.65, and loss tangent is 0.001, as shown in Figure 8.9. The permittivity is approximately a function of the dimension h that is defined in Figure 8.9b, and meanwhile, the permeability can be considered as 1. The resulted refractive index ranges from $n = 1.07$ to 1.87 at 10 GHz, satisfying the requirement of index distribution in Figure 8.6. IMLs are not involved in this design because the cell-to-cell change in dimension is minor, and therefore the impedance is matched gradually in the whole cloak over the entire measurement frequency range.

Since only nonresonant metamaterials are involved, the compact ground-plane cloak is expected to obtain broadband and low-loss properties. Similarly, a near-field scanning system is built up within a planar waveguide, with the wave polarization restricted to transverse electric. The near-field distributions, as well as the far-field patterns of the ground-plane cloak, are measured at X band. Figure 8.10a plots the field distribution when an EM wave is incident on an empty ground plane at an angle of 45° with respect to the normal. Once an uncloaked triangular bump is located on the ground, as drawn in Figure 8.10b, the reflected beam shows strong scatterings. To hide the bump, the designed ground-plane cloak is added in Figure 8.10c. Again, a single reflected beam is observed at 10 GHz. Corresponding far-field patterns for the three scenarios in Figure 8.10a–c are also plotted in

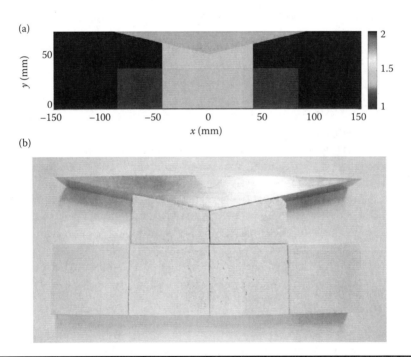

Figure 8.8 (a) Dielectric 2D map of a ground-plane cloak composed of 4 × 2 blocks. The dielectric blocks are rectangles of dimension 34.25 by 30 mm. The fabricated cloak is shown in (b). An aluminum triangle of height 16 mm and base 144 mm on the ground plane is hidden by the ground-plane cloak. (The source of the material D. Bao et al. All-dielectric invisibility cloaks made of $batio_3$-loaded polyurethane foam. *New J. Phys.*, 13: 103023, Copyright 2011, IOP Publishing & Deutsche Physikalische Gesellschaft. CC BY-NC-SA, is acknowledged.)

Nearly Isotropic Inhomogeneous Metamaterials ■ 247

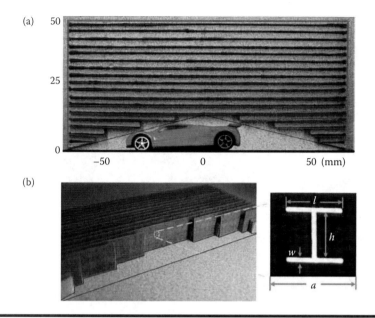

Figure 8.9 (a) Photograph of the fabricated metamaterial ground-plane cloak and the concealed car model. (b) The design of nonresonant elements. The dimensions of the metamaterial unit cells are $a = 3$ mm, $w = 0.2$ mm, $l = h + 2w$ mm, and h varying from 0 to 2.2 mm. (From H. F. Ma et al. Compact-sized and broadband carpet cloak and free-space cloak. *Opt. Express*, 17: 19947, 2009. With permission of Optical Society of America.)

Figure 8.10g–i. For the empty ground plane and the cloaked bump, the beam profiles show a very similar single peak; while for the bump alone, a multipeak far-field pattern is observed. The cloaking behavior is also confirmed in measurements from 11 to 13 GHz, as shown in Figure 8.10d–f. In this way, the broadband properties of the compact ground-plane cloak is demonstrated in experiment.

Realization of the 2D compact ground-plane cloak is a major step toward the real applications of invisibility cloaks. Furthermore, a 3D device can be easily achieved by rotating/extending the 2D model to an axis. We will study how to design and realize a 3D ground-plane cloak in Section 8.5.

8.3 2D Ground-Plane Illusion-Optics Devices

The invisibility cloak works as a "radar-cheating" device because it conceals a real object as if there is nothing existing. More generally, all devices that confuse the detecting radar could be considered as illusion devices. In Section 6.5, the concept of illusion-optics devices has been introduced and some examples have been

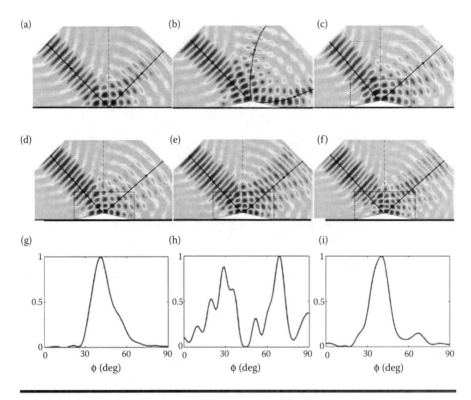

Figure 8.10 Measured electric field mapping of (a) the ground plane, (b) triangular metallic bump, and (c) ground-plane cloaked bump when collimated beam is incident at 10 GHz. The rays display the wave propagation direction, and the dashed line indicates the normal of the ground in free space and that of the ground-plane cloak in the transformed space. (d–f) Collimated beam incident on the ground-plane cloaked bump at 11, 12, and 13 GHz, respectively. Far-field patterns when collimated beam is incident at 10 GHz on (g) the ground plane, (h) the triangular metallic bump, (i) the ground-plane cloaked bump. (From H. F. Ma et al. Compact-sized and broadband carpet cloak and free-space cloak. *Opt. Express*, 17: 19947, 2009. With permission of Optical Society of America.)

given. Here, we focus on a special class of illusion devices, the 2D ground-plane illusion-optics devices, which can be realized using nearly isotropic metamaterials with gradient refractive indices.

Assuming that there is an arbitrarily shaped bump (e.g., a trapezoidal bump here) on the ground-plane, marked as *Region C* in Figure 8.11c. An illusion device is expected to misguide the incident wave and consequently make the detecting radar locate the object at a wrong position (e.g., *Region A* in Figure 8.11a) and with a wrong shape (a small triangular pit here). The space in Figure 8.11a is defined as the virtual space and the one in Figure 8.11c is defined as the physical space.

To achieve a nearly isotropic physical space, which serves as the target illusion device, quasi-conformal-mapping transformation optics (QCTO) is also adopted. The difference between the design of a ground-plane cloak and a ground-plane illusion device is that the former one requires only one step of QCTO and the latter one requires a two-step composite transformation. Instead of making a mapping directly from the virtual space to the physical space, an intermediate space as shown in Figure 8.11b is inserted. In the first step, a similar QCTO procedure as presented in Section 8.1 is carried out from the virtual space to the intermediate space, and the rectangular structure in Figure 8.11b becomes equivalent to the irregular one in the virtual space in Figure 8.11a. In the second step, another QCTO is utilized from the intermediate space to the final physical space, and this produces a different bump on the bottom side of the rectangle, as illustrated in Figure 8.11c. Finally, from the TO theory, we know that the bump in Figure 8.11c with surrounding materials will be identified as a pit in Figure 8.11a by EM waves due to the equivalence between the two spaces.

In 2D circumstance, the material parameters in the physical space can be calculated using Equations 8.1, 8.22, and 8.23 twice. Mei et al. presented a 2D nearly isotropic illusion device in Reference 18. In this work, the transformation region is 250 by 100 mm, the irregular pit in Figure 8.11a is a triangular pit and the bump in Figure 8.11c is a trapezoidal bump. The obtained refractive index distribution in the physical space after the two consecutive mappings is given in Figure 8.11d. Again, all

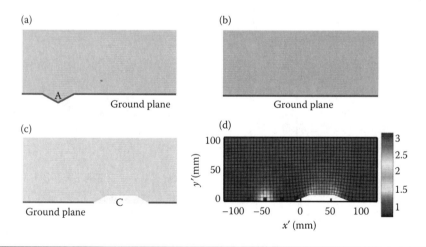

Figure 8.11 Quasi-conformal mappings between (a) the virtual space, (b) the intermediate space, and (c) the physical space. (d) The actual refractive index distribution in the physical space after the two consecutive mappings. (Reprinted with permission from Mei, Z. L., J. Bai, and T. J. Cui. *J. Electromagn. Waves Appl.*, 24, 2561. Copyright 2010, Taylor & Francis.)

250 ■ *Metamaterials*

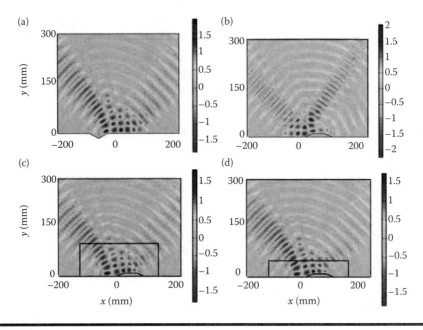

Figure 8.12 The electric field distribution under a Gaussian beam incidence, in which the incident angle is $\pi/4$, the waist width is two wavelengths, and the waist center is located at the origin. (a) With a bare triangular pit. (b) With a bare trapezoidal bump. (c) With the trapezoidal bump plus the illusion device. (d) Same as (c) but the device is halved in height and the GRIN materials are adjusted to have a loss tangent of 0.01. The working frequency is 10 GHz. (Reprinted with permission from Mei, Z. L., J. Bai, and T. J. Cui. *J. Electromagn. Waves Appl.*, 24, 2561. Copyright 2010, Taylor & Francis.)

the parameters below 1 can be approximately set to 1 at the price of relatively acceptable performance degradation. In addition, since most refractive index is close to 1 in the outer part of the device, a size reduction is possible and the device becomes more compact.

The performance of the 2D illusion device has been demonstrated by numerical calculations using Comsol Multiphysics and homemade MATLAB codes. Figure 8.12 shows distributions of the total electric fields in the near-field region for different configurations. In Figure 8.12a, the incident wave, a Gaussian beam from the upper-left corner, illuminates the ground plane with a bare triangular pit. Most reflected field assembles around the 45° direction. In Figure 8.12b, the same incident wave illuminates the ground plane with a bare trapezoidal bump. The field distribution is totally different. Most reflected field assembles in the 30° and 60° direction. However, once the trapezoidal bump is surrounded by the designed illusion device, as shown in Figure 8.12c, the overall field distribution coincides with that of the bare triangular pit again. The field distribution outside the device looks the same as that of the triangular pit, though there are differences inside the device. In other words, one is able to change a trapezoidal bump into a triangular pit using the illusion device. Furthermore, a more practical and compact illusion device is tested. The height of the illusion device is halved and a loss tangent of 0.01 is added to the composing materials. The simulation result is given in Figure 8.12d. It is observed that the full illusion device and the halved one produce very similar field distributions, as we expected.

It should be pointed out that, in practice, various combinations of *Region A* in Figure 8.11a and *Region C* in Figure 8.11c are possible, leading to devices with different functions other than the one reported in Reference 18. For instance, if *Region A* and *Region C* have the same shape and different size, the illusion device works as a "shrinking device" or an "enlarging device." If *Region A* and *Region C* lie at different positions on the ground plane, the illusion device obtains an additional shift effect and as a result a target is wrongly detected with a different profile and at a different position. Furthermore, if *Region A* is replaced with two subregions filled with different dielectrics, it is possible to change the scattering pattern of a metallic bump in *Region C* into two dielectric blocks. More functions of the illusion device will be presented in Chapter 9 when we talk about 3D spacial illusion-optics devices.

8.4 2D Planar Parabolic Reflector

For the reason that the optical transformation is inversible, there are many underlying applications other than the invisibility cloak. From an engineering point of view, the quasi-conformal mapping can result in nearly isotropic and inhomogeneous metamaterial devices with low cost and readiness for mass production. In these designs, instead of using a physical space with distorted coordinate systems to mimic a virtual space with orthogonal coordinate systems, one is able to use a physical space with orthogonal coordinate systems to mimic a virtual space with distorted coordinate systems, so as to obtain specific functions. For example, in antenna systems, many widely used devices have curved surfaces, such as parabolic reflectors and convex lenses. Although improvements have been reported to modify the existing parabolic reflectors and convex lenses, the required metamaterials are usually anisotropic, resonant, and consequently difficult to apply [19,20]. Here, in this section, we are going to use the technique of quasi-conformal mapping to design an equivalent device that operates in the same manner as a parabolic reflector but has a planar profile. In the virtual space perceived by the EM waves, the parabolic reflector has a curved surface, contains homogeneous and isotropic materials (the air), and can be described with distorted coordinate systems. By using appropriate quasi-conformal mapping grids, these distorted coordinate systems are mapped to the physical space, where the new device is

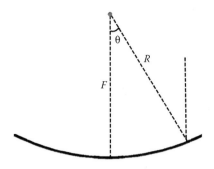

Figure 8.13 Geometry of the parabolic surface.

planar, contains inhomogeneous but nearly isotropic materials, and is easy to be realized.

To design a planar parabolic reflector, let us suppose that a metallic parabolic reflector is placed in the free space. The geometry of the parabolic surface is given in Figure 8.13, where

$$R = \frac{2F}{1 + \cos\theta}. \quad (8.27)$$

To carry out the optical transformation, a curved parabolic reflector is located in the virtual space, represented as the PEC surface in Figure 8.14a. In the physical space shown in Figure 8.14b, there is a flat PEC surface instead, covered by transformation media. The flat PEC surface and the transformation media together compose a planar parabolic reflector. The two spaces have the same boundaries to the air. Their south boundaries (defined in Figure 8.14) are different because the PEC surfaces are different. Note that in Figure 8.14, the focal point is outside the transformation region, and hence, in the physical space, a source at the focal point is not embedded in the transformation media.

Near-orthogonal grids are generated in both the physical space and the virtual space for quasi-conformal mapping, as plotted in Figure 8.14. Both spaces are discretized into small cells, and in each cell there is a set of local coordinates defined. In Section 8.1, we talked about how to apply optical transformation to design a 2D ground-plane cloak. Here, a similar designing procedure is employed to create an all-dielectric planar parabolic reflector. Because the original parabolic reflector is symmetric, we can simply design a 2D planar reflector and straightforwardly obtain a 3D model by rotating the 2D model to the axis.

An example of the 2D planar parabolic reflector has been reported in Reference 21. The planar reflector is designed based on a conventional parabolic reflector with an aperture of 180 mm and a focal length of 109 mm. The permittivity distribution is calculated and plotted in Figure 8.15a. Applying the two steps

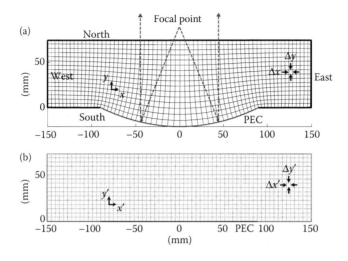

Figure 8.14 (Section view) (a) The virtual space with distorted coordinates. A parabolic reflector is placed in the free space, as illustrated by the curved PEC. (b) The physical space with orthogonal coordinates. The parabolic reflector is replaced with a flat PEC. (From W. Tang et al. Discrete coordinate transformation for designing all-dielectric flat antennas. *IEEE Trans. Antenn. Propag.*, 58: 3795, © 2010 IEEE.)

of simplification, as we mentioned in the previous section, the resolution of the permittivity map is reduced and the complexity is gradually depressed. Finally, a relatively low resolution sampling map of 16 by 3 blocks as shown in Figure 8.15d is achieved. The size of each block is 18.75 mm, which is half a wavelength in free space at 8 GHz.

Instead of commercial softwares, the finite-difference time-domain (FDTD) method-based in-house codes have been used in Reference 21 to verify the performance of the planar parabolic reflector [22–25]. First, a plane wave with E polarization comes along the $+y$ direction at 8 GHz, in order to compare the low-resolution planar reflector shown in Figure 8.15d with the parabolic reflector in terms of the focal length. Figure 8.16a and b illustrates the real part of the E_z field when the planar reflector and the conventional parabolic reflector are applied, respectively. Their focal lengths are measured as the distances from the center of the PEC surface to the narrowest envelope marked with the dashed gray line in (a) and (b). The focal lengths in (a) and (b) are 102.6 and 102.7 mm, respectively, very close to each other. In addition, slightly different reflections are observed on two sides of the reflectors. These reflections are mainly caused by the neglecting of less-than-unity permittivities during simplifications. In Figure 8.16c and d, the electric field distributions when these two reflectors are fed by a small horn are

254 ■ *Metamaterials*

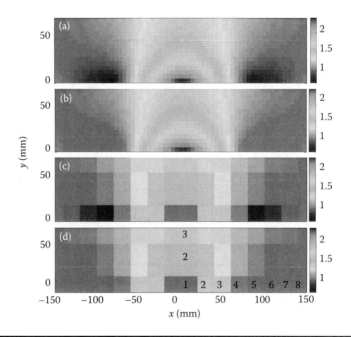

Figure 8.15 (a) Relative permittivity map consisting of 64 × 16 blocks. (b) Relative permittivity map consisting of 64 × 16 blocks, without less-than-unity values. (c) Relative permittivity map consisting of 16 × 3 blocks. (d) Relative permittivity map consisting of 16 × 3 blocks, without less-than-unity values. (From W. Tang et al. Discrete coordinate transformation for designing all-dielectric flat antennas. *IEEE Trans. Antenn. Propag.*, 58: 3795, © 2010 IEEE.)

presented, respectively. Very similar field distributions indicate that the planar reflector maintains the function of transforming an incident cylindrical wave to a plane wave and consequently obtains highly directive beams. Such beam-forming performance are also observed at 4, 6, 10, and 12 GHz. Therefore, the planar parabolic is numerically demonstrated to operate within a wide frequency band.

Experimental verification of a similar broadband planar parabolic reflector has been completed by Mei et al [26] in the X band. The reflector is designed using quasi-conformal optical transformation, simplified in a similar way as we discussed in Section 8.2, and realized with nonresonant I-shaped metamaterial unit cells. The manufactured prototype possesses an aperture of 200 mm and a focal length of 75 mm, and is composed of 4 by 4 mm small squares whose refractive index varies from 1 to 1.47. Finally, the fabricated reflector has a total size of 200 by 80 mm^2 and a height of 12 mm (3 unit cells in the vertical direction), as shown in Figure 8.17a. The metamaterial sample is made of I-shaped structures printed on F4B substrate (whose permittivity is 2.65 with the loss tangent 0.001 at 10 GHz)

Nearly Isotropic Inhomogeneous Metamaterials ■ 255

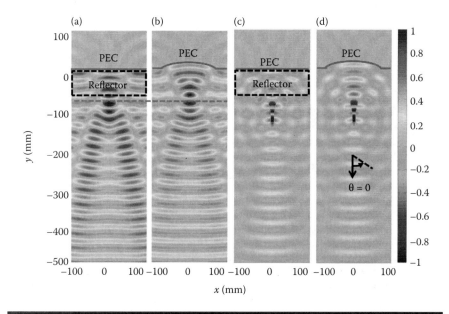

Figure 8.16 Real part of the E_z field at 8 GHz. (a) A plane wave along the y direction illuminates a low-resolution planar parabolic reflector shown in Figure 8.15d. The focal length (measured from the dashed gray line to the center of the PEC) is 102.6 mm. (b) The same plane wave illuminates the conventional parabolic reflector. The focal length is 102.7 mm. (c) A small horn antenna is applied at the focal point to feed the planar parabolic reflector. (d) A small horn antenna is applied at the focal point to feed the conventional parabolic reflector. (From W. Tang et al. Discrete coordinate transformation for designing all-dielectric flat antennas. *IEEE Trans. Antenn. Propag.*, 58: 3795, © 2010 IEEE.)

using the lithography technology, and the supporting frame is a hand-made foam structure with permittivity close to one.

To measure the performance of the planar reflector, a near-field scanning system in Figure 8.17b, which is very similar to the one used in the experiment for a 2D ground-plane cloak, is used to map the electric field distribution within a planar waveguide. The measurement results are illustrated in Figure 8.18. For the original parabolic reflector, the near-field distribution at 10 GHz is given in Figure 8.18a. The incident wave from a line source is transformed to a quasi-plane wave effectively. A similar field distribution is observed in Figure 8.18b when the planar reflector is located instead of the original one. The broad bandwidth of the planar reflector is illustrated by the field maps taken at 8.5 (not shown here for the sake of clarity) and 11.5 GHz (Figure 8.18c) in the figure, which show similar performances as those of the map at 10 GHz. To quantitatively investigate the measured performance, far-field radiation patterns have been calculated using the measured data and the results

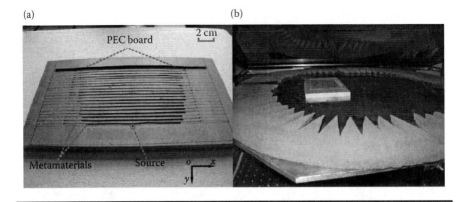

Figure 8.17 (a) Photograph of the fabricated planar antenna. (b) Experimental setup for the near-field measurement. (The source of the material Z. L. Mei, J. Bai, and T. J. Cui. Experimental verification of a broadband planar focusing antenna based on transformation optics. *New J. Phys.*, 13: 063028. Copyright 2011, IOP Publishing & Deutsche Physikalische Gesellschaft. CC BY-NC-SA, is acknowledged.)

are given in Figure 8.18d. First of all, at 10 GHz, radiation patterns for the simulated planar reflector (see the black solid curve) and the measured one (see the solid curve with star) agree well in the peak radiation direction, and only small differences are found near the side lobes, which suggests the correctness of the fabrication technique. Second, relatively large differences are found between the measured parabolic reflector (gray solid curve with diamond) and the planar one. This is mainly due to the simplification of parameters and reduction of size. In fact, this difference represents an essential trade-off between the performance and the structural complexity in practical applications. The measured far-field radiation patterns at 8.5 GHz (see the dash-dotted curve) and 11.5 GHz (dashed curve) again confirm the broadband performance of the planar reflector.

In addition, broadband all-dielectric beam-steerable planar reflectors using quasi-conformal optical transformation have also been reported [27]. Instead of moving or tilting the feed/reflector, one is able to manipulate the reflected emission by tuning the permittivity of dielectrics. This method has a merit of maintaining the profile of the feed-reflector combined system; therefore, it is potentially applicable for mounting reflectors on the platform such as the surface of an aircraft.

8.5 3D Ground-Plane Invisibility Cloak

In previous sections in this chapter, we introduced some transformation-optics-based devices. They are able to be constructed using nearly isotropic metamaterials.

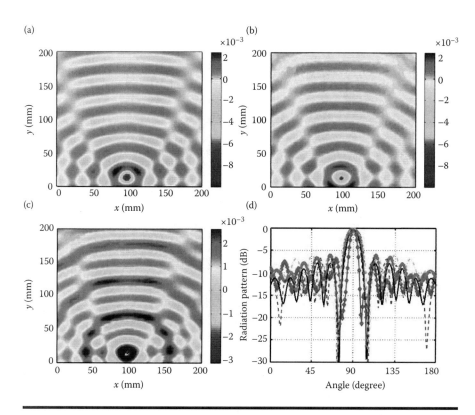

Figure 8.18 Measured results for the parabolic reflector and planar reflector at different frequencies. (a–c) Measurement results using the 2D mapper for the parabolic reflector at 10 GHz, and for the fabricated planar reflector at 10 and 11.5 GHz, respectively. (d) Normalized far-field radiation patterns for reflectors at different frequencies. Gray solid curve with diamond: the parabolic reflector; black solid curve: simulation data for the fabricated planar reflector at 10 GHz; dash-dotted curve: measured result at 8.5 GHz; solid curve with star: measured result at 10 GHz; dashed curve: measured result at 11.5 GHz. The near-field distribution for 8.5 GHz is not shown in the figure since it is similar to those given. (The source of the material Z. L. Mei, et al. Experimental verification of a broadband planar focusing antenna based on transformation optics. *New J. Phys.*, 13: 063028. Copyright 2011, IOP Publishing & Deutsche Physikalische Gesellschaft. CC BY-NC-SA, is acknowledged.)

However, the quasi-conformal mapping between the virtual space and the physical space is only validated in 2D coordinates. As a result, the above-designed devices, such as the ground-plane cloaks and illusion devices, are all within 2D limits.

In view of the difficulty to realize full-parameter free-space transformation-optics-based devices, axisymmetric 3D devices derived from 2D ones have been taken into consideration. For example, when rotated to the vertical axis, a 2D ground-plane cloak in Figure 8.3 becomes a 3D one. The 3D cloak is located on the ground plane, as sketched in Figure 8.1a, and objects placed inside the secured region beneath the cloak become undetectable to EM waves. It is easy to understand that such axisymmetric 3D cloaks are effective to differently polarized incident waves providing the cloak is composed of isotropic dielectrics. Nevertheless, it had been challenging to build 3D isotropic metamaterials because most electrically or magnetically resonant particles are polarization-sensitive, as we pointed out in Sections 3.1 and 3.2.

The first practical implementation of a fully 3D broadband and low-loss ground-plane cloak at microwave frequencies, which can conceal a 3D object located under a curved conducting plane from all viewing angles by imitating the reflection of a flat conducting plane, was reported by Ma and Cui in 2010 [28]. In this work, the 3D cloak was realized by drilling inhomogeneous holes in multilayered dielectric plates. We studied resonant particles in Chapter 3 and remarked that metamaterials composed of dielectric spheres in host media are isotropic but difficult to fabricate and integrate. In practice, dielectric cylinder arrays have been alternatives to compose nearly isotropic metamaterials at microwave frequencies. This method has been used in this 3D ground-plane cloak design, with differently sized subwavelength air hole arrays drilled on ordinary dielectric substrates for different effective permittivities.

To obtain the 3D metamaterial cloak, first of all, the refractive index distribution for the 2D compact ground-plane cloak presented in Reference 13 is given again and shown in Figure 8.19d. Note that in the 2D circumstance, the incident waves are restricted to the *xoz* plane and the electric field is along the *y* axis (TE-polarized incidence). Then, a 3D cylindrical ground-plane cloak is generated by rotating the 2D one around the *z* axis. Next, the composing unit cell for isotropic metamaterials are decided based on two considerations. First, since the span of refractive indices in Figure 8.19d is not large (from 1 to 1.63), nonresonant metamaterials are competent. Second, dielectric particles are less sensitive to polarization when compared with dielectric-metal ones. Therefore, the 3D ground-plane cloak is decided to be realized by drilling inhomogeneous holes in multilayered dielectric plates.

Photographs of the fabricated 3D cloak sample are illustrated in Figure 8.19a–c. On the bottom of the cloak, there exists a cone-shaped region (see Figure 8.19b), the secured region, where conducting objects can be hidden. The dielectric plates are made of polytetrafluoroethylene and glass fiber (F4B), with the relative permittivity of 2.65 and loss tangent of 0.001. It has been demonstrated in Reference 13 that the drilled-hole unit cells shown in Figure 8.19a–c remain almost the same characteristics with three orthogonal polarizations of incident waves and, therefore, the metamaterials can be approximately regarded as isotropic. By varying the diameter of the drilled hole (D), the effective indices of refraction change gradually and effectively fulfill the required values. In total, the 3D cloak contains 1310

Nearly Isotropic Inhomogeneous Metamaterials ■ 259

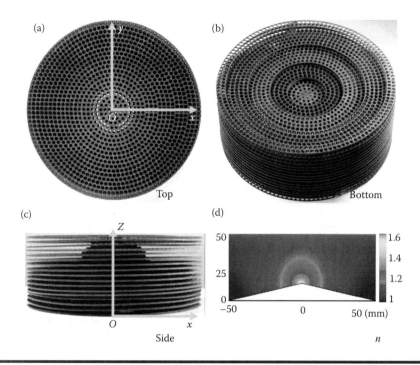

Figure 8.19 3D microwave ground-plane cloak and its refractive index distribution. (a) Top view of the cloak. (b) Bottom view of the cloak. (c) Side view of the cloak. (d) The distribution of refractive index in the x–z plane. The material parameters of the whole cloak are obtained by rotating the x to z plane pattern around the z axis. (Reprinted by permission from Macmillan Publishers Ltd. *Nat. Commun.* H. F. Ma and T. J. Cui. Three-dimensional broadband ground-plane cloak made of metamaterials. 1: 21. Copyright 2010.)

inhomogeneous holes whose diameters vary from 0.6 mm (for the central hole) to 2.7 mm (for the boundary holes).

The experimental demonstration is carried out in a fully anechoic chamber. To test the performance of designed ground-plane cloak, scattering behaviors of the flat ground plane, the ground plane with the cone-shaped perturbation, and the ground plane with the perturbation covered by the cloak are measured for comparison. In the experimental setup, in the near field, a carefully designed metamaterial lens is applied to launch a narrow-beam plane wave as the incidence. The gradient index MTMs lens is loaded inside a transmitting horn antenna (refer to the one shown in Figure 3.16c) to improve the radiation from the conventional horn antenna and obtain high gains in both E-plane and H-plane. Meanwhile, on the other side of the anechoic chamber, a normal X-band rectangular horn antenna is used as the receiver to measure the far-region scattered fields. In the experiment, the incident angle is

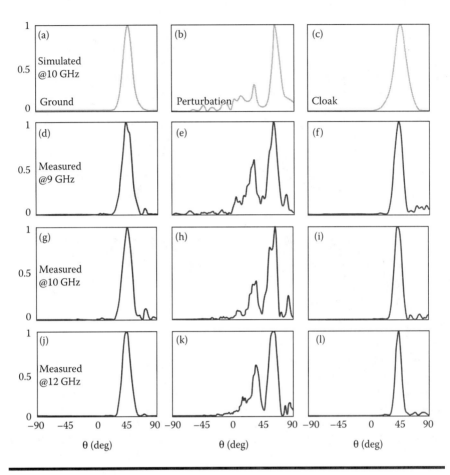

Figure 8.20 Simulated and measured electric fields in the far region. The fields are shown at different frequencies under the incidence of parallel-polarized electric field emitted from the metamaterial lens antenna. θ is the incidence angle, whereas the electric field amplitudes are normalized to unity. (a–c) Simulated results for the three situations at 10 GHz. (d–f) Measured results for the three situations at 9 GHz. (g–i) Measured results for the three situations at 10 GHz. (j–l) Measured results for the three situations at 12 GHz. (Reprinted by permission from Macmillan Publishers Ltd. *Nat. Commun.* H. F. Ma and T. J. Cui. Three-dimensional broadband ground-plane cloak made of metamaterials. 1: 21. Copyright 2010.)

chosen to be −45° in the *xoz* plane, and the receiving direction varies from −90° to 90° when the whole system is rotated around the *y* axis.

Two polarizations of incident waves have been considered in the measurement. Figure 8.20 shows the simulated and measured results for the parallel polarization when the incident electric field is parallel to the *xoz* plane. Expected performance has been observed from the results. First, measurements (Figure 8.20g–i) agree well with the simulations (Figure 8.20a–c), proving the equivalence of the drilled-hole metamaterials and the continuous materials. Second, the measured scattering property of the ground plane with both perturbation and cloak (Figure 8.20i) is nearly the

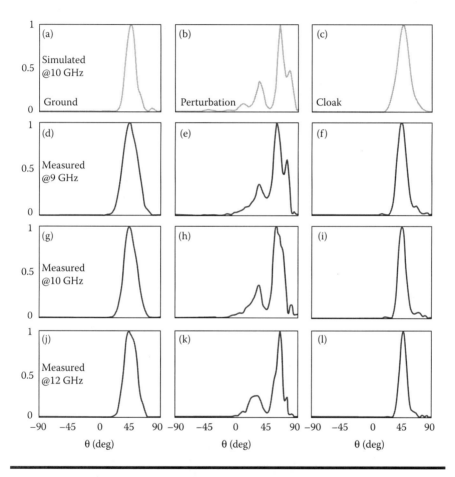

Figure 8.21 The simulated and measured electric fields in the far region. The fields are shown at different frequencies under the incidence of perpendicular-polarized electric field emitted from the metamaterial lens antenna. θ is the incidence angle, whereas the electric field amplitudes are normalized to unity. (a–c) Simulated results for the three situations at 10 GHz. (d–f) Measured results for the three situations at 9 GHz. (g–i) Measured results for the three situations at 10 GHz. (j–l) Measured results for the three situations at 12 GHz. (Reprinted by permission from Macmillan Publishers Ltd. *Nat. Commun.* H. F. Ma and T. J. Cui. Three-dimensional broadband ground-plane cloak made of metamaterials. 1: 21. Copyright 2010.)

same as that of ground plane (Figure 8.20g). A single-peak reflection at the mirror-reflecting direction of the incident wave is detected. In contrast, the scattering property of the ground plane with perturbation alone (Figure 8.20h) is significantly distinct. Multipeak reflections are detected at different directions. Third, the cloak is demonstrated to operate in a broad frequency band from 9 to 12 GHz, as plotted in Figure 8.20d–f and Figure 8.20j–l. This broadband property is guaranteed by nonresonant property of the metamaterials.

Figure 8.21 shows the simulated and measured results for the perpendicular polarization when the incident electric field is perpendicular to the *xoz* plane. Similarly good cloaking performance is observed from 9 to 12 GHz.

Metamaterials fabricated with multilayered dielectric plates by drilling inhomogeneous holes have been proved as a good candidate for 3D near-isotropic inhomogeneous devices by a series of work. Characteristics of such kind of metamaterial, as well as the detailed designing method, have been studied in Reference 29. A half-spherical Luneburg lens and a half Maxwell fisheye lens have been fabricated and measured as another two examples. These two gradient-index lenses are designed based on geometric optics. In the next section, we will design a 3D near-isotropic inhomogeneous lens using quasi-conformal mapping.

8.6 3D Flattened Luneburg Lens

Aside from the 3D ground-plane cloak, nearly isotropic inhomogeneous metamaterials have also been adopted for other applications, among which, microwave lenses have become an important part because they serve widely in wireless communication systems. Two categories of metamaterial lenses have been mainly reported. The first category is based on uniform material lenses such as the conventional convex or concave lenses (arrays). Transformation optics have been involved to change profiles or improve performance of existing lenses, and isotropic or anisotropic inhomogeneous metamaterials have been utilized to realize the design, as have been reported in References 19–21 and 30. The second category of lenses are composed of inhomogeneous materials decided by geometric optics, for instance, the gradient MTM lens [31], the Luneburg lens, and the Maxwell fisheye lens. Refractive index inside these lenses varies gradually along the radius. It is easy to understand that metamaterials may be directly applied to construct these lenses because they can easily fulfill the required refractive index range, as has been reported in References 32 and 33.

Figure 8.22 sketches a spherical Luneburg lens that generates perfect geometrical images of two given concentric spheres onto each other. In other words, if one focal point lies at infinity, the other one will be on the opposite surface of the lens. This property can be adopted to create highly directive beams if one locates a point source on the surface of the lens. Furthermore, if the point source is moved on the surface of the lens, the directive beam becomes steerable in large scanning angles, as is

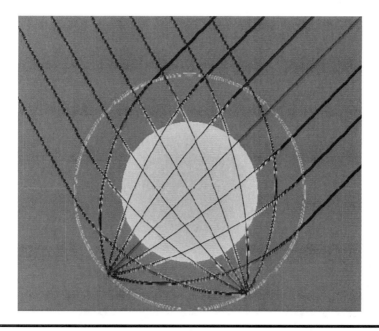

Figure 8.22 Ray tracing for a spherical Luneburg lens.

implied in Figure 8.22. A simple Luneburg lens can be achieved if the refractive index distribution follows:

$$n = \sqrt{2 - \frac{r^2}{R^2}}, \qquad (8.28)$$

where R is the radius of the sphere. Clearly, the Luneburg lens is a 3D gradient lens.

One consideration from the engineering point of view is that the spherical geometry of the Luneburg lens inevitably brings in difficult fabrication. In addition, the spherically focal surface cannot be well matched to the planar feeding source or detector array. Therefore, transformation optics has been proposed to change the profile of the Luneburg lens so as to obtain a flattened focal plane [34,35]. Since a Luneburg lens may encounter differently polarized waves, an isotropic lens is more competent in practice. In Reference 35, the technique of quasi-conformal mapping was applied to create the refractive index map of the flatten Luneburg lens. During the transformation, the virtual space is a Luneburg lens with the refractive index distribution described in Equation 8.28 in the free space, as shown in Figure 8.23a. A part of the spherical surface of the Luneburg lens with an open angle θ is flattened and hence the physical space becomes a rectangular region shown in Figure 8.23b. Near-orthogonal grids are generated in both the physical space and the virtual space to guarantee quasi-conformal mapping, so as to result in a lens made of isotropic all-dielectric materials. In the work reported by Kundtz and Smith, copper strips on an FR4 substrate were used as polarizable particles to realize the distribution of dielectric constant.

3D demonstration of flattened Luneburg lens using nearly isotropic inhomogeneous metamaterials was delivered by Ma and Cui in 2010 [36]. A 2D simplified refractive index distribution is first generated using the method of quasi-conformal mapping that has been studied in the previous section of this chapter. Note that the range of refractive index (from 0.5 to 2.2) in Figure 8.23b is restricted to a narrower one (from 1.1 to 2.1) in Figure 8.23c because a small portion in the physical space with less-than-one refractive index is replaced by free space. In this way, only dielectric materials are needed.

A novel 3D Luneburg lens is generated by rotating the 2D profile in Figure 8.23c around the z axis, as shown in Figure 8.23d. Similar to the 3D ground-plane cloak, this lens is also composed of nonresonant metamaterials which are fabricated with multilayered dielectric plates. In each layer, different inhomogeneous holes are drilled to realize the refractive index map given in Figure 8.23c. Figure 8.23d illustrates the photograph of the final sample of the 3D lens. Note that two kinds of dielectric plates, the FR4 dielectric and the F4B dielectric, are involved in the fabrication in order to fulfill all the required refractive indices. The detailed designing and fabrication process of the 3D lens is given in Reference 36 and its supplementary files.

Experimental demonstration for this 3D flattened Luneburg lens has been taken in both the near-field and far-field regions. A feeding source, a Ku-band coax-to-waveguide device, is placed at different positions on the flattened focal surface. In the near-field measurement, a coaxial probe is placed in front of the lens to detect the electric fields in the near-field region. In order to get the best measurement results, the probe should be parallel to the polarization direction of electric fields. To change the polarization in the experiment, the feeding coax-to-waveguide device is adjusted accordingly. The measured plane is on the xoz plane with $y = 0$ mm (x, y, z axes are defined in Figure 8.23), which contains the optical axis. According to the theory behind the design, it is expected that when a source is placed at different positions on the flattened focal surface, the lens will radiate directive beams at different angles on the planes containing the optical axis. On the planes not containing the optical axis, the beams will have some distortions. This prediction has been proved by the measured near-field results shown in Figure 8.24. When the feed is located at the lens center, polarized vertically, at 12.5 GHz, the flattened Luneburg lens generates very good planar wave fronts in the z direction (Figure 8.24a). When the feed is 10 mm off the center, the beam is steered to the angle of 20° off the z axis (Figure 8.24d). When the feed is further moved to be 30 mm off the center, the beam is steered to the angle of 50° off the z axis (Figure 8.24g). The lens has been tested to work in a broad frequency band from 12.4 to 18 GHz, and measured results at 15 and 18 GHz can be found in Figure 8.24b, e, h and c, f, i, respectively.

The far-field measurement is carried out in the anechoic chamber and far-field radiations in both horizontal and vertical polarizations are detected and recorded. Measured results in Figure 8.25 have verified the steerable highly directive beams created by the lens. At 12.5, 15, and 18 GHz, the radiation angles changes gradually from 0° to about 50° with respect to the normal direction (z axis) when the feeding source moves from the center of the focal plane to the position 10 mm off the center.

The 3D flattened Luneburg lens is realized using nearly isotropic inhomogeneous metamaterials. It is significantly advantageous when compared with conventional uniform material lens and the spherical Luneburg lens. For instance, it has zero focal distance, flattened focal surface, broad scanning angle, and low aberrations. Furthermore, if a planar array of feeding sources is placed on the flattened focal surface and is properly controlled, the radiation beam will be scanned in a

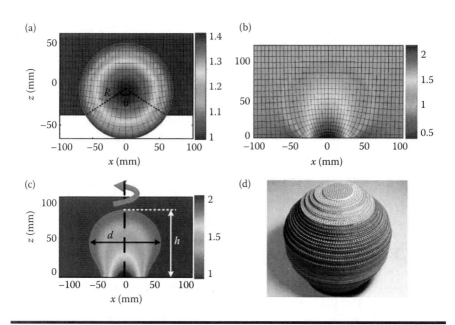

Figure 8.23 Quasi-conformal mapping of 2D flattened Luneburg lens and generation of 3D flattened Luneburg lens. (a) The 2D Luneburg lens and its transformation space, in which $R = 70\,mm$ and $\theta = 120°$. (b) The 2D flattened Luneburg lens and its refractive index distribution. (c) The simplified refractive index distribution in the xoz plane. The 3D lens is generated by rotating the profile around the z axis. (d) The photograph of the fabricated 3D lens. (Reprinted by permission from Macmillan Publishers Ltd. *Nat. Commun.* H. F. Ma and T. J. Cui. Three-dimensional broadband and broad-angle transformation-optics lens. 1: 124. Copyright 2010.)

266 ■ *Metamaterials*

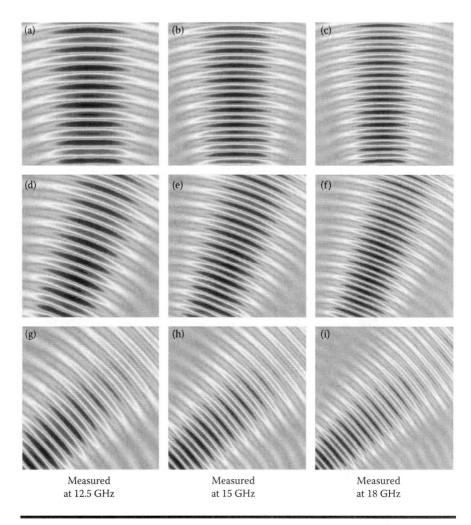

Measured at 12.5 GHz	Measured at 15 GHz	Measured at 18 GHz

Figure 8.24 Measurement results of near-electric fields outside the 3D lens. The measurements are shown at 12.5, 15, and 18 GHz. (a–c) The feeding source is located at the center of the focal plane ($x = 0$ and $y = 0$). (d–f) The feeding source is 10 mm off the center of the focal plane ($x = -10$ mm and $y = 0$). (g–i) The feeding source is 30 mm off the center of the focal plane ($x = -30$ mm and $y = 0$). (Reprinted by permission from Macmillan Publishers Ltd. *Nat. Commun.* H. F. Ma and T. J. Cui. Three-dimensional broadband and broad-angle transformation-optics lens. 1: 124. Copyright 2010.)

range of 50° around the z axis. This feature may be applied in a novel compact antenna array system.

This chapter has mainly focused on the application of the nearly isotropic inhomogeneous metamaterials in antenna and microwave engineering. Optical

Nearly Isotropic Inhomogeneous Metamaterials ■ 267

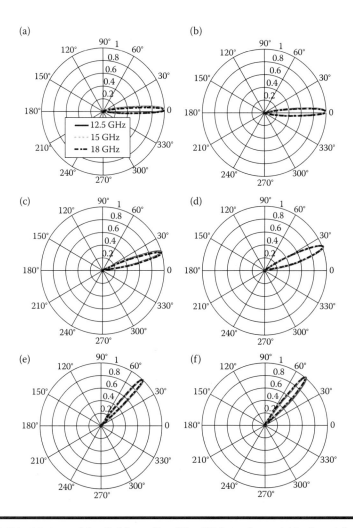

Figure 8.25 Measured far-field results of the 3D lens. (a, c, e) Horizontal polarizations and (b, d, f) vertical polarizations at 12.5, 15, and 18 GHz. (a, b) The feeding source is located at the center of the focal plane ($x = 0$ and $y = 0$). (c, d) The feeding source is 10 mm off the center of the focal plane ($x = -10$ mm and $y = 0$). (e, f) The feeding source is 30 mm off the center of the focal plane. (Reprinted by permission from Macmillan Publishers Ltd. *Nat. Commun.* H. F. Ma and T. J. Cui. Three-dimensional broadband and broad-angle transformation-optics lens. 1: 124. Copyright 2010.)

transformation has been adopted in its discretized form through the method of quasi-conformal mapping, aiming to provide an all-dielectric approach to create new devices with controllable shapes and novel properties. The designing method studied in this chapter can help the engineering community to better understand the new research field of optical transformation, and to easily conceive and design broadband devices using metamaterials.

References

1. D. Schurig, J. J. Mock, B. J. Justice, S. A. Cummer, J. B. Pendry, A. F. Starr, and D. R. Smith. Metamaterial electromagnetic cloak at microwave frequencies. *Science*, 314: 977, 2006.
2. D. Schurig, J. B. Pendry, and D. R. Smith. Calculation of material properties and ray tracing in transformation media. *Opt. Express*, 14: 9794, 2006.
3. W. Jiang, J. Chin, Z. Li, Q. Cheng, R. Liu, and T. J. Cui. Analytical design of conformally invisible cloaks for arbitrarily shaped objects. *Phys. Rev. E*, 77: 66607, 2008.
4. W. Cai, U. Chettiar, A. Kildishev, and V. Shalaev. Optical cloaking with metamaterials. *Nat. Photon.*, 1: 224, 2007.
5. A. Alu and N. Engheta. Multifrequency optical invisibility cloak with layered plasmonic shells. *Phys. Rev. Lett.*, 100: 113901, 2008.
6. D. Gaillot, C. Croenne, D. Lippens et al. An all-dielectric route for terahertz cloaking. *Opt. Express*, 16: 3986, 2008.
7. J. Li and J. Pendry. Hiding under the carpet: A new strategy for cloaking. *Phys. Rev. Lett.*, 101: 203901, 2008.
8. J. F. Thompson, B. K. Soni, and N. P. Weatherill. *Handbook of Grid Generation*. CRC Press, Boca Raton, FL, 1999.
9. P. Knupp and S. Steinberg. *Fundamentals of Grid Generation*. CRC Press, Boca Raton, FL, 1994.
10. R. Holland. Finite-difference solution of Maxwell's equations in generalized nonorthogonal coordinates. *IEEE Trans. Nucl. Sci.*, 30: 4589, 1983.
11. Y. Hao and C. Railton. Analyzing electromagnetic structures with curved boundaries on Cartesian FDTD meshes. *IEEE Trans. Microw. Theory Tech.*, 46: 82, 1998.
12. R. Liu, C. Ji, J. J. Mock, J. Y. Chin, T. J. Cui, and D. R. Smith. Broadband groundplane cloak. *Science*, 323: 366, 2009.
13. H. F. Ma, W. X. Jiang, X. M. Yang, X. Y. Zhou, and T. J. Cui. Compact-sized and broadband carpet cloak and free-space cloak. *Opt. Express*, 17: 19947, 2009.
14. D. Bao, K. Rajab, Y. Hao, E. Kallos, W. Tang, C. Argyropoulos, Y. Piao, and S. Yang. All-dielectric invisibility cloaks made of $batio_3$-loaded polyurethane foam. *New J. Phys.*, 13: 103023, 2011.
15. E. Kallos, C. Argyropoulos, and Y. Hao. Ground-plane quasicloaking for free space. *Phys. Rev. A*, 79: 63825, 2009.
16. C. E. Shannon. Communication in the presence of noise. *Proc. IRE*, 37: 10, 1949.
17. Nyquist-Shannon sampling theorem. http://en.wikipedia.org/wiki/Nyquist-Shannon_sampling_theorem.
18. Z. L. Mei, J. Bai, and T. J. Cui. Illusion devices with quasi-conformal mapping. *J. Electromagn. Waves Appl.*, 24: 2561, 2010.

19. F. Kong, B. Wu, J. Kong, J. Huangfu, S. Xi, and H. Chen. Planar focusing antenna design by using coordinate transformation technology. *Appl. Phys. Lett.*, 91: 253509, 2007.
20. D. Roberts, N. Kundtz, and D. Smith. Optical lens compression via transformation optics. *Opt. Express*, 17: 16535, 2009.
21. W. Tang, C. Argyropoulos, E. Kallos, W. Song, and Y. Hao. Discrete coordinate transformation for designing all-dielectric flat antennas. *IEEE Trans. Antenn. Propag.*, 58: 3795, 2010.
22. A. Taflove and S. C. Hagness. *Computational Electrodynamics: The Finite-Difference Time-Domain Method*, 3rd Edition. Artech House, Boston, 2005.
23. Y. Hao and R. Mittra. *FDTD Modelling of Metamaterials: Theory and Applications*. Artech House, MA, 2009.
24. Y. Zhao, C. Argyropoulos, and Y. Hao. Full-wave finite-difference time-domain simulation of electromagnetic cloaking structures. *Opt. Express*, 16: 6717, 2008.
25. C. Argyropoulos, Y. Zhao, and Y. Hao. A radially-dependent dispersive finite-difference time-domain method for the evaluation of electromagnetic cloaks. *IEEE Trans. Antenn. Propag.*, 57: 1432, 2009.
26. Z. L. Mei, J. Bai, and T. J. Cui. Experimental verification of a broadband planar focusing antenna based on transformation optics. *New J. Phys.*, 13: 063028, 2011.
27. R. Yang, W. Tang, and Y. Hao. Wideband beam-steerable flat reflectors via transformation optics. *Antenn. Wireless Propag. Lett.*, 10(4): 1290, 2011.
28. H. F. Ma and T. J. Cui. Three-dimensional broadband ground-plane cloak made of metamaterials. *Nat. Commun.*, 1: 21, 2010.
29. H. F. Ma, B. G. Cai, T. X. Zhang, Y. Yang, W. X. Jiang, and T. J. Cui. Three-dimensional gradient-index materials and their applications in microwave lens antennas. *IEEE Trans. Antenn. Propag.*, 61(5): 2561, 2013.
30. K. L. Morgan, D. E. Brocker, S. D. Campbell, D. H. Werner, and P. L. Werner. Transformation-optics-inspired anti-reflective coating design for gradient index lenses. *Opt. Lett.*, 40(11): 2521, 2015.
31. D. R. Smith, J. J. Mock, A. F. Starr, and D. Schurig. Gradient index metamaterials. *Phys. Rev. E*, 71: 036609, 2005.
32. H. F. Ma, X. Chen, X. M. Yang, H. S. Xu, Q. Cheng, and T. J. Cui. A broadband metamaterial cylindrical lens antenna. *Chin. Sci. Bull.*, 55: 2066, 2010.
33. Z. L. Mei, J. Bai, T. M. Niu, and T. J. Cui. A half Maxwell fish-eye lens antenna based on gradient-index metamaterials. *IEEE Trans. Antenn. Propag.*, 60(1): 398, 2012.
34. D. Schurig. An aberration-free lens with zero f-number. *New J. Phys.*, 19: 115034, 2008.
35. N. Kundtz and D. R. Smith. Extreme-angle broadband metamaterial lens. *Nat. Mater.*, 9: 129, 2010.
36. H. F. Ma and T. J. Cui. Three-dimensional broadband and broad-angle transformation-optics lens. *Nat. Commun.*, 1: 124, 2010.

Chapter 9
Anisotropic Inhomogeneous Metamaterials

Anisotropic inhomogeneous metamaterials play a very important role in inhomogeneous metamaterials. In Chapter 6, we have introduced the design methods of inhomogeneous metamaterial devices. Here, we will focus on some experimental work in anisotropic inhomogeneous metamaterials, especially transformation-optics devices.

9.1 Spatial Invisibility Cloak

In recent years, the experiments on anisotropic inhomogeneous metamaterials have achieved important progress. Transformation optics allows anisotropic metamaterials to be tailor-made according to practical needs. In this chapter, we will review some typical experiments on transformation optics and metamaterials devices, for both time-varied fields and static fields. All these devices have unique and exciting properties.

Invisibility means something that indeed exists but cannot be observed or perceived by others. This idea has long captivated the popular imagination. The most attractive and fascinating device based on anisotropic inhomogeneous metamaterials is the invisibility cloak [1]. Thus we first introduce the realization of cloak in this chapter. Since 2006, invisibility has become a practical matter for the scientific community, with the proposal of transformation optics. The first experimental demonstration of invisibility cloak in free space was reported by Smith's group from Duke University [2]. To design a cylindrical cloak, coordinate transformation in circularly cylindrical coordinates (r, φ, z) should be constructed. For example, a general linear transformation for designing the cloak is

$$r' = \frac{b-a}{b}r + a, \quad \varphi' = \varphi, \quad z' = z, \tag{9.1}$$

where r, φ, and z are the radial, angular, and vertical coordinates in the physical system, respectively, and r', φ', and z' are the coordinates in the actual system. a is radius of the cloaked object, that is, the inner radius of the cloak, and b is the outer radius of the cloak. From the transformation optics theory, the expressions for the permittivity and permeability components of the invisibility cloak are expressed as

$$\varepsilon_r = \mu_r = \frac{r-a}{r}, \quad \varepsilon_\varphi = \mu_\varphi = \frac{r}{r-a}, \quad \varepsilon_z = \mu_z = \left(\frac{b}{b-a}\right)^2 \frac{r-a}{r}. \tag{9.2}$$

All components are gradient functions of radius, as can be seen from Equation 9.2, and hence, the invisibility cloak would be an anisotropic and inhomogeneous device. In the first cloaking experiment, the invisible effect was verified for TE polarized waves. In such a case, the electric field is polarized along the z axis (cylinder axis), and only ε_z, μ_r, and μ_φ are of interest. Moreover, the following simplified material parameters

$$\varepsilon_z = \left(\frac{b}{b-a}\right)^2, \quad \mu_r = \left(\frac{r-a}{r}\right)^2, \quad \mu_\varphi = 1 \tag{9.3}$$

have the same dispersion relationship as that of Equation 9.2. Since the wave trajectory inside the cloak is only determined by the dispersion relation, in the geometric limit, rays will follow the same paths in media defined by Equation 9.2 or 9.3, and refraction angles into or out of the media are also the same [2]. The difference is that the reflectance will be nonzero using the simplified material properties.

From Equation 9.3, for the simplified cloak, the electric component ε_z is constant, μ_φ is one, and only μ_r varies radially throughout the structure. The SRRs, which provide a magnetic response, can satisfy the need of the parameters if they are positioned with their axes along the radial direction (see Figure 9.1). By tuning the geometric parameters of SRRs, the desired ε_z and μ_r can be obtained. It is fortunate that the frequencies of the electric and magnetic resonances for SRRs can be controlled and shifted by two geometrical parameters of the structures, respectively. To obtain the effective material parameters ε_z and μ_r from the scattering parameters, a standard retrieval procedure was performed [3]. The first free-space cloak consists of 10 concentric cylinders, each of which has a height of three unit cells, as shown in Figure 9.1. Evenly spaced set of cylinder radii was chosen so that an integral number

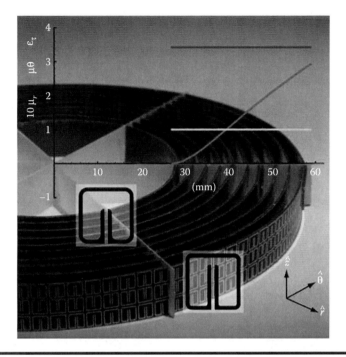

Figure 9.1 Photograph of the first free-space invisibility cloak in microwave band. (From D. Schurig et al. Metamaterial electromagnetic cloak at microwave frequencies. *Science*, 314: 977, 2006. Reprinted with permission of AAAS.)

of unit cells fits exactly around the circumference of each cylinder, necessitating a particular ratio of radial-to-circumferential unit cell size.

For the experimental confirmation, the metamaterial cloak was measured in a parallel plate waveguide, now known as 2D field mapping system. The simulation and measurement of the cloak that surrounds a 25-mm-radius copper cylinder were performed. By comparison, it is shown that the cloak reduces both the backscattering (reflection) and forward scattering (shadow), as shown in Figure 9.2. The wave fronts on the boundary of the cloak match the wave fronts outside the cloak, which essentially correspond to those of empty space. The results of the first fabricated invisibility cloak provided an experimental display of the EM cloaking mechanism and demonstrated the feasibility of implementing media specified by the transformation optics method with metamaterial technology. It has opened up the realization of complex inhomogeneous and anisotropic EM or optical devices made of metamaterials. Many later studies on metamaterial experiments are inspired by this work.

Plasmonic cloaking is a scattering-cancellation technique based on the local negative polarizability of metamaterials. The first experimental realization was reported

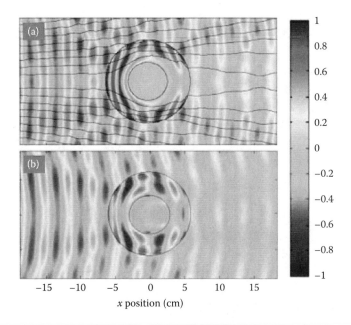

Figure 9.2 Simulated and experimented near-field distributions of the first microwave invisibility cloak. (a) Simulation, (b) experiment. (From D. Schurig et al. Metamaterial electromagnetic cloak at microwave frequencies. *Science*, 314: 977, 2006. Reprinted with permission of AAAS.)

and measured at microwave frequencies by Engheta's group [4]. To create a metamaterial plasmonic shell capable of cloaking a dielectric cylinder, they embedded an array of metallic fins into a high-permittivity fluid. In 2012, an alternative approach of invisibility cloaking was experimentally demonstrated, which was also based on scattering cancellation and implemented with anisotropic metamaterials [5]. In their design, nonsuperluminal propagation of EM waves is inherited, and the working bandwidth is relatively broad. Such a scheme may be used in future practical implementation of cloaking devices at large scales in free space. More recently, Smith's group designed and experimentally verified a 2D unidirectional cloak, which could reduce the scattering of an object with the size of ten wavelengths. In their approximation-free design, the performance characteristics promised by transformation optics was regained [6].

9.2 D.C. Circuit Invisibility Cloak

Besides the time-varying fields, the cloaking phenomenon also exists in the steady current fields. The first D.C. electric invisibility cloak was realized by Yang et al. [7].

By exploiting the connection between electric conductivity in conducting materials and resistors in the circuit theory, a D.C. cloak was fabricated based on the resistor network. The working principle is as follows: when the D.C. cloak is wrapped on the object, it can smoothly guide the electric currents around the object and keep the perturbations only inside the cloak. Outside the cloak, the current lines return to their original direction as if nothing happens.

When Sir J. B. Pendry proposed the idea of transformation optics, he also pointed out that the transformation optics theory applies equally well to the static fields [1]. In the D.C. limit, the transformation optics is simplified as transformation electrostatics [8], which is exactly based on the invariance of Laplace equations. For an anisotropic conducting material, the conductivity can be defined by a symmetric and positively definite matrix-valued function, $\sigma = \sigma_{ij}(x)$. In the absence of sources, the electric potential V satisfies the Laplace equation

$$(\nabla \cdot \sigma \nabla) V = 0, \tag{9.4}$$

from which the electric field is expressed as $\vec{E} = -\nabla V$. Suppose that the coordinate transformation between a virtual space and a real space (or physical space) is $x' = x'(x)$. In the new coordinate system, the Laplace equation 9.4 is form invariant, and hence the conductivity can be derived in the virtual space as $\sigma = A\sigma A^T/|A|$, in which A is the Jacobian transformation matrix with $A_{ij} = \partial x_i/\partial x_{j'}$.

Using the same spatial transformation as Smith's cloak in 2D case [2], the invisibility cloak was designed in electrostatics. The conductivity tensor for the invisibility cloak is written as

$$\sigma_r = \frac{r-a}{r}\sigma_0, \quad \sigma_\varphi = \frac{r}{r-a}\sigma_0, \quad \sigma_z = \left(\frac{b}{b-a}\right)^2 \frac{r-a}{r}\sigma_0, \tag{9.5}$$

in which a and b represent the inner and outer radii of the cloak, respectively. The conducting materials defined in Equation 9.5 implies that the realization of D.C. cloak requires anisotropic and inhomogeneous conductivities.

Based on the circuit theory and resistor networks, Yang et al. proposed a method to realize the required radial and tangential components of the conductivity tensor for the D.C. cloak. Their idea is illustrated in Figure 9.3. Figure 9.3a shows a continuous conducting material plate with the conductivity σ and thickness h, which may extend to infinity in the radial direction. To make an equivalence of the material to a resistor network, the continuous material is first discretized using the polar grids, as shown in Figure 9.3a. According to Ohm's law, each elementary cell in the grid can be implemented by two resistors

$$R_r = \frac{\Delta r}{\sigma_r r \Delta \varphi h}, \quad R_\varphi = \frac{r \Delta \varphi}{\sigma_\varphi \Delta r h}, \tag{9.6}$$

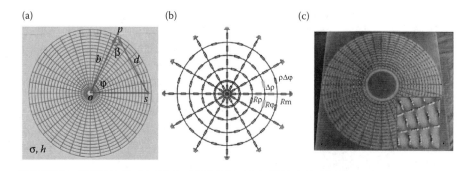

Figure 9.3 (a) Illustration of a D.C. invisibility cloak. (b) The equivalent resistor network of a continuous conducting materials. (c) Photograph of the D.C. invisibility cloak. (Reprinted with permission from F. Yang et al., *Phys. Rev. Lett.*, 109: 053902. Copyright 2012 by the American Physical Society.)

where Δr and $\Delta \varphi$ are step lengths in the radial and tangential directions, respectively. Thus, the anisotropic conductivity tensor can be implemented easily using different resistors in different directions, as illustrated in Figure 9.3b. To make simulations and measurements, the infinitely large material should be tailored to have a suitable size. Like the PMLs in the time-varying problems, matching resistors are added in the outer ring to emulate an infinite material. Using the theorem of uniqueness, the matching resistors can be easily obtained as

$$R_m = \frac{d(\ln r_0 - \ln d)}{\sigma bh \cos\beta \Delta\varphi}, \tag{9.7}$$

in which r_0 is the distance between the ground and the source point S, and all other variables are noted in Figure 9.3a.

In the experimental setup, the conductivity of background material was chosen as 1 S/m. The background material together with the D.C. cloak is discretized into 20×36 cells using the polar grid. The photo of fabricated D.C. cloak using resistor network is illustrated in Figure 9.3c. The cloaked region covers an area of 5 layers, and the cloak occupies another 5 layers. The remaining 10 layers are used for the background material. The simulation and measurement results of the resistor networks agree excellently with each other. According to Equation 9.5, the conductivity becomes singular in the inner boundary of the D.C. cloak, that is, $\sigma_\rho = 0$ and $\sigma_\varphi = \infty$. Such a singularity can be handled conveniently in the steady current field. The corresponding resistors to the singular conductivity are given by $R_\rho = \infty$ and $R_\varphi = 0$, respectively, which can be realized using the short and open circuits in the resistor network.

Recently, an ultrathin and nearly perfect D.C. electric cloak has been proposed based on new strategy of transformation electrostatics [9], which is made of homogeneous and anisotropic conducting material. This D.C. invisibility cloak was composed of an ultrathin resistor network layer with the thickness of one unit cell, which is the smallest size of artificial metamaterials. Although it is ultrathin, the D.C. electric cloak has nearly perfect cloaking behavior.

Beginning with the design of a cylindrical elliptical D.C. electric cloak, a 2D coordinate transformation is constructed in the elliptical coordinate system (ξ, η)

$$\xi' = (\xi_2 - \xi_1)\xi/\xi_2, \quad \eta' = \eta, \tag{9.8}$$

in which ξ_1 and ξ_2 are the coordinate parameters of the inner and outer boundaries of the elliptical cloak. The inner and outer lengths of major axes for the elliptical D.C. cloak are R_1 and R_2, and the inner and outer elliptical boundaries share the same focal length $2p$, then a closed form representation can be obtained: $\xi_i = \ln(R_i/p + \sqrt{(R_i/p)^2 - 1})$, $i = 1, 2$. Hence, the conductivity tensor of the planar elliptical D.C. electric cloak was derived from the transformation electrostatics,

$$\sigma_{\xi'} = \frac{\xi_2 - \xi_1}{\xi_2}\xi_0, \quad \sigma_{\eta'} = \frac{\xi_2}{\xi_2 - \xi_1}\xi_0, \tag{9.9}$$

in which ξ_0 is the conductivity of the homogeneous background material. The above equation clearly implies that both ξ and η components of the conductivity tensor are constants.

If the focal length $2p$ is chosen as a very small value, the elliptical D.C. cloak is nearly a circular D.C. cloak with the outer radius R_2 and inner radius R_1. In such a case, σ_ξ becomes σ_r and σ_η becomes σ_φ, indicating the radian and angular components of the conductivity. The conductivity tensor of the nearly circular D.C. electric cloak is given by

$$\sigma_{r'} = k\xi_0, \quad \sigma_{\varphi'} = \xi_0/k, \tag{9.10}$$

in which $k = (\xi_2 - \xi_1)/\xi_2 = \ln(R_2/R_1)/\ln(2R_2/p)$. It is very interesting to see that this is a nearly perfect D.C. cloak since it crushes the cloaked object to nearly a point (i.e., a very short line segment with length $2p$). From Equation 9.10, the material parameters of such a D.C. electric cloak $\sigma_{r'}$ and $\sigma_{\varphi'}$ are constants, which are only related to the inner and outer radii of the cloak, and hence the D.C. cloak is anisotropic but homogeneous. Particularly, when the focal length $2p$ is close to 0, k is nearly zero. Then, the radial conductivity tends to 0, which means insulation in the radial direction; while the azimuth conductivity tends to very large, which means good conducting in the azimuth direction. Hence, the nearly perfect D.C. cloak is a combination of such two conditions and is anisotropic.

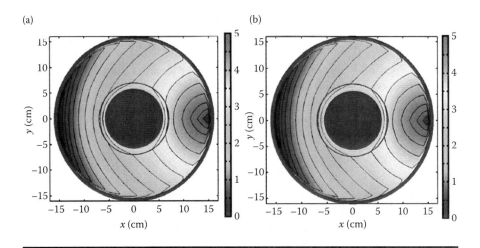

Figure 9.4 Simulated and experimented electrical potential distributions of the ultrathin D.C. invisibility cloak. (a) Simulation, (b) experiment. (Reprinted with permission from W. X. Jiang et al. An ultrathin but nearly perfect direct current electric cloak. *Appl. Phys. Lett.*, 102: 014102. Copyright 2013, American Institute of Physics.)

In numerical simulations and experiments, an insulating shell is chosen as the inner boundary of D.C. electric cloak owing to the nearly zero radial conductivity. The homogeneous feature of the proposed design makes it possible to realize ultrathin D.C. electric cloak. Figure 9.4a and b demonstrates the simulated and tested electrical potential distributions of one-layer electric D.C. cloak excited by a point source, showing excellent cloaking performance. It is obvious that the potential distributions outside the cloak in both simulation and experiment are almost the same as that of the point source in homogeneous conducting material. Hence, the cloaking behavior is nearly perfect even though the thickness of D.C. cloak is only one layer, which is the thinnest case in artificial materials.

Next, we discuss an exterior D.C. cloak, which was reported in Reference 10. To design an exterior cloak in two dimensions, a larger annulus ($b < r < c$, cloaked region) is folded onto a smaller one ($a < r < b$, complementary region). At the same time, the larger circle with radius c is linearly changed into a smaller one with radius a. The linear transformation function, which involves the folding operation, is shown as follows

$$\rho' = f(\rho) = \begin{cases} \dfrac{a}{c}\rho & 0 < \rho < c \\ -\dfrac{b-a}{c-b}(\rho - b) + b & b < \rho < c \end{cases}, \quad \varphi' = \varphi, \quad z' = z \quad (9.11)$$

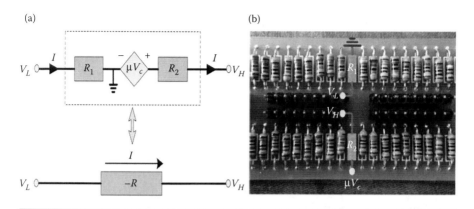

Figure 9.5 (a) The negative-resistor model to realize negative conductivity and resistance. (b) The realization of negative conductivity and resistance using resistors and active devices. (F. Yang et al.: A negative conductivity material makes a dc invisibility cloak hide an object at a distance. *Adv. Func. Mat.* 23: 4306, 2013. Copyright Wiley-VCH Verlag GmbH & Co. KGaA. Reproduced with permission.)

in the cylindrical coordinates, in which a and b represent the inner and outer radii of the cloak, respectively, and c is the outer radius of the cloaked region. Assume that the conductivities of the background material and cloaked region are σ_0 and σ_m, respectively, which may be inhomogeneous. Then the conductivity of the invisibility cloak can be calculated by using transformation optics in direct current.

Because the design of exterior cloaks involves the folded geometry, negative material parameters are included. In the D.C. case, negative conductivity, or equivalently, negative resistance are needed. A negative resistor model is presented in Figure 9.5a, which is composed of two resistors and a controlled voltage source. Here, I represents the current running through the negative resistor, μV_C stands for the electromotive force of the controlled voltage source, μ is the controlling ratio, and V_C is the voltage of excitation in the network. Resistors R_1 and R_2 are used to match the current and voltage of the port, while V_H and V_L denote the high and low voltages, respectively, and $-R$ is the resistance of the ideal negative resistor. Since the current I running through $-R$ is $(V_H - V_L)/R$, and the proposed model is used to replace the ideal negative resistor, the resistances of R_1 and R_2 are obtained as

$$R_1 = \frac{V_L}{I}, \quad R_2 = \frac{\mu V_C - V_H}{I}. \tag{9.12}$$

Because the network to fabricate is composed of linear components, the current and voltage at each node in the network change proportionally with the excitation signal, making R_1 and R_2 unchanged. Hence the model can be used to mimic an ideally negative resistor. In the design process, the negative conductivity is first calculated, and then mapped to the negative resistors on a polar grid. Next, a circuit simulation is performed on the resulted resistor network with ideal negative resistors, which gives the voltage and current distributions at each node, including V_H, V_L, and I mentioned above. Finally, Equation 9.12 is used to give the corresponding resistors in the circuit model. We remark that μ is a fixed value depending on the active sources available at hand in this process. Figure 9.5b illustrates the experimental realization of the negative resistor, in which a unit cell is shaded in gray colors.

Based on the above principle, the D.C. exterior cloak was fabricated and tested. The simulation and experiment results of the exterior cloak based on resistor network are demonstrated in Figure 9.6. To calculate the current distribution using the gradient of voltage, the voltage at each node in the resistor network is first obtained. Figure 9.6a illustrates the potential and current distributions of the point source in the background containing the objects and D.C. exterior cloak simultaneously. The measured results are presented in Figure 9.6b, which have excellent agreements to those of continuous materials shown in Figure 9.6a, confirming the correctness of the equivalent resistor network experimentally verifies the functionality of the D.C. exterior cloak.

The first experiment on active cloaking and an active illusion was recently presented for the Laplace equation [11]. By dynamically changing the controlled

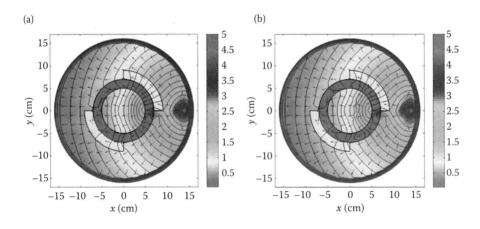

Figure 9.6 (a) The simulated voltage and current distributions when the designed D.C. exterior cloak is placed close to the objects. (b) The measured voltage and current distributions when the designed D.C. exterior cloak is placed close to the objects. (F. Yang et al.: A negative conductivity material makes a dc invisibility cloak hide an object at a distance. *Adv. Func. Mat.* 23: 4306, 2013. Copyright Wiley-VCH Verlag GmbH & Co. KGaA. Reproduced with permission.)

sources, the center protected region can be cloaked or disguised as different objects. The measurement results agree very well with numerical simulations.

9.3 Spatial Illusion-Optics Devices

Spatial illusion devices, as the extension of invisibility cloak, are of great interest because it can potentially transform an actual perception into the predesigned perception. Such concept devices may empower unprecedented applications in the advanced material science, camouflage, cloaking, optical and/or microwave cognition, and defense security in the future. Based on the transformation optics theory and engineering capability of metamaterials, some experiments of illusion devices have been proposed and performed. In this section, we will review three functional spatial illusion devices, which are capable of creating another one or multiple virtual images at the original object's position under the illumination of EM waves. The scattering signature of the object may be perceived as a smaller virtual target, or a completely different target, or multiple virtual targets with different geometries and compositions. The illusion materials, which are being inhomogeneous and anisotropic, have been realized using varying unit cells working off resonance. Experimental demonstrations of illusion have validated the theory of scattering metamorphosis and opened a novel avenue to the wave-dynamic illusion, cognitive deception, manipulate strange light (or matter) behaviors, and novel optical and microwave devices. Unlike invisible cloaking, the illusion device not only conceals the radar signature of the real object, but also generates the image of a virtual object with arbitrarily predesigned size and arbitrary material parameters. From 2006, transformation optics, together with metamaterials, has received much attention since it can be used to reveal a large range of novel and unusual phenomena. The invisibility cloak, one of the applications of transformation optics, can transform an object into nothing. Optical illusion devices, the other application, can transform an object into another one with a different shape and material makeup. Illusion devices are as important as the invisibility cloaks, which can fool the detector (or viewer) into making the wrong decision.

9.3.1 Shrinking Devices

First, we introduce a shrinking device, which can compress an object with arbitrary material properties to a virtual smaller one with desired size and material parameters [12]. In other words, the illusion device shrinks the size of any object and makes it look like another smaller one with different material parameters. It is also interesting that the illusion device may generate a virtual object with extremely large parameters, which does not exist in nature.

A 2D case is considered with the electric fields polarized in the z direction. To design a shrinking device with inner and outer radii a and b, a transformation is carried out in cylindrical coordinates,

$$r = k_1(r' - c) + a, \quad \varphi = \varphi', \quad z = z', \tag{9.13}$$

in which c is the radius of the virtual object, and $k_1 = (b-a)/(b-c)$. The material properties of the shrinking device are calculated by transformation optics,

$$\mu_r = \frac{r - a + k_1 c}{r}\mu_0, \quad \mu_\varphi = \frac{r}{r - a + k_1 c}\mu_0, \quad \varepsilon_z = \frac{r - a + k_1 c}{k_1^2 r}\varepsilon_0, \tag{9.14}$$

in which, (ε_0, μ_0) are the EM parameters of background material. When the radius of virtual cylinder approaches zero, the shrinking device becomes a perfect invisibility cloak. For the object inside the shrinking shell, EM properties are assumed to be described by $\varepsilon_{in}(r') = \varepsilon_{in1} + j\varepsilon_{in2}$ and $\varepsilon_{in}(r') = \varepsilon_{in1} + j\varepsilon_{in2}$, which may be inhomogeneous. The virtual object will be a smaller inhomogeneous object and the EM space will be defined by the following mappings:

$$r = k_2 r', \quad \varphi = \varphi', \quad z = z', \tag{9.15}$$

where $k_2 = a/c$ is the shrinking coefficient of the illusion object. Given that the real dielectric object is composed of many isotropic and inhomogeneous media, the material parameters of virtual object are also calculated by transformation optics,

$$\mu_{rv} = \mu_{\varphi v} = \mu_{in}(r), \quad \varepsilon_{zv} = k_2^2 \varepsilon_{in}(r), \tag{9.16}$$

in which $\varepsilon_{in}(r)$ and $\mu_{in}(r)$ are permittivity and permeability of the actual inhomogeneous object, respectively. From Equation 9.16, the material properties of virtual object are determined by the real object and the shrinking coefficient.

Equation 9.14 implies that all three components of material parameters are gradient functions of radius, which are very difficult to operate and optimize using current technology of metamaterials. The material parameters of shrinking device can be reduced by dispersion relations,

$$\mu_r = \frac{(r - a + k_1 c)^2}{r^2}\mu_0, \quad \mu_\varphi = \mu_0, \quad \varepsilon_z = \frac{1}{k_1^2}\varepsilon_0. \tag{9.17}$$

The reduced parameters provide the same wave trajectory inside the shrinking device. In such a case, only one component changes along the r direction, which makes it much more feasible to realize the shrinking device. To achieve the required material parameters, we select the SRR structures, which are positioned with their axes along the radial direction, as shown in Figure 9.7. The SRR can provide magnetic responses that can be tailored.

Figure 9.7 Photograph of the first free-space illusion device, a shrinking device, in microwave band. (Reprinted with permission from W. X. Jiang et al. Shrinking an arbitrary object as one desires using metamaterials. *Appl. Phys. Lett.*, 98: 204101. Copyright 2011, American Institute of Physics.)

The photograph of the shrinking device is shown in Figure 9.7, which consists of eight concentric rings. Each layer is four unit cells tall (12 mm). The desired permittivity and permeability can be obtained from split rings by selecting appropriate geometrical parameters. The effective medium parameters ϵ_z and μ_r are retrieved from the reflection and transmission coefficients.

In the experiment, the electric field distributions of the shrinking device were scanned in a parallel-plate waveguide system. A monopole probe is fixed inside the planar waveguide as the feeding source. The shrinking performance has been demonstrated numerically and experimentally. A 30-mm-wide copper square, which is regarded as an object with an extremely high loss, is placed in the center. Figure 9.8a shows the simulation results of electric fields inside and outside the reduced-parameter shrinking device with a 30-mm-wide copper square object in the center. The measured electric fields are illustrated in Figure 9.8b. Obviously, these two results have excellent agreements to each other. In other words, the illusion device shrinks the square metallic object as desired with good performance.

The shrinking device can be extended to a scaling device, which can shrink or enlarge an object using the metamaterials [13]. The EM scattering properties of such scaling devices have been rigorously analyzed using the Eigen-mode expansion method. A scaling device can make a dielectric or metallic object look smaller or larger. The rigorous analysis shows that the scattering coefficients of the scaling devices are exactly the same as those of the equivalent virtual objects.

9.3.2 Material Conversion Devices

In this section, we introduce another metamaterial-based illusion device, the material conversion device [14], which can change the scattering signal of a metal target

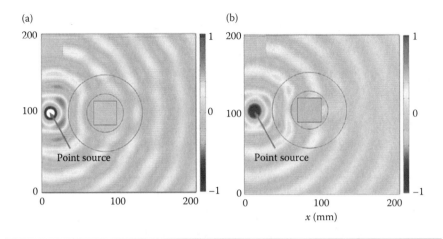

Figure 9.8 Near-field distribution of the shrinking device. (a) Simulated result, (b) experimental result. (Reprinted with permission from W. X. Jiang et al. Shrinking an arbitrary object as one desires using metamaterials. *Appl. Phys. Lett.*, 98: 204101. Copyright 2011, American Institute of Physics.)

into a dielectric one whose material parameters can be predesigned. As is known, the scattering characteristics of metal and dielectric targets are significantly different. However, when the metal target is covered by the anisotropic inhomogeneous material conversion device, its scattering characteristics and EM image are exactly the same as those of a predesigned dielectric object. The SRRs are used to fabricate the conversion device at microwave frequencies.

For feasible implementations, we consider a 2D metallic cylinder with infinite length along the z direction as the actual object. In general, the conductivity of metal is always very large, hence the metal is regarded as a perfect conductor at microwave band. Covered by an illusion device, the radar image of the metallic cylinder will be a dielectric cylinder with predesigned size and material parameters. We assume the radius of the actual metallic cylinder to be ρ_1 and the required radius of the virtual dielectric cylinder to be ρ_0. The metallic cylinder is enclosed by a conversion device with inner radius ρ_1 and outer radius ρ_3. Such an illusion device can be designed using transformation optics in which the real space (or the physical space) is the metallic cylinder with the conversion device, and the virtual space is the dielectric cylinder with the predesigned material parameters. To design the illusion device, we define a transformation from the virtual space to the physical space in the cylindrical coordinates as

$$\rho = \begin{cases} k_1 \rho' + \rho_1, & 0 \leq \rho' \leq \rho_0, \\ k_2(\rho' - \rho_0) + \rho_2, & \rho_0 \leq \rho' \leq \rho_3 \end{cases} \quad \varphi = \varphi', \quad z = z', \quad (9.18)$$

in which ρ_2 is the radius of an interior surface inside the illusion device ($\rho_1 < \rho_2 < \rho_3$), and $k_1 = (\rho_2 - \rho_1)/\rho_0$ and $k_2 = (\rho_3 - \rho_2)/(\rho_3 - \rho_0)$ are external and internal transformation coefficients, respectively. The preceding mapping implies that the virtual space includes two parts: an inner cylinder ($0 \leq \rho \leq \rho_0$) with the predesigned material parameters (ε_{vi}, μ_{vi}) and an outer cylindrical shell ($\rho_0 \leq \rho \leq \rho_3$) with the predesigned material parameters (ε_{vo}, μ_{vo}). Correspondingly, the real material of the illusion device in the physical space also contains two parts: an inner shell ($\rho_1 \leq \rho \leq \rho_2$) and an outer shell ($\rho_1 \leq \rho \leq \rho_3$). From transformation optics, the material parameters of the illusion device can be calculated. Due to the nature of planar-waveguide apparatus used in the experiment, only components ε_z, μ_ρ, and μ_φ are relevant and required. If the products of $\varepsilon_z\mu_\rho$ and $\varepsilon_z\mu_\varphi$ are kept the same under TE polarization, the dispersion relations of the transformation media remain unchanged. The rescaled material parameters of the conversion device were chosen as

$$\mu_\rho = \left(\frac{\rho - \rho_1}{\rho}\right)^2, \quad \mu_\varphi = 1, \quad \varepsilon_z = \frac{1}{k_1^2} \text{ for the inner part,} \qquad (9.19)$$

$$\mu_\rho = \left(\frac{\rho - \rho_2 + k_2\rho_0}{\rho}\right)^2, \quad \mu_\varphi = 1, \quad \varepsilon_z = \frac{1}{k_2^2} \text{ for the outer part.} \qquad (9.20)$$

Such rescaled material parameters can provide the same wave trajectory inside the illusion device.

To implement the rescaled material parameters using artificial metamaterials, we have to choose and design appropriate unit cells of metamaterials. In both the inner and outer parts of the radar illusion device, ε_z and μ_φ are constants, while μ_ρ varies in the radial direction, as shown in Equations 9.19 and 9.20. Hence we select the artificial structures with adjustable magnetic responses, such as SRR, as unit cells. To achieve the required material parameters, the split ring structures are positioned with their axes parallel to the radial direction, as shown in Figure 9.9.

The layout of the radar illusion device makes use of unit cells that are diagonal in the cylindrical basis, as shown in Figure 9.9. In the inner part of the illusion device, there are five 3-mm-thick layers; in the outer part, there are six 3-mm-thick layers. Hence the whole conversion device consists of 11 concentric rings of low-loss printed circuit boards that are made of the F4B dielectric substrate, which has a thickness of 0.25 mm and a relative permittivity of 2.65 with the loss tangent of 0.001. The height of the illusion device is 12 mm, which is four unit cells tall.

In the designing of the unit cells of material conversion device, the inner part and the outer part are considered, respectively. The reflection and transmission coefficients ($S11$, $S21$) of the split-ring unit cells are first acquired. The effective-medium parameters ε_z and μ_ρ are then retrieved from these scattering coefficients by using the standard procedure. The material conversion device was measured in a parallel-plate waveguide mapping system, which restricts the EM fields to the 2D

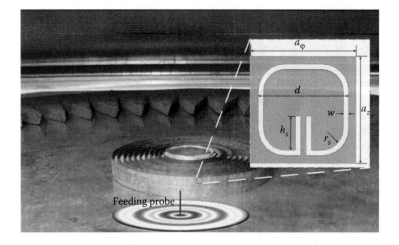

Figure 9.9 Experimental setup and photograph of the material conversion device. (Reprinted with permission from W. X. Jiang and T. J. Cui, *Phys. Rev. E*, 83: 026601. Copyright 2011 by the American Physical Society.)

TE-polarization mode. Distance between the upper and lower parallel aluminum plates is set to 13.5 mm. The sample of the conversion device is placed on the center of the lower plate, which is mounted on a step motor to translate in two dimensions. A monopole probe is fixed inside the waveguide as the feeding source, as shown in Figure 9.9. To measure the electric fields on a plane above the conversion device being tested, a sensitive detecting probe is inserted through a hole in the center of the upper plate.

Figure 9.10a demonstrates the simulated electric field distribution of the 15-mm-radius copper cylinder covered with the conversion device, which is described by the rescaled parameters in Equations 9.19 and 9.20 and discretized by 11 layers. Practical effects are incorporated into the simulations to correspond with the experimental conditions. Figure 9.10b gives the measured E field distributions of the illusion device with a copper cylinder located at the center. Obviously, the measured and simulated results agree perfectly. Therefore, the experimental realization also faithfully produces the desired illusion effect. Hence the illusion device indeed makes the EM image of the metallic cylinder look like a dielectric cylinder.

9.3.3 Virtual Target Generation Devices

Recently, a distinguished virtual target generation device based on anisotropic inhomogeneous metamaterials in electromagnetics was reported by a joint group from Southeast University in China, National University of Singapore, and University of Birmingham in UK [15]. With such a device, the scattering signature of an arbitrary

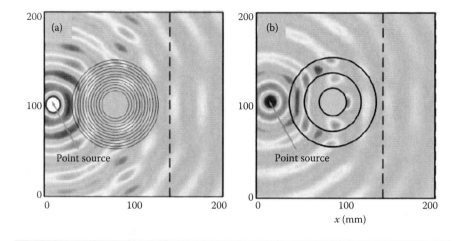

Figure 9.10 Simulation and experiment results of the material conversion device. (a) Simulation, (b) experiment. (Reprinted with permission from W. X. Jiang and T. J. Cui, *Phys. Rev. E*, 83: 026601. Copyright 2011 by the American Physical Society.)

object will be concealed and replaced with that of multiple isolated virtual objects. Furthermore, in both the near-field pattern and far-field scattering cross section, the multiple illusion objects can be arbitrarily predesigned and precontrolled. For example, the function of the target generation device to discuss is transforming a perfectly conducting object into three objects, including a shrunk metallic object in the original position and two "virtual" objects (dielectric or metallic). The target generation device shrinks the original metallic object and produces two additional illusion images in the radar signature. In other words, radar only perceives that there are three stand-alone objects, though only a single object is physically present at the center.

The above functionality can be directly derived using the transformation optics theory. In the actual space, the radius of the object (or the inner radius of the ghost device) is a, and the outer radius of the ghost device is c. In the virtual space, there are three distinct objects: a shrunk object and two wing-ghosts. Here, we just design two wing-ghosts as an example. Actually, we can design an arbitrary number of virtual objects. The radius of shrunk object is a' and the inner and outer radii of wing-ghosts are b' and c, respectively. The general 3D optical transformation for the above functionality is written in the spherical coordinates as

$$(r, \theta, \varphi) = (k(r' - a') + a, \theta', \varphi'), \tag{9.21}$$

in which $k = (c - a)/(c - a')$. Then the constitutive parameters in the region from $r = a$ to $r = c$ except for the two wing-ghost areas (see region I in Figure 9.11) and

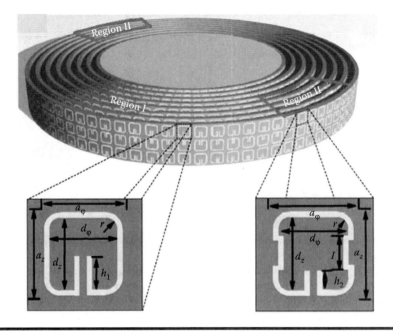

Figure 9.11 Schematic of the virtual target generation device. (W. X. Jiang et al.: Creation of ghost illusions using wave dynamics in metamaterials. *Adv. Func. Mat.* 23: 4028, 2013. Copyright Wiley-VCH Verlag GmbH & Co. KGaA. Reproduced with permission.)

the two wing-ghost areas (region II in Figure 9.11) in the spherical coordinates can be obtained by transformation optics.

For the 2D case, the cross section of the cylindrical ghost device is the same. The cylindrical device is characterized by its diagonal-parameter tensors in the cylindrical coordinates (r, φ, z), and the same material properties hold for the diagonal components (μ_r, μ_φ, μ_z, ε_r, ε_φ, ε_z) of full parameters. In order to be realized by artificial materials, the full parameter can be simplified under the TE polarization. The simplified parameters of the 2D ghost-illusion device for z-polarized electric fields are written as

$$(\mu_r, \mu_\varphi, \varepsilon_z) = \left(\left(\frac{r - a + ka'}{k} \right)^2, 1, \frac{1}{k^2} \right), \quad (9.22)$$

$$(\mu_r, \mu_\varphi, \varepsilon_z) = \left(\left(\frac{r - a + ka'}{k} \right)^2, 1, \frac{\varepsilon_{virtual}}{k^2} \right), \quad (9.23)$$

in which $\varepsilon_{virtual}$ is the relative permittivity of the ghost-wing targets, which is a predesigned and known parameter.

The geometrical sizes of the virtual device are set by $a' = 0.41\lambda_0$, $a = 0.95\lambda_0$, $b' = 1.08\lambda_0$, $c = 1.75\lambda_0$, $\varepsilon_{virtual} = 2.23$, and the free-space wavelength $\lambda_0 = 30$ mm. In region I, the permittivity maintains 2.82 and the permeability ranges from 0.065 to 0.226, which are realized by artificially structured materials, the conventional SRRs etched on a dielectric substrate; while in region II, the permittivity maintains 6.29 and the permeability ranges from 0.226 to 0.355, which are realized by a class of modified SRRs shown in Figure 9.11. Such modified SRR structures can raise the permittivity remarkably.

The designed target generation device illustrated in Figure 9.11 is synthesized with eight concentric layers of low-loss PCBs. On each layer, SRRs of 35-μm-thick copper were coated on one side of the 0.25-mm-thick substrate (F4B) with the relative permittivity $\varepsilon = 2.65$ and loss tangent 0.001. Although we apply different structures in Regions I and II, the height of each layer (10.8 mm) and the distance between adjacent layers (3 mm) remain constant. The inner and outer radii of the ghost device are 28.5 and 52.5 mm, respectively. Each layer carries three rows of SRRs, whose length and height are 3.14 and 3.6 mm. The designed SRRs create the required radial permeability μ_r and z direction permittivity ε_z at 10 GHz. The PCB rings are adhered to a 2-mm-thick hard-foam board that was cut with concentric circular rabbets to fit the concentric layers exactly.

The target generation device is also effective for dielectric objects. For example, we can simply choose the object as free space (or air) with the same geometrical parameters. Such a ghost device can virtually transform the void (i.e., air) into the EM images of three distinct objects. It is noted that the ghost device not only shrinks the central air area, but also changes the material properties virtually.

The distance between the source and the center of the ghost device is $2.43\lambda_0$. Figure 9.12a illustrates the experimental setup together with the resultant 2D fields in the near zone. The electric field depends on x and y coordinates, but does not vary in the z direction, which is the direction of polarization parallel to the antenna's orientation. In the experiment, the distance between the source and the center of the ghost device is set as $3.93\lambda_0$. Figure 9.12b–d shows the simulation and measurement results of the ghost device. As illustrated in Figure 9.12b, the measured electric fields inside the ghost device are perturbed by anisotropic metamaterials, but the scattered waves outside are in very good agreement with the numerical results for a layered ghost-device profile, as shown in Figure 9.12d. For more details, Figure 9.12c compares the simulated and measured electric field intensity along a preselected line, located at $3.8\lambda_0$ (114 mm) away from the illusion device. In this experiment, the illusion device has just 8 layers including two regions, but a good illusion performance has been observed. The simplified prototype preserves the functionality of the ideal virtual target generation device, indicating the robustness of the design for practical applications.

In the above simulation and experiment results, one situation of the target generation device is discussed, which transforms a metallic object in the real space into a smaller metallic object with two stand-alone dielectric wings in the virtual space.

290 ■ *Metamaterials*

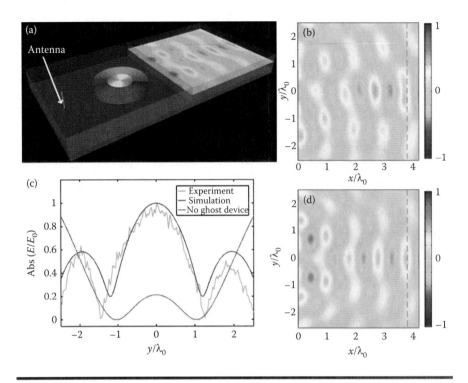

Figure 9.12 (a) Experimental setup of the virtual target generation device. (b) The simulation result of the illusion device. (c) The comparison of the simulated and experimental field distributions along an observing line. (d) The experimental result of the illusion device. (W. X. Jiang et al.: Creation of ghost illusions using wave dynamics in metamaterials. *Adv. Func. Mat.* 23: 4028, 2013. Copyright Wiley-VCH Verlag GmbH & Co. KGaA. Reproduced with permission.)

Actually, more types of ghost scattering can be produced under other design configurations. For example, when we choose $a' = 0$ mm (i.e., the radius of metallic cylinder is zero in the virtual space) with other parameters unchanged, then the device will produce only two ghost-wing objects. In other words, this ghost device virtually camouflages the initial EM scattering of a metallic cylinder, as if there existed only two dielectric objects. It is very important that there will be no scattering center in the original position of the actual target in this case. When the two wings of the ghost device are metallic, they are also effective for the dielectric and metallic objects. Furthermore, if we divide region II into several smaller areas or consider more blocks of region II, the ghost device will transform the signature of one object to an equivalence of a cluster of multiple different objects. Hence, many interesting illusion performance can be manipulated via the proposed ghost devices. Introducing more blocks of region II can also make the device rotationally symmetric and more independent of the incident direction.

In this section, we have discussed another kind of transformation devices, the illusion device, which can enable an object to possess arbitrarily predesigned scattering properties. The principle can be precisely controlled for producing the desired scattering metrics of the illusion device. Metamaterial structures are employed to realize the 2D prototype, so that the illusion signatures are produced at nonresonant frequency with low-loss feature thereby. The metamorphosis of EM signature has been verified by numerical investigations and experiments, which demonstrate good ghost-illusion performance in wave dynamics. All these devices are scalable in different frequency regions, provided that the future nanofabrication technique can produce the required subwavelength elements in the infrared or visible light.

9.4 Circuit Illusion-Optics Devices

In 2010, a 2D illusion optics analog, "the invisible gateway," was first experimentally demonstrated based on an inhomogeneous transmission-line medium [16]. The so-called "invisible gateway" is actually an open channel, but it appears to be blocked for waves in a selected frequency band. The scheme of "invisible gateway" is illustrated in Figure 9.13a. A PEC wall divides space into two regions. A channel is then opened in the PEC wall with a negative-index material (NIM, $\varepsilon' = \mu' = -1$) filled inside the trapezoidal region and an air channel. From the viewpoint of transformation optics, the NIM will project the adjacent PEC boundary into another optically equivalent PEC boundary, the dashed line in Figure 9.13a. Hence, the wall with an air channel will appear as a continuous PEC wall for the outside detectors at the designed frequency band. In another word, EM wave incident from one side of the wall cannot propagate to the other side while other entities are allowed to pass through the open channel, and hence the effect of invisible gateway is illustrated.

An inductor–capacitor (LC) network has been used to mimic the configuration by the authors. The unit cell to mimic the air region is shown in Figure 9.13b, where the series inductors and shunt capacitors can act as composing units for an isotropic medium with effective positive permittivity and permeability. The NIM is mimicked by the dual LC configuration, as shown in Figure 9.13c, where the position of L and C are interchanged. The experimental device, shown in Figure 9.13d, is fabricated on a grounded FR4 substrate with thickness of 1mm and dielectric constant ε_r of 4.3. There were 61 grid nodes in the x direction and 41 grids in the y direction for the whole structure. The distance between adjacent nodes is 6 mm. The totally measured region is about 390 mm × 270 mm. The magnified views of the grid nodes are also shown in the insets of Figure 9.13d. The nodes in the positive-index medium (PIM) region consist of four surface-mounted inductors in series and one capacitor in shunt to the ground by a via-hole, as shown in the schematic plot in Figure 9.13b. The left-handed unit cell consists of four surface-mounted capacitors

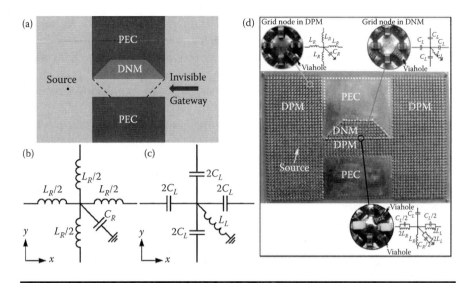

Figure 9.13 (a) Schematic of the EM invisible gateway. (b, c) The *LC* circuit unit cell to mimic PIMs and NIMs. (d) The fabricated circuit illusion device. (Reprinted with permission from C. Li et al., *Phys. Rev. Lett.*, 105: 233906. Copyright 2010 by the American Physical Society.)

in series and one inductor is shunt to the ground by a via-hole, as illustrated by the schematic plot in Figure 9.13c. Grounded PEC plates are used to mimic the PEC walls in Figure 9.13a. In their design, the outer boundaries of the PIM regions are truncated by being connected with Bloch impedances, which are applied to mimic the PML. The series branches of the node consists of the parallel connection of $2L_R$ and $C_L/2$, while the shunt branch consists of the parallel connection of $C_R/2$ and $2L_L$, as shown in the lower inset of Figure 9.13d. Hence, a precise equivalence to the continuity condition of tangential electric and magnetic field at the interface between two different media can be achieved.

The measured voltage distribution at the designed frequency $f = 51\,\text{MHz}$ is shown in Figure 9.14a. For comparison, Figure 9.14b shows the measured results when the NIM is replaced by the PEC, which verifies that the circuit device has the similar illusion effect of the invisible gateway. In the frequency band around the designed frequency, the chip inductors and capacitors have nonresonant property and the values of L_R, L_L and C_R, C_L change gradually when the frequency changes. Hence, the illusion property can be extended to the neighboring frequencies from 46 to 56 MHz.

Recently, a strategy has been proposed to design effective localized transformation optics devices [17], which was a kind of circuit illusion device. For the global invisibility cloak, the mathematical principle is to compress the space in a circle with

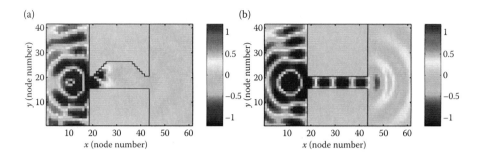

Figure 9.14 Measured node voltage distribution of the circuit illusion device, (a) the gateway with NIMs, (b) the gateway with NIMs replaced by PEC. (Reprinted with permission from C. Li et al., *Phys. Rev. Lett.*, 105: 233906. Copyright 2010 by the American Physical Society.)

radius r_1 to an annular region ($r_0 \leq r \leq r_1$). The object encloaked in the real space is invisible and cannot be perceived by outer detectors. Unlike the global invisibility cloak, a localized invisibility cloak makes the enclosed object partially visible and partially invisible. To generate such a localized invisibility cloak, the physical space is first divided into m parts along the azimuth direction and n parts along the radial direction in the cylindrical coordinates, so that it contains $m \times n$ subregions, as shown in Figure 9.15a. Then the visible subregions and invisible subregions are chosen by desire. The spatial transformation for the localized invisibility cloak is the same as that for the global cloak, and the parameters of invisibility media can be noted as ($\bar{\varepsilon}, \bar{\mu}$).

The $m \times n$ subregions in the real object (Figure 9.15a) will be mapped to the corresponding $m \times n$ subregions in the localized cloaking shell (Figure 9.15b). If we want to make a subregion visible, we can fill in the transformed medium with ($\varepsilon_f \bar{\varepsilon}, \mu_f \bar{\mu}$) in the corresponding subregion in the cloaking shell, where (ε_f, μ_f) is the material property of the original object; if we want a subregion invisible, the material in the corresponding subregion will be exactly the invisibility media. As a result, we can make any subregions be invisible or visible at will by using the similar strategy. In a similar way, other localized transformation devices can be designed, such as the localized optical illusion.

In the D.C. limit, the localized transformation optics is reduced to the localized transformation electrostatics. For example, the localized D.C. invisibility cloak can be designed and then the material properties in concealing subregions can be calculated. The localized D.C. invisibility cloak was emulated by circuit theory [17]. To demonstrate the localized cloaking effect experimentally, a localized D.C. invisibility cloak was fabricated using a resistor network on printed circuit board. The measurement results (equipotential lines) of the whole metal object with a D.C. localized

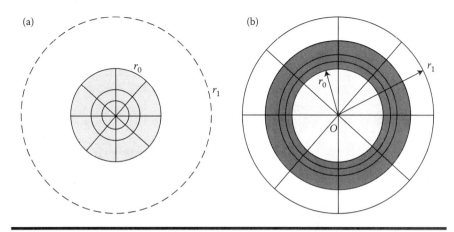

Figure 9.15 Design scheme of the localized transformation optics device. The original object is divided into many subregions (shaded part in (a)), which are mapped to the cloaking-shell regions (shaded part in (b)). (Reprinted with permission from W. X. Jiang et al. Localized transformation optics devices. *Appl. Phys. Lett.*, 103: 214104. Copyright 2013, American Institute of Physics.)

cloak and the predesigned partially visible object are illustrated in Figure 9.16. It can be observed that the measured equipotential lines of the metal object enclosed by the localized D.C. cloak are exactly the same as those of part of the object, verifying the functionality of localized cloaking.

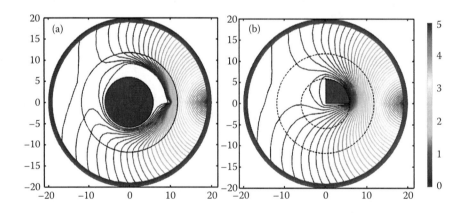

Figure 9.16 (a) The measured equipotential lines when the metallic cylinder is coated by the localized D.C. cloak. (b) The measured equipotential lines for the predesigned locally invisible object. (Reprinted with permission from W. X. Jiang et al. Localized transformation optics devices. *Appl. Phys. Lett.*, 103: 214104. Copyright 2013, American Institute of Physics.)

More recently, another straightforward 2D D.C. illusion device was reported based on the transformation optics theory [18], which can transform a metallic object with radius a in the conducting background into a magnified dielectric one with radius b, as shown in Figure 9.17a and b. Such a device was designed and fabricated using the resistor network. The illusion device is composed of complicated materials with anisotropic and inhomogeneous conductivities. In the design, the metallic cylinder was set as pure copper with $\sigma = 5.8 \times 10^7$ S/m, the background medium has a conductivity of $\sigma = 1$ S/m, and the virtual dielectric object has a conductivity of $\sigma = 10$ S/m. To emulate an infinite background, matching resistors are used around the circle with radius of 15 cm.

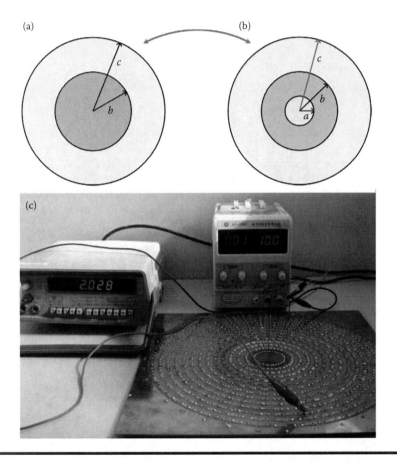

Figure 9.17 Illustration of a D.C. illusion device. (a) Virtual space, (b) real space, (c) photograph of the experimental setup for the D.C. illusion device. (Reprinted with permission from M. Liu et al. DC illusion and its experimental verification. *Appl. Phys. Lett.*, 101: 051905. Copyright 2012, American Institute of Physics.)

The D.C. illusion device was realized using the printed circuit board (PCB) technology and metal-film resistors. Due to space limitation, tangential resistors in the inner-most three layers are surface-mounted resistors. In the experiment, a D.C. power supply with the magnitude of 10 V is used to excite the resistor network, and then a 4-1/2 multimeter is utilized to measure the voltage at each node one by one. The experimental setup is shown in Figure 9.17c, which also gives the photo of the fabricated device. The measured voltage distribution fits the simulations very well, which shows that a good D.C. illusion has been achieved.

In conclusion, we have reviewed some experimental work on anisotropic and inhomogeneous devices, such as harmonic invisibility cloak, D.C. invisibility cloak, free-space illusion devices, and circuit illusion devices. The success of all the above novel devices with complex material parameters shows that metamaterials can indeed be employed to realize detailed and exacting specifications, including gradients and nonrectangular geometry, such as cylindrical shape.

References

1. J. B. Pendry, D. Schurig, and D. R. Smith. Controlling electromagnetic fields. *Science*, 312: 1780, 2006.
2. D. Schurig, J. J. Mock, B. J. Justice, S. A. Cummer, J. B. Pendry, A. F. Starr, and D. R. Smith. Metamaterial electromagnetic cloak at microwave frequencies. *Science*, 314: 977, 2006.
3. D. R. Smith, S. Schultz, P. Markos, and C. M. Soukoulis. Determination of effective permittivity and permeability of metamaterials from reflection and transmission coefficients. *Phys. Rev. B*, 65: 195104, 2002.
4. B. Edwards, A. Alu, M. G. Silveirinha, and N. Engheta. Experimental verification of plasmonic cloaking at microwave frequencies with metamaterials. *Phys. Rev. Lett.*, 103: 153901, 2009.
5. S. Xu, X. Cheng, S. Xi, R. Zhang, H. O. Moser, Z. Shen, Y. Xu et al. Experimental demonstration of a free-space cylindrical cloak without superluminal propagation. *Phys. Rev. Lett.*, 109: 223903, 2012.
6. N. Landy and D. R. Smith. A full-parameter unidirectional metamaterial cloak for microwaves. *Nat. Mater.*, 12: 25, 2013.
7. F. Yang, Z. L. Mei, T. Y. Jin, and T. J. Cui. dc electric invisibility cloak. *Phys. Rev. Lett.*, 109: 053902, 2012.
8. T. J. Cui and Z. L. Mei. Transformation Electrostatics. 2015. http://www.e-fermat.org.
9. W. X. Jiang, C. Y. Luo, Z. L Mei, and T. J. Cui. An ultrathin but nearly perfect direct current electric cloak. *Appl. Phys. Lett.*, 102: 014102, 2013.
10. F. Yang, Z. L. Mei, X. Y. Yang, T. Y. Jin, and T. J. Cui. A negative conductivity material makes a dc invisibility cloak hide an object at a distance. *Adv. Func. Mat.*, 23: 4306, 2013.
11. Q. Ma, Z. L. Mei, S. K. Zhu, T. Y. Jin, and T. J. Cui. Experiments on active cloaking and illusion for Laplace equation. *Phys. Rev. Lett.*, 111: 173901, 2013.
12. W. X. Jiang, T. J. Cui, X. M. Yang, H. F. Ma, and Q. Cheng. Shrinking an arbitrary object as one desires using metamaterials. *Appl. Phys. Lett.*, 98: 204101, 2011.

13. W. X. Jiang, B. B. Xu, Q. Cheng, T. J Cui, and G. X. Yu. Design and rigorous analysis of transformation-optics scaling devices. *J. Opt. Soc. Am. A*, 30: 1698, 2013.
14. W. X. Jiang and T. J. Cui. Radar illusion via metamaterials. *Phys. Rev. E*, 83: 026601, 2011.
15. W. X. Jiang, C.-W. Qiu, T. Han, S. Zhang, and T. J. Cui. Creation of ghost illusions using wave dynamics in metamaterials. *Adv. Func. Mat.*, 23: 4028, 2013.
16. C. Li, X. Meng, X. Liu, F. Li, G. Fang, H. Chen, and C. T. Chan. Experimental realization of a circuit-based broadband illusion-optics analogue. *Phys. Rev. Lett.*, 105: 233906, 2010.
17. W. X. Jiang, S. Ge, C. Luo, and T. J. Cui. Localized transformation optics devices. *Appl. Phys. Lett.*, 103: 214104, 2013.
18. M. Liu, Z. L. Mei, X. Ma, and T. J. Cui. DC illusion and its experimental verification. *Appl. Phys. Lett.*, 101: 051905, 2012.

Chapter 10
Conclusions and Remarks

Since the first wire medium with negative permittivity was artificially realized in 1996 [1], together with the first negative permeability realized in 1999 [2] in verifying the fabulous negative refraction [3] in experiments [4], metamaterials have achieved tremendous developments in the past 20 years. It is impossible to include all excellent advances on metamaterials in a single book. Hence, this book summarizes only a small part of the vast progresses of metamaterials, which is conducted in the State Key Laboratory of Millimeter Waves, Southeast University, China. In this book, we concentrate only on the metamaterials based on the effective medium theory in the microwave frequencies.

10.1 Summary of the Book

The whole book can be divided into two parts: the fundamentals and applications of metamaterials. The fundamental part was described by the effective medium theory for artificial metamaterial structures and the designs of metamaterial particles, that is, meta-atoms. Based on such two fundamental issues, three kinds of metamaterials have been designed and fabricated in microwave frequencies—homogeneous metamaterials, random metamaterials, and inhomogeneous metamaterials, from which a number of new physical phenomena have been verified and several functional devices have been developed for practical applications.

In homogeneous metamaterials, four typical cases have been introduced: single-negative (either negative permittivity or negative permeability), double-negative, zero-refractive-index, and double-positive metamaterials. In each situation, special behaviors of EM waves have been investigated, such as the evanescent wave amplification on the interface of two single-negative metamaterials, strong wave localization on a four-quadrant double-negative and double-positive open cavity, wave tunneling through a tiny channel of isotropic zero-index metamaterials, high directivity or omnidirectional radiations of EM waves by anisotropic zero-index metamaterials, and polarization conversions by anisotropic double-positive metamaterials. In all scenarios, special institutive parameters are used to realize the unusually physical phenomena.

In random metamaterials, two approaches have been used to design the special arrangements of meta-atoms: making randomly gradients of refractive indexes and random reflection phases. The significant feature of random metamaterials is to change the EM reflections to diffusion, which has found applications in practical engineering to reduce the RCSs of targets. Recently, a rapid and efficient method has been proposed to design the random metamaterials using the optimization algorithm and far-field scattering pattern analysis, which can reach broadband and broad-angle RCS reductions [5]. This concept has also been extended from microwave to terahertz frequencies [6], reaching efficient terahertz diffusions.

In inhomogeneous metamaterials, three methodologies have been used to obtain different complexities: the GO method, quasi-conformal mapping method, and transformation optics. Based on the GO method, gradient-refractive-index metamaterials have been designed and fabricated, from which an EM black hole and a number of metamaterial lens antennas have been developed. Due to the significant improvement to antenna performance, such metamaterial lenses have been applied in real engineering. From the quasi-conformal mapping, nearly isotropic inhomogeneous metamaterials have been realized, reaching the exciting broadband ground-plane invisibility cloaks and flattened Luneburg lens. Using the transformation optics, we have designed anisotropic inhomogeneous metamaterials, which have been used to construct unusual EM functions such as spatial invisibility cloaks and illusion-optics devices.

In summary, the metamaterials based on the effective medium theory have been well developed in the past two decades, in both fundamental theories and practical engineering, as stated in this book.

10.2 New Trends of Metamaterials

Metamaterial is a fast-developing research area, and a number of new versions have emerged in recent years, such as metasurfaces, coding and digital metamaterials, programmable metamaterials, and planar plasmonic metamaterials. Here, we provide a brief introduction to such topics, although they are beyond the scope of this book.

10.2.1 Planar Metamaterials: Metasurfaces

Recently, a planar version of metamaterials, metasurfaces with gradient phase changes, has received great interests in both microwave and optical frequencies due to their good functional performance, planar geometries, and easy fabrications [7–14]. By designing periodically gradient phase shifts from 0 to 2π along one direction on a metasurface using differently oriented V-shaped nano antennas, anomalous reflections, and refractions have been demonstrated when the EM waves are incident through the metasurface, which are governed by the Fermat principle [7]. Such anomalous reflections and refractions are referred to as generalized Snell's laws. Based on the generalized Snell's laws, the gradient-phase metasurfaces can be used to realize a lot of interesting physical phenomena, including optical vortex [7], light bending [8], photonic spin Hall effect [9], polarization traffic control [10], polarization rotation [11], dual-polarity plasmonic lens [12], and efficient conversion from spatial waves to SPPs [13].

The other kind of metasurface is holographic metasurface, in which the functions to be realized are written on the holographic interference pattern. Using holographic metasurfaces, one is capable of controlling EM radiations in microwave frequencies to achieve special performance that cannot be obtained by conventional approach. For instance, a combined theory of holography and leaky wave has been presented to produce multibeam radiations by generating special surface interference patterns [14]. When the frequency changes, it was observed that the radiation beams could accomplish 2D scans regularly by using the proposed holographic metasurface [14]. This is the first time to realize 2D scans of antennas by changing the frequency. Metasurfaces will play more important roles in controlling the EM waves (in both reflected and transmitted modes) and designing planar devices with special functionalities in the optical and microwave frequencies.

10.2.2 Coding Metamaterials and Programmable Metamaterials

All metamaterials discussed in this book are described by the macroscopic effective medium parameters, which can be regarded as analog metamaterials. As a counterpart of analog metamaterials, digitally coding metamaterials have been proposed recently [15]. Instead of using effective medium parameters to describe metamaterials, it was suggested to describe metamaterials with the aid of coding sequences of "0" and "1" particles, which are unit cells with 0 and π phase responses [15]. By coding "0" and "1" particles (in 1-bit code) or "00," "01," "10," and "11" particles (in 2-bit code) with controlled sequences, one could manipulate EM waves and realize different functionalities. Here, the four particles in 2-bit coding have phase responses of 0, $\pi/2$, π, and $3\pi/2$, respectively. To digitally control the coding metamaterials, a unique metamaterial particle was proposed, which can reach either "0" or "1" response controlled by a biased diode. Using the field-programmable gate array (FPGA), programmable metamaterial has been presented [15], which realized lots of distinct functionalities (e.g., high-gain radiations, multibeam radiations, beam scans, and RCS reductions) by the FPGA programming. Coding metamaterials have been extended from microwave to terahertz frequencies, producing abnormal terahertz reflections and diffusions [16,17].

Coding metamaterials and programmable metamaterials open a new venue to control the EM terahertz waves instantly. It is expected to build up new-concept radar systems, communication systems, and imaging systems using such kinds of metamaterials.

10.2.3 Plasmonic Metamaterials

As discussed in Section 1.2.3, SPPs are highly localized surface waves existing on the interface of two media with opposite permittivity, and hence they only naturally exist in the optical frequency, in which metals have negative permittivity. To produce spoof SPPs in the microwave and terahertz frequencies, plasmonic metamaterials have been proposed by making artificial structures on the metal surface [18]. However, such plasmonic metamaterials are usually 3D and hence inconvenient to be integrated for engineering applications. Recently, a planar plasmonic metamaterial was presented to guide the spoof SPPs in microwave and terahertz frequencies, which is composed of ultrathin metallic strip with subwavelength corrugations [19]. Numerical simulations and experiments demonstrate that the ultrathin plasmonic waveguide sustains highly localized SPPs along two orthogonal directions in broadband by keeping good modal shape and propagating long distance with low bending loss. It was further shown that the surface plasmon waves can propagate around curved surfaces with arbitrary shapes, resulting in conformal surface plasmons (CSPs) [19]. The flexible and ultrathin SPP waveguide can be bent, folded, and even twisted to mold the flow of CSPs, producing a number of passive plasmonic devices (e.g., SPP bends, splitters, filters, polarizers, and resonators) [20]. Similarly, a planar ultrathin textured metallic disk with subwavelength scale has been illustrated to support spoof localized surface plasmons (LSPs) in the microwave and terahertz frequencies [21,22]. Multipolar plasmonic resonances were observed in both numerical simulations and experiments, including the dipole, quadrupole, hexapole, octopole, decapole, dodeca-pole, and quattuordec-pole modes [21]. Such spoof LSP resonances are sensitive to the disk's geometry and local dielectric environments, and hence the ultrathin textured metallic disk has potential applications as the plasmonic sensor in the microwave and terahertz frequencies.

Another advantage of the planar plasmonic metamaterials is the easy integration with semiconductor devices to produce active SPP components and circuits in the microwave frequency. Using a commercially available subwavelength-scale amplifier chip, the first SPP amplifier was developed, which could directly amplify the SPP waves by 20 dB in wide frequency band [23]. Both measured transmission coefficients and near-field distribution have verified the SPP amplification. Based on the nonlinear property of a subwavelength-scale FET chip, higher-order harmonics of

SPPs were generated [24]. By properly designing the SPP waveguides connecting to the field effect transistor (FET) chip, the second harmonic is amplified and the higher-order harmonics are suppressed, which results in an SPP multiplier device. The above idea can be directly extended to produce SPP mixers. To bridge the traditional transmission line and SPP waveguide, an efficient conversion between the spatial waves and SPP modes has been presented [25]. According to this conversion as well as passive and active SPP and LSP devices, integrated spoof SPP/LSP circuits and system could be realized in the near future.

References

1. J. B. Pendry, A. J. Holden, W. J. Stewart, and I. Youngs. Extremely low frequency plasmons in metallic mesostructures. *Phys. Rev. Lett.*, 76: 4773, 1996.
2. J. B. Pendry, A. J. Holden, D. J. Robbins, and W. J. Stewart. Magnetism from conductors and enhanced nonlinear phenomena. *IEEE Trans. Microw. Theory Tech.*, 47: 2075, 1999.
3. V. G. Veselago. The electrodynamics of substances with simultaneously negative values of ϵ and μ. *Sov. Phys. Usp.*, 10: 509, 1968.
4. R. Shelby, D. R. Smith, and S. Schultz. Experimental verification of a negative index of refraction. *Science*, 292: 77, 2001.
5. K. Wang, J. Zhao, Q. Cheng, D. Dong, and T. J. Cui. Broadband and broad-angle low-scattering metasurface based on hybrid optimization algorithm. *Sci. Rep.*, 4: 5935, 2014.
6. D. S. Dong, J. Yang, Q. Cheng, J. Zhao, L. H. Gao, S. J. Ma, S. Liu, H. B. Chen, Q. He, and W. W. Liu. Terahertz broadband low-reflection metasurface by controlling phase distributions. *Adv. Opt. Mater.*, 3(10): 1405, 2015.
7. N. Yu, P. Genevet, M. A. Kats, F. Aieta, J. Tetienne, F. Capasso, and Z. Gaburro. Light propagation with phase discontinuities: Generalized laws of reflection and refraction. *Science*, 334: 333, 2011.
8. X. Ni, N. K. Emani, A. V. Kildishev, A. Boltasseva, and V. M. Shalaev. Broadband light bending with plasmonic nanoantennas. *Science*, 335: 427, 2012.
9. X. Yin, Z. Ye, J. Rho, Y. Wang, and X. Zhang. Photonic spin Hall effect at metasurfaces. *Science*, 339: 1405, 2013.
10. J. Lin, M. J. P. Balthasar, Q. Wang, G. Yuan, N. Antoniou, X. Yuan, and F. Capasso. Polarization-controlled tunable directional coupling of surface plasmon polaritons. *Science*, 340: 331, 2013.
11. E. Miroshnichenko and Y. S. Kivshar. Polarization traffic control for surface plasmons. *Science*, 340: 283, 2013.
12. X. Chen, L. Huang, H. Muhlenbernd, G. Li, B. Bai, Q. Tan, G. Jin, C. W. Qiu, S. Zhang, and T. Zentgraf. Dual-polarity plasmonic metalens for visible light. *Nat. Commun.*, 3: 1198, 2012.
13. S. Sun, Q. He, S. Xiao, Q. Xu, X. Li, and L. Zhou. Gradient-index meta-surface as a bridge linking propagating waves and surface waves. *Nat. Mater.*, 11: 426, 2012.
14. Y. B. Li, X. Wan, B. G. Cai, Q. Cheng, and T. J. Cui. Frequency-controls of electromagnetic multi-beam radiations and beam scanning by metasurfaces. *Sci. Rep.*, 4: 6921, 2014.

15. T. J. Cui, M. Q. Qi, X. Wan, J. Zhao, and Q. Cheng. Coding metamaterials, digital metamaterials, and programmable metamaterials. *Light: Sci. Appl.*, 3: e218, 2014.
16. L. Gao, Q. Cheng, J. Yang, S. Ma, J. Zhao, S. Liu, H. Chen et al. Broadband diffusion of terahertz waves by multi-bit coding metasurfaces. *Light: Sci. Appl.*, 4: e324, 2015.
17. L. Liang, M. Qi, J. Yang et al. Anomalous terahertz reflection and scattering by flexible and conformal coding metamaterial. *Adv. Opt. Mater.*, 3(10): 1373, 2015.
18. J. B. Pendry, L. Martĭn-Moreno, and F. J. Garcia-Vidal. Mimicking surface plasmons with structured surfaces. *Science*, 305: 847, 2004.
19. X. Shen, T. J. Cui, D. Martin-Canob, and F. J. Garcia-Vidal. Conformal surface plasmons propagating on ultrathin and flexible films. *Proc. Natl. Acad. Sci.*, 110: 40, 2013.
20. X. Shen and T. J. Cui. Planar plasmonic metamaterial on a thin film with nearly zero thickness. *Appl. Phys. Lett.*, 102: 211909, 2013.
21. X. Shen and T. J. Cui. Ultrathin plasmonic metamaterial for spoof localized surface plasmons. *Laser Photon. Rev.*, 8: 137, 2014.
22. P. A. Huidobro, X. Shen, J. Cuerda, E. Moreno, L. Martin-Moreno, F. J. Garcia-Vidal, T. J. Cui, and J. B. Pendry. Magnetic localized surface plasmons. *Phys. Rev. X*, 4: 021003, 2014.
23. H. C. Zhang, S. Liu, X. Shen, L. H. Chen, L. Li, and T. J. Cui. Broadband amplification of spoof surface plasmon polaritons in the microwave frequency. *Laser Photon. Rev.*, 9: 83, 2015.
24. H. C. Zhang, Y. Fan, J. Guo, X. Fu, L. Li, C. Qian, and T. J. Cui. Second-harmonic generation of spoof surface plasmon polaritons using nonlinear plasmonic metamaterials. *ACS Photonics*, 3(1): 139, 2016.
25. H. F. Ma, X. Shen, Q. Cheng, W. X. Jiang, and T. J. Cui. Broadband and high-efficiency conversion from guided waves to spoof surface plasmon polaritons. *Laser Photon. Rev.*, 8: 146, 2014.

Index

A

Anisotropic inhomogeneous metamaterials, 9, 11, 178, 271–296
   circuit illusion-optics devices, 291–296
   D.C. circuit invisibility cloak, 274–281
   design of, 176–177
   spatial illusion-optics devices, 281–291
      material conversion devices, 283–286
      shrinking devices, 281–283
      virtual target generation devices, 286–291
   spatial invisibility cloak, 271–274
Anisotropic zero-index metamaterial (AZIM), 4
   directivity enhancement to Vivaldi antennas using compact, 125–126
   highly directive radiation by line source in, 119–122
   spatial power combination for omnidirectional radiation via radial, 122–125
Ansoft HFSS, 69, 76, 116, 117, 129, 199
Antiresonant phenomenon, 25
Averaged refractive index, 237–238

B

Babinet principle, 40, 75
Bessel function, 124
Bianisotropy, 36
Bloch–Floquet theorem, 78, 84, 104, 105

C

Carpet cloak, *see* Ground-plane cloak
CC–LL bilayer structure, 92–96
Circuit illusion-optics device, 291–296
Coding metamaterials, 301–302
Compact-sized ground-plane cloaks, 241, 243–244, 245
Complementary electric-LC resonator (CELCR), 76, 77, 98–101
Complementary particles, 38–42
Complementary split-ring resonator (CSRR), 40, 41–42, 75, 76, 115–119
Composite right-/lefthander (CRLH) TL metamaterial, 30, 55–57, 83
Concentrators, 180–182
Conformal surface plasmons (CSPs), 302
Control antenna, 136–138
Crystals, 1, 29

D

D.C. circuit invisibility cloak, 274–281
D.C. exterior cloak, 280
D.C. illusion device, 11, 295–296
D.C. particles, 57–62
Dielectric-metal resonant particles, 36–38, 39, 43, 51
Dielectric particles, 43–48
Discrete coordinate transformation (DCT), 236, 253
Double-negative (DNG) metamaterial, 2, 38, 39, 45, 79–84
   free-space LHM super lens based on fractal-inspired, 107–112
   localization of EM waves using LHM–RHM open cavities, 101–107
Double-positive (DPS) metamaterial, 38, 83, 85–89, 102
   microstrip antennas by magneto-dielectric metamaterials loading, 132–139
   TL representation of, 88–89
   transmission polarizer based on anisotropic, 127–132

E

Edge-coupled split ring resonator (EC-SRR), 68–69, 70, 79, 80–81, 85

305

Effective medium parameters, 21–24, 100, 101, 129, 136, 137, 283, 301
Effective medium theory, 17
　effective medium parameters, retrieval methods of, 21–24
　general effective medium theory, 24–28
　Lorentz–Drude models, 17–21
Effective permittivity, 47
　frequency response of, 116
　of I-shaped resonator arrays, 33
　and permeability, 25, 27, 45–46, 53, 78, 92, 100
ELCR, 72, 73, 74, 79–81, 128–130
Electrically resonant particles, 29, 30–34, 35–36, 40
Electric permittivity ε, 23, 30, 38, 79, 85
Electromagnetic black hole, 225–230
　experiments, 228–230
　full-wave simulations (continuous medium), 227–228
　metamaterials utilized, 228
　ray tracing performance, 227
　refractive index profile, 226–227
Embedded meander line (EML), 87, 88, 132, 134–136, 138, 139
EM waves, 41, 147, 169, 170, 195, 300, 301
　localization of, using LHM–RHM open cavities, 101–107
　random metamaterials and, 147
ε-negative (ENG) metamaterial, 37, 38, 68, 71, 73, 74, 77–78, 79, 81, 85, 92
ENZ metamaterials, EM tunneling through waveguide channel filled with, 112–119
Evanescent-wave amplification (EWA), in MNG–ENG bilayer slabs, 89–97
Experiments
　electromagnetic black hole, 228–230
　3D half Luneburg lens, 222–223
　3D planar gradient-index lens, 217–218
　2D half Maxwell Fisheye lens, 210–212
　2D Luneburg lens, 204–207
　2D planar gradient-index lenses, 201

F

Faraday's law of induction, 26
Fermat, Pierre de, 170, 301
Fermat's principle, 49, 170, 191
Field-programmable gate array (FPGA), 301, 302
Finite difference time domain (FDTD), 25, 253

Free space LHM super lens, fractal-inspired DNG metamaterials and, 107–112
Full-wave simulations (continuous medium)
　electromagnetic black hole, 227–228
　3D half Luneburg lens, 220
　3D Maxwell Fisheye lens, 224–225
　3D planar gradient-index lens, 214–215
　2D half Maxwell Fisheye lens, 208–209
　2D Luneburg lens, 204
　2D planar gradient-index lenses, 198–199
Full-wave simulations (discrete medium)
　2D planar gradient-index lenses, 200–201

G

General effective medium theory, 24–28
Geometrical optics (GO) method, 9–10, 171–173, 191, 202, 213, 262
GO method, *see* Geometrical optics (GO) method
Gradient-index inhomogeneous metamaterials
　electromagnetic black hole, 225–230
　　experiments, 228–230
　　full-wave simulations (continuous medium), 227–228
　　metamaterials utilized, 228
　　ray tracing performance, 227
　　refractive index profile, 226–227
　gradient-refractive-index (GRIN), 191–194
　GRIN metamaterials, 194–197
　　hole-array metamaterial, 194–195
　　I-shaped metamaterial, 195
　　waveguide metamaterial, 195–197
　3D half Luneburg lens, 218–223
　　experiments, 222–223
　　full-wave simulations, 220
　　metamaterials utilized, 221–222
　　ray tracing performance, 219
　　refractive index profile, 219
　3D Maxwell Fisheye lens, 223
　　full-wave simulations and experiments, 224–225
　　ray tracing performance, 223–224
　　refractive index profile, 223
　3D planar gradient-index lens, 212–218
　　experiments, 217–218
　　full-wave simulations (continuous medium), 214–215
　　metamaterials utilized, 215–217
　　refractive index profile, 213–214
　2D half Maxwell Fisheye lens, 207–212

experiments, 210–212
full-wave simulations (continuous medium), 208–209
metamaterials utilized, 209–210
ray tracing performance, 208
refractive index profile, 207–208
2D Luneburg lens, 201–207
experiments, 204–207
full-wave simulations (continuous medium), 204
metamaterials utilized, 204
ray tracing performance, 202–203
refractive index profile, 202
2D planar gradient-index lenses, 197–201
derivation of refractive index profile, 197–198
experimental realization, 201
full-wave simulations (continuous medium), 198–199
full-wave simulations (discrete medium), 200–201
hole-array metamaterials, 199–200
Gradient-refractive-index (GRIN) materials, 191–192
comparison, 193
discretization, 192, 193
fabrication, 193
flowchart for design, analysis, and realization of, 192–193
geometry–EM parameter relation, 193
index profile, 192
measurement, 193
metamaterials, 194–197
hole-array metamaterial, 194–195
I-shaped metamaterial, 195
waveguide metamaterial, 195–197
simulation, 193
unit cell study, 193
Ground-plane cloak, 46, 173–176, 234–235, 240–249, 257–259, 261
3D ground-plane cloak, 10, 50–51, 247, 258, 262, 264
2D ground-plane cloak, 10, 50, 51, 240, 241, 252, 255

H

Half Maxwell fisheye lens (HMFE), 191, 207, 208, 262; *see also* 3D Maxwell Fisheye lens; 2D half Maxwell fisheye lens
Hamiltonian system of equations, 202–203, 219, 226

High-performance antennas, 182–185
Hole-array metamaterial, 194–195, 199–200
Holographic metasurface, 301
Homogeneous metamaterials, 1, 2–7, 67, 299–300
DNG metamaterials, *see* Double-negative (DNG) metamaterial
DPS metamaterials, *see* Double-positive (DPS) metamaterial
left-handed materials, 2–3
negative-epsilon materials, 4–6
negative-mu materials, 6–7
SNG metamaterials, *see* Single-negative (SNG) metamaterial
zero-index metamaterials, *see* Zero-index metamaterials
zero-refractive-index materials, 3–4

I

Illusion-optics devices, 11
inhomogeneous metamaterials and, 185–188
spatial, 281–291
material conversion devices, 283–286
shrinking devices, 281–283
virtual target generation devices, 286–291
2D ground-plane, 247–251
Impedance matching layers (IML), 212, 217, 240, 241, 245
Inductor–capacitor (LC) network, 30, 291
Inhomogeneous metamaterials, 1, 9–11, 49, 299, 300
anisotropic, *see* Anisotropic inhomogeneous metamaterials
EM devices based on, 178–188
concentrators, 180–182
high-performance antennas, 182–185
illusion-optics devices, 185–188
invisibility cloaks, 178–180
geometrical optics (GO) method, 9–10, 171–173
gradient-index, *see* Gradient-index inhomogeneous metamaterials
isotropic, *see* Isotropic inhomogeneous metamaterials
nonperiodic arrays of meta-atoms, 169–171
optical transformation, 176–177
quasi-conformal mapping method, 10–11, 173–176
transformation optics, 11

Invisibility cloak, 11, 173, 176, 300
    conductivity tensor for, 275
    D.C. circuit, 57, 58, 274–281, 293
    inhomogeneous metamaterials and, 178–180
    spatial, 271–274
    3D ground-plane, 256–262
    2D compact ground-plane, 241–247
    2D ground-plane, 233–241
"Invisible gateway", 291, 292
I-shaped metamaterial, 195–197, 204
Isotropic inhomogeneous metamaterials, 233
    design of, 171–176
    3D flattened Luneburg lens, 262–268
    3D ground-plane invisibility cloak, 256–262
    2D compact ground-plane invisibility cloak, 241–247
    2D ground-plane illusion-optics devices, 247–251
    2D ground-plane invisibility cloak, 233–241
    2D planar parabolic reflector, 251–256

J

Jacobian matrix, 175, 237, 275

K

Ku-band coax-to waveguide device, 220, 264

L

Lai's illusion, 185–186
Lambertian reflection, 7–9
LC particles, 51–57
Left-handed materials (LHMs), 2–3, 38, 48, 53, 79, 110
    RHM open cavities, localization of EM waves using, 101–107
    super lens based on fractal-inspired DNG metamaterials, 107–112
Localized transformation optics devices, 292, 293, 294
Lorentz–Drude models, 17–21, 25, 34
Luneburg lens, 11, 47, 191, 195, 263, 264
    3D flattened, 262–268
    3D half, 10, 47, 218–223, 224
        experiments, 222–223
        full-wave simulations, 220
        metamaterials utilized, 221–222
        ray tracing performance, 219
        refractive index profile, 219
    2D, 10, 201–207
        experiments, 204–207
        full-wave simulations (continuous medium), 204
        metamaterials utilized, 204
        ray tracing performance, 202–203
        refractive index profile, 202

M

Magnetically resonant particles, 29, 34–36, 40, 48–49
Magnetic permeability μ, 20, 21, 23, 26, 30, 178
Magneto-dielectric metamaterials loading, microstrip antennas and, 132–139
Man-made atoms, 29, 30
Material conversion devices, 283–286
Maxwell equation, 2–3, 4–5, 25, 26, 120, 177, 236
Maxwell's fisheye (MFE) lenses, 9–10, 191, 262
    3D, 223–225
        full-wave simulations and experiments, 224–225
        ray tracing performance, 223–224
        refractive index profile, 223
    2D half, 207–212
        experiments, 210–212
        full-wave simulations (continuous medium), 208–209
        metamaterials utilized, 209–210
        ray tracing performance, 208
        refractive index profile, 207–208
Meta-atoms, 1, 2, 7, 9, 29, 30, 67, 169–171, 300
Metamaterials, 1
    advantages of, 1
    anisotropic inhomogeneous, *see* Anisotropic inhomogeneous metamaterials
    classification of, xxv, xxvi, 1
    coding, 301–302
    features of, xxv
    gradient-index inhomogeneous, *see* Gradient-index inhomogeneous metamaterials
    homogeneous, *see* Homogeneous metamaterials
    inhomogeneous, *see* Inhomogeneous metamaterials
    isotropic inhomogeneous, *see* Isotropic inhomogeneous metamaterials
    new trends of, 300–303
    nonresonant, 2
    planar, 300–301
    plasmonic, 302–303

programmable, 301–302
random, *see* Random metamaterials
resonant, 1–2
utilized
    electromagnetic black hole, 228
    3D planar gradient-index lens, 215–217
    3D half Luneburg lens, 221–222
    2D half Maxwell Fisheye lens, 209–210
    2D Luneburg lens, 204
Metasurfaces, 7, 8, 300–301
    holographic, 301
    with random distribution of reflection phase, 150–152
    RCS reduction by, 163–166
Microstrip antennas, magneto-dielectric metamaterials loading and, 132–139
MLR, 74, 75, 85, 126
μ-negative (MNG) metamaterials, 37, 38, 68, 70, 97–101
MNG–ENG bilayer slabs, evanescent-wave amplification (EWA) in, 89–97

N

Natural materials, 1, 17, 68
Negative-epsilon materials, 4–6
Negative index metamaterials (NIM), 32
Negative-mu materials, 6–7
Negative resistor model, 61, 279
Newton's second law of motion, 17–18
Nonbianisotropic SRRs (NB-SRRs), 36
Noncrystals, 1, 29
Nonresonant metamaterials, 1, 2, 240, 246, 258, 264
Nonresonant particles, 48–51
Nonuniform rational B-spline (NURBS), 179
NRI TL-MTM, 104, 105, 106
Nyquist–Shannon sampling theorem, 236, 244

O

Ohm's law, 58, 275
1D randomly gradient index metamaterial, 148–150, 153–154, 157, 158, 159–160
Optical black hole, 225–226
Optical path length, 170
Optical transformation, 176–177, 182, 183, 233–235, 251, 252

P

Parallel capacitor model, 194–195
Pendry, J. B., 48, 175, 178, 275

Planar metamaterials, 40, 300–301
Plasmonic cloaking, 273–274
Plasmonic metamaterials, 302–303
Printed circuit board (PCB), 36–37, 240, 293, 296
PRI TL-MTM, 103–104, 105
Programmable metamaterials, 301–302

Q

Quasi-conformal mapping method, 10–11, 173–176, 233, 235–236, 244, 249, 251, 263–264, 300
Quasi-conformal-mapping transformation optics (QCTO), 249
Quasicrystal, 1, 9, 29

R

Radial anisotropic zero-index metamaterial (RAZIM), 122–125
Randomly distributed gradients of refractive index, 147, 152–156
    amount of subregions or length of coating, 157
    experimental verification of diffuse reactions, 159–163
    influence of impedance mismatch, 157–158
    influence of random distribution mode, 158–159
Randomly gradient index metamaterial, 147–150
Random metamaterials, 1, 7–9, 299, 300
    diffuse reflections by metamaterial with randomly distributed gradients, 152–163
    EM waves and, 147
    metasurface with random distribution of reflection phase, 150–152
    randomly gradient index metamaterial, 147–150
    RCS reduction by metasurface with random distribution of reflection phase, 163–166
Ray tracing performance, 208
    electromagnetic black hole, 227
    3D half Luneburg lens, 219
    3D Maxwell Fisheye lens, 223–224
    2D half Maxwell Fisheye lens, 208
    2D Luneburg lens, 202–203
RCS reduction by metasurface, with random distribution of reflection phase, 163–166

Reflection phase
    metasurface with random distribution of, 150–152
    RCS reduction by metasurface with random distribution of, 163–166
Refractive index profile, 207–208
    derivation of, 197–198
    electromagnetic black hole, 226–227
    3D half Luneburg lens, 219
    3D Maxwell Fisheye lens, 223
    3D planar gradient-index lens, 213–214
    2D half Maxwell Fisheye lens, 207–208
    2D Luneburg lens, 202
Resonant metamaterials, 1–2, 49, 226
Retrieval methods, of effective medium parameters, 21–24
Right-handed material (RHM), 52–54, 79, 101–107

S

Sampling, 244
Shechtman, Daniel, 1
Shrinking devices, 251, 281–283
Single-negative (SNG) metamaterials, 37–38, 68–79, 85
    anisotropic MNG metamaterials, partial focusing by, 97–101
    evanescent-wave amplification in MNG–ENG bilayer slabs, 89–97
Snell's law, 3, 38, 79, 301
Spatial illusion-optics devices, 281–291
    material conversion devices, 283–286
    shrinking devices, 281–283
    virtual target generation devices, 286–291
Spatial invisibility cloak, 271–274
Specular reflection, 7, 8
Split-ring-resonators (SRRs), 6, 11, 19–21, 23, 24, 28, 34–36, 40–41, 121, 124, 272, 282, 289
Super crystal, 29, 62, 67
Super noncrystal, 29, 62, 147, 166
Super quasicrystal, 9, 29, 62, 169
Surface plasmon polaritons (SPPs), 2, 4, 6–7, 302–303

T

"Ten Breakthroughs of Science", xxv, xxvi
3D flattened Luneburg lens, 47, 262–268
3D ground-plane cloak, 10, 50–51, 247, 258, 262, 264
3D ground-plane invisibility cloak, 256–262
3D half Luneburg lens, 218–223
    experiments, 222–223
    full-wave simulations, 220
    metamaterials utilized, 221–222
    ray tracing performance, 219
    refractive index profile, 219
3D Maxwell fisheye lens, 223–225
    full-wave simulations and experiments, 224–225
    ray tracing performance, 223–224
    refractive index profile, 223
3D planar gradient-index lens, 212–218
    experiments, 217–218
    full-wave simulations (continuous medium), 214–215
    metamaterials utilized, 215–217
    refractive index profile, 213–214
Transformation optics, 11, 176–177, 178, 182, 184–185, 262, 271, 275, 281–282, 285, 293–295, 300
Transmission polarizer based on anisotropic DPS metamaterials, 127–132
2×2 covariant metric g, 238–239
2D compact ground-plane invisibility cloak, 241–247
2D ground-plane cloak, 10, 50, 51, 240, 241, 252, 255
2D ground-plane illusion-optics devices, 247–251
2D ground-plane invisibility cloak, 233–241
2D half Maxwell fisheye lens, 207–212
    experiments, 210–212
    full-wave simulations (continuous medium), 208–209
    metamaterials utilized, 209–210
    ray tracing performance, 208
    refractive index profile, 207–208
2D illusion device, performance of, 250–251
2D Luneburg lens, 201–207
    experiments, 204–207
    full-wave simulations (continuous medium), 204
    metamaterials utilized, 204
    ray tracing performance, 202–203
    refractive index profile, 202
2D near-field mapping apparatus, 98, 140–141, 162
2D NRI TL metamaterial, 83–84
2D planar gradient-index lenses, 197–201
    derivation of refractive index profile, 197–198

experimental realization, 201
full-wave simulations (continuous medium), 198–199
full-wave simulations (discrete medium), 200–201
hole-array metamaterials, 199–200
2D planar parabolic reflector, 251–256

V

Vector network analyzer (VNA), 21, 22, 95, 96, 111, 141, 164–165
Veselago, V. G., 2, 3, 79, 107
Virtual target generation devices, 286–291
Vivaldi antennas, directivity enhancement to, using compact AZIMs, 125–126

W

Waveguided-metamaterial (WG-MTM), 42, 51, 74–77, 85, 88, 195–197
Waveguide method, 23, 24
WG-MDM antenna, 134–139
Wire media, 71–72

Z

Zero-index metamaterials (ZIMs), 3–4, 38, 85, 112, 300
　directivity enhancement to Vivaldi antennas using compact AZIMs, 125–126
　electromagnetic tunneling through waveguide channel, 112–119
　highly directive radiation byline source in anisotropic, 119–122
　spatial power combination for omnidirectional radiation via radial AZIM, 122–125
Zero-refractive-index materials, 3–4, 299